T0282501

LONDON MATHEMATICAL SOCIETY STUDENT TEXTS

Managing Editor: Professor C.M. SERIES, Mathematics Institute.
University of Warwick, Coventry CV4 7AL, United Kingdom

London Mathematical Society Student Texts 44

Classical Invariant Theory

Peter J. Olver
School of Mathematics
University of Minnesota

CAMBRIDGE
UNIVERSITY PRESS

PUBLISHED BY THE PRESS SYNDICATE OF THE UNIVERSITY OF CAMBRIDGE
The Pitt Building, Trumpington Street, Cambridge, United Kingdom

CAMBRIDGE UNIVERSITY PRESS
The Edinburgh Building, Cambridge CB2 2RU, UK   http://www.cup.cam.ac.uk
40 West 20th Street, New York, NY 10011–4211, USA   http://www.cup.org
10 Stamford Road, Oakleigh, Melbourne 3166, Australia

First published 1999

Typeface Computer Modern 10/12 pt.   *System* TEX [AU]

*A catalog record of this book is available from
the British Library.*

Library of Congress Cataloging in Publication data
Olver, Peter J.
Classical invariant theory / Peter J. Olver
p.   cm. – (London Mathematical Society student texts ; 44)
Includes bibliographical references and indexes.
ISBN 0 521 55243 5 (hc). – ISBN 0 521 55821 2 (pbk.)
1. Invariants.   I. Title.   II. Series.
QA201.O48   1999   98-33722
512.5 – dc21   CIP

0 521 55243 5 hardback
0 521 55821 2 paperpack

Transferred to digital printing 2003

To my parents — Grace E. Olver and Frank W.J. Olver

As all roads lead to Rome so I find in my own case at least that all algebraic inquiries, sooner or later, end at the Capitol of Modern Algebra over whose shining portal is inscribed the Theory of Invariants.

— Sylvester, quoted in [**72**; p. 143].

The *theory of invariants* came into existence about the middle of the nineteenth century somewhat like Minerva: a grown-up virgin, mailed in the shining armor of algebra, she sprang forth from Cayley's Jovian head. Her Athens over which she ruled and which she served as a tutelary and beneficent goddess was *projective geometry*.

— Weyl, [**230**].

Like the Arabian phoenix arising out of its ashes, the theory of invariants, pronounced dead at the turn of the century, is once again at the forefront of mathematics.

— Kung and Rota, [**135**].

Contents

Introduction

Classical invariant theory is the study of the intrinsic or geometrical properties of polynomials. This fascinating and fertile field was brought to life at the beginning of the last century just as the theory of solubility of polynomials was reaching its historical climax. It attained its zenith during the heyday of nineteenth-century mathematics, uniting researchers from many countries in a common purpose, and filling the pages of the foremost mathematical journals of the time. The dramatic and unexpected solution to its most fundamental problem — the finitude of the number of fundamental invariants — propelled the young David Hilbert into the position of the most renowned mathematician of his time. Following a subsequent decline, as more fashionable subjects appeared on the scene, invariant theory sank into obscurity during the middle part of this century, as the abstract approach entirely displaced the computational in pure mathematics. Ironically, though, its indirect influence continued to be felt in group theory and representation theory, while in abstract algebra the three most famous of Hilbert's general theorems — the Basis Theorem, the Syzygy Theorem, and the Nullstellensatz — were all born as lemmas (*Hilfsätze*) for proving "more important" results in invariant theory! Recent years have witnessed a dramatic resurgence of this venerable subject, with dramatic new applications, ranging from topology and geometry, to physics, continuum mechanics, and computer vision. This has served to motivate the dusting off of the old computational texts, while the rise of computer algebra systems has brought previously infeasible computations within our grasp. In short, classical invariant theory is the closest we come in mathematics to sweeping historical drama and romance. As a result, the subject should hold a particular fascination, not only for the student and practitioner, but also for any mathematician with a desire to understand the culture, sociology, and history of mathematics.

I wrote this introductory textbook in the hope of furthering the recent revival of classical invariant theory in both pure and applied mathematics. The presentation is not from an abstract, algebraic standpoint, but rather as a subject of interest for applications in both mathematics and other scientific fields. My own training is in differential equations and mathematical physics, and so I am unashamed to restrict my attention to just real and complex polynomials. This approach allows me to directly employ differentiation and other analytical tools as the need arises. In this manner, the exposition at times resembles that of the classical texts from the last century, rather than that of more modern treatments that either presuppose an extensive training in the methods of abstract algebra or reduce the subject to a particular case of general tensor analysis. Nevertheless, a fair amount of more recent material and modern developments is covered, including several original results that have not appeared in print before. I have designed the text so that it can be profitably read by students having a fairly minimal number of mathematical prerequisites.

Notes to the Reader

The purpose of this book is to provide the student with a firm grounding in the basics of classical invariant theory. The text is written in a non-abstract manner and makes fairly low demands on the prospective reader. In addition, a number of innovations — in methodology, style, and actual results — have been included that should attract the attention of even the most well-seasoned researcher. We shall concentrate on the basic theory of binary forms, meaning polynomials in a single variable, under the action of the projective group of linear fractional transformations, although many of the methods and theoretical foundations to be discussed have far wider applicability. The classical constructions are all founded on the theory of groups and their representations, which are developed in detail from the beginning during the exposition.

The text begins with the easiest topic of all: the theory of a single real or complex quadratic polynomial in a single variable. Although completely elementary, this example encapsulates the entire subject and is well worth reviewing one more time — although an impatient reader can entirely omit this preliminary chapter. As any high school student knows, the solution to the quadratic equation relies on the associated discriminant. Less obvious is the fact that the discriminant is (relatively) unchanged under linear fractional transformations. Hence it forms the first (both historical and mathematical) example of an *invariant* and so can be used for classification of canonical forms. The text starts in earnest in Chapter 2, which provides an overview of the basics of classical invariant theory within the context of binary forms. Here we meet up with the basic definitions of invariants and covariants, and investigate how the geometry of projective space governs the correspondence between homogeneous and inhomogeneous polynomials, as well as their transformation properties under, respectively, linear and projective transformations. The motivating examples of cubic and quartic polynomials are discussed in detail, including complete lists of invariants, covariants, and canonical forms. The Fundamental Theorem of Algebra guarantees the existence of a complete system of (complex) roots, whose geometrical configuration is governed by the invariants. Two particularly important invariants are the classical resultant, which indicates the existence of common roots to a pair of polynomials, and the discriminant, which indicates multiple roots of a single polynomial. The chapter concludes with a brief introduction to the Hilbert Basis Theorem, which states that every system of polynomials has only a finite number of polynomially independent invariants, along with remarks on the classification of algebraic relations or syzygies among the invariants.

With this preliminary survey as our motivating guide, the next two chapters provide a grounding in the modern mathematical foundations of the subject, namely, transformation groups and representation theory. Chapter 3 is a self-contained introduction to groups and their actions on spaces. Groups originally arose as the symmetries of a geometric or algebraic object; in our case the object is typically a polynomial. The chapter includes a discussion of the equivalence problem — when can two objects be transformed into each other by a suitable group element — and the allied concept of a canonical form. Chapter 4 concentrates on the theory of linear group actions, known as representations. For general transformation groups, the associated multiplier representations act on the functions defined on the space; the linear/projective actions on polynomials form a very particular instance of this general construction. The invariant functions arise as fixed points for such representations, and so the focus of classical invariant theory naturally falls within this general framework.

The next three chapters describe the core of the classical constructive algebraic theory of binary forms. The most important operations for producing covariants are the "transvection" processes, realized as certain bilinear differential operators acting on binary forms, or, more generally, analytic functions. According to the First Fundamental Theorem of classical invariant theory, all of the invariants and covariants for any system of polynomials or, more generally, functions, can be constructed through iterated transvectants and, in the inhomogeneous case, scaling processes. Thus, a proper grounding in these basic techniques is essential. Traditionally, such invariant processes are based on the symbolic method, which is the most powerful computational tool for computing and classifying invariants. However, no aspect of the classical theory has been as difficult to formalize or as contentious. The point of view taken here is nonstandard, relying on the construction of covariants and invariants as differential polynomials. Taking inspiration from work of Gel'fand and Dikii, [**77**], in solitons and the formal calculus of variations, I introduce a transform that mimics the Fourier transform of classical analysis and maps differential polynomials into algebraic polynomials. The transform is, in essence, the symbolic method realized in a completely natural manner, applicable equally well to polynomials and more general functions. The chapter concludes with proofs of the First Fundamental Theorem, which states that every covariant has symbolic form given by a polynomial in certain "bracket factors", and the Second Fundamental Theorem, which completely classifies the syzygies among the brackets. Although the determination of a complete Hilbert basis for the covariants of a general binary form turns out to be an extremely difficult problem, which has been solved only for forms of low degree, I shall prove a result due to Stroh and Hilbert that constructs an explicit rational basis for a form of arbitrary degree.

Chapter 7 introduces a graphical version of the symbolic method that can be used to simply and pictorially analyze complicated invariant-theoretic identities for binary forms. Each symbolic expression has an equivalent directed graph, or "digraph" counterpart, whereby algebraic identities among the symbolic forms translate into certain graphical operations that bear much similarity to basic operations in knot theory, [**124**], and thereby lead to a significant simplification with visual appeal. As an application, I show how to implement Gordan's method for constructing a complete system of fundamental invariants and covariants for binary forms, illustrated by the cubic and quartic examples.

At this point, we have covered the classical algebraic techniques underlying the theory of binary forms. Since the group of linear/projective transformations depends analytically on parameters, it is an example of a Lie transformation group. The theory of Lie groups includes a wide range of powerful calculus-based tools for the analysis of their invariants. Chapter 8 begins with a very brief introduction to Lie groups, including the general Frobenius Theorem that completely determines the local structure of the orbits and the fundamental invariants for regular actions. Here, invariants are classified up to functional dependence, rather than polynomial or rational dependence as was done in the more algebraic aspects of the theory; the number of fundamental invariants depends solely on the dimension of the group orbits. Even better, there is an explicit computation algorithm, which relies just on the Implicit Function Theorem, for constructing the invariants of regular Lie group actions. This method, known as "normalization", has its origins in Élie Cartan's theory of moving frames, [**33, 93**], which was developed for studying the geometry of curves and surfaces. Surprisingly, the normalization method has not been developed at all in the standard literature; the construction relies on a new theory of moving frames for general transformation group actions recently established by the author in collaboration with Mark Fels, [**69, 70**]. Applications to the classification of joint invariants and differential invariants for interesting transformation groups are provided.

In the theory of moving frames, the determination of symmetries, the complete solution to the equivalence problem, and the construction of canonical forms rely on the analysis of suitable differential invariants. In the case of planar curves, there is a single basic differential invariant — the group-theoretic *curvature* — along with a group-invariant *arc length* element. Higher order differential invariants are obtained by repeatedly differentiating curvature with respect to arc length. The first two fundamental differential invariants trace out the *signature set* which uniquely characterizes the curve up to group transformations. A direct application of the moving frame method leads to a remarkable theorem that the equivalence and symmetry of a binary form relies on merely *two* classical rational covariants! This result, first established in [**167**], reduces the entire complicated algebraic Hilbert basis to a simple pair of rational covariants whose functional dependencies completely encode the geometric properties of the binary form. I present a number of striking new consequences of this result, including a new bound on the number of discrete symmetries of polynomials. These innovative techniques are of much wider applicability and clearly deserve further development in the multivariate context.

While Chapter 8 develops "finite" Lie theory, the following chapter is concerned with Lie's powerful infinitesimal approach to invariance. Each Lie-theoretic object has an infinitesimal counterpart, and the replacement of complicated group-theoretic conditions by their infinitesimal analogs typically linearizes and significantly simplifies the analysis. The infinitesimal version of a Lie group is known as a Lie algebra, which contains the infinitesimal generators of the group action, realized as first order differential operators (or vector fields). Assuming connectivity, a function is invariant under the group if and only if it is annihilated by the infinitesimal generators, allowing methods from the theory of partial differential equations to be applied to the analysis of invariants. In the context of binary forms, the infinitesimal generators were, in fact, first recognized by Cayley, [**41**], to play an important role in the theory. I show how one can use these to build up general invariants from simpler "semi-" and "isobaric" invariants through an inductive procedure based on invariance under subgroups. The chapter culminates in a proof of the Hilbert Basis Theorem that relies on a particular differential operator that converts functions into invariants.

The final chapter is included to provide the reader with an orientation to pursue various generalizations of the basic methods and theories to multivariate polynomials and functions. Unfortunately, space has finally caught up with us at this point, and so the treatment is more superficial. Nevertheless, I hope that the reader will be sufficiently motivated to pursue the subject in more depth.

I have tried to keep the prerequisites to a minimum, so that the text can be profitably read by anyone trained in just the most standard undergraduate material. Certainly one should be familiar with basic linear algebra: vectors, matrices, linear transformations, Jordan canonical form, norms, and inner products — all of which can be found in any comprehensive undergraduate linear algebra textbook. Occasionally, I employ the tensor product construction. No knowledge of the general theory of polynomial equations is assumed. An introductory course in group theory could prove helpful to the novice but is by no means essential since I develop the theory of groups and their representations from scratch. All constructions take place over the real or complex numbers, and so no knowledge of more general field theory is ever required. One certainly does not need to take an abstract algebra course before starting; indeed, this text may serve as a good motivation or supplement for such a course!

In Chapters 8 and 9, I rely on multivariable differential calculus, at least as far as the Implicit Function Theorem, and the basic theory of first order systems of ordinary differential equations. In particular, the reader should be familiar with the solution to linear systems of differential equations, including matrix exponentials and their computation via Jordan canonical forms. I do not require any experience with Lie group theory or differential geometry, although the reader may wish to consult a basic text on manifolds, vector fields, and Lie groups to supplement the rather brief exposition here. (Chapter 1 of my own book [**168**] is particularly recommended!) Some of the more difficult results are stated without proof, although ample references are provided. I should remark that although the transform method adopted in Chapter 6 is inspired by the Fourier transform, no actual knowledge of the analytical Fourier theory is required.

Inevitably, the writing of an introductory text of moderate size requires making tough choices on what to include and what to leave out. Some of my choices are unorthodox. (Of course, if all choices were "orthodox", then there wouldn't be much point writing the book, as it would be a mere reworking of what has come before.) The most orthodox choice, followed in all the classical works as well as most modern introductions, is to concentrate almost entirely on the relatively modest realm of binary forms, relegating the vast hinterlands of multivariate polynomials and functions to an all too brief final chapter that cannot possibly do them justice. Of course, one motivation for this tactic is that most of the interesting explicit results and methods already make their appearance in the binary form case. Still, one tends to leave with the wish that such authors (including the present one) had more to say of substance in the multidimensional context.

Less orthodox choices include the reliance on calculus — differential operators, differential equations, differential invariants — as a framework for the general theory. Here we are in good company with the classics — Clebsch, [**49**], Gordan, [**89**], Grace and Young, [**92**], and even Hilbert, [**107**]. Post–Noetherian algebraists will no doubt become alarmed that I have regressed, in that the calculus-based tools are only valid in characteristic zero, or, more specifically, for the real and complex numbers, while "true" invariant theory requires that all fields be treated as equals, which means throwing out such "antiquated" analytical tools. My reply (and I speak here as the semi-applied mathematician I am) is that the primary physical and geometrical applications of invariant theory, which, after all, motivated its development, remain either real or complex, and it is here that much of the depth, beauty, and utility of the subject still resides. Another, more provocative, response is that the more interesting generalization of the classical techniques is not necessarily to fields of nonzero characteristic, but rather to more general associative and non-associative algebras, starting with the quaternions, octonions, Clifford algebras, quantum groups, and so on. One retains calculus (the quaternion calculus is a particularly pretty case) but gives up commutativity (and even possibly associativity). The development of a non-commutative classical invariant theory remains, as far as I know, completely unexplored.

The most original inclusion is the application of the Cartan theory of moving frames to the determination of symmetries and a solution to the equivalence problem for binary forms. Most of the constructions and results in this part of the text are new but can be readily comprehended by an advanced undergraduate student. This connection between geometry and algebra, I believe, opens up new and extremely promising vistas in both subjects — not to mention the connections with computer vision and image processing that served as one of my original motivations.

An unorthodox omission is the combinatorial and enumerative techniques that receive a large amount of attention in most standard texts. This was a difficult decision, and a topic I really did want to include. However, as the length of the manuscript crept up and up, it became clear that something had to go, and I decided this was it. The combinatorial formulae that count the number of invariants, particularly those based on Hilbert and Molien series and their generating functions, are very pretty and well worth knowing; see [200, 204], for instance. However, as far as practical considerations go, they merely serve as indicators of what to expect and are of less help in the actual determination and classification of invariants. Indeed, in all the examples presented here, enumeration formulae are never used, and so their omission will not leave any gaps in the exposition. But the reader is well advised to consult other sources to rectify this omission.

The text is designed for the active reader. As always, one cannot learn mathematics by merely reading or attending lectures — one needs to *do* mathematics in order to absorb it. Thus, a large number and variety of exercises, of varying degrees of difficulty, are liberally interspersed throughout the text. They either illustrate the general theory with additional interesting examples or supply further theoretical results of importance that are left for the reader to verify. The student is strongly encouraged to attempt most exercises while studying the material.

I have also included many references and remarks of historical and cultural interest. I am convinced that one cannot learn a mathematical subject without being at least partially conversant with its roots and its original texts. Modern reformulations of classical mathematics, while sometimes (but not always) more digestible to the contemporary palate, often shortchange the contributions of the original masters. Worse yet, such rewritings can actually be harder for the novice to digest, since they tend to omit the underlying motivations or significance of the results and their interconnections with other parts of mathematics and applications. I am a firm believer in the need for a definite historical consciousness in mathematics. There is no better way of learning a theorem or construction than by going back to the original source, and a text (even at an introductory level) should make significant efforts to uncover and list where the significant ideas were conceived and brought to maturation. On the other hand, I do not pretend that my list of references is in any sense complete (indeed, the sheer volume of the nineteenth-century literature precludes almost any attempt at completeness); nevertheless, it includes many obscure but vital papers that clearly deserve a wider audience. I hope the reader is inspired to continue these historical and developmental studies in more depth.

The text has been typeset using the author's own O$T_{\!E}$X system of macros. Details and software can be found at my web site:

`http://www.math.umn.edu/~olver` .

The figures were drawn with the aid of MATHEMATICA. Comments, corrections, and questions directed to the author are most welcome.

A Brief History

Classical invariant theory's origins are to be found in the early-nineteenth-century investigations by Boole, [**24**], into polynomial equations. The subject was nurtured by that indefatigable computer Cayley, to whom we owe many of the fundamental algorithms. Any reader of Cayley's collected works, [**36**], which include page after page of extensive explicit tables, cannot but be in awe of his computational stamina. (I often wonder what he might have accomplished with a functioning computer algebra system!) While the British school, led by Cayley and the flamboyant Sylvester, joined by Hermite in France, was the first to plow the virgin land, the actual flowering and maturation of the theory passed over to the Germans. The first wave of German experts includes Aronhold, the progenitor of the mystical "symbolic method", Clebsch, whose contributions metamorphosed into basic formulae in representation theory with profound consequences for quantum physics, and, most prominently, Gordan, the first among equals. Gordan's crowning achievement was his computational procedure and proof of the fundamental Basis Theorem that guarantees only a finite number of independent invariants for any univariate polynomial. The classical references by Clebsch, [49], Faà di Bruno, [67], and Gordan, [89], describe the resulting invariant theory of binary forms. A very extensive history of the nineteenth-century invariant theory, including copious references, was written by F. Meyer, [151]. Modern historical studies by Fisher, [72], and Crilly, [55], also document the underlying sociological and cultural implications of its remarkable history.

Despite much effort, extending Gordan's result to polynomials in two or more variables proved too difficult, until, in a profound stroke of genius, David Hilbert dramatically unveiled his general Basis Theorem in 1890. Hilbert's first, existential proof has, of course, had an incomparable impact, not just in classical invariant theory, but in all of mathematics, since it opened the door to the abstract algebraic approach that has characterized a large fraction of twentieth-century mathematics. Its immediate impact was the discreditation of the once dominant computational approach, which gradually fell into disrepute. Only in recent years, with the advent of powerful computer algebra systems and a host of new applications, has the computational approach to invariant theory witnessed a revival.

Nevertheless, the dawn of the twentieth century saw the subject in full florescence, as described in the marvelous (and recently translated) lectures of Hilbert, [107]. The texts by Grace and Young, [92], and Elliott, [65], present the state of the computational art, while Weitzenböck, [229], reformulates the subject under the guiding light of the new physics and tensor analysis. So the popular version of history, while appealing in its drama, is not entirely correct; Hilbert's paper did not immediately kill the subject, but rather acted as a progressive illness, beginning with an initial shock, and slowly consuming the computational body of the theory from within, so that by the early 1920's the subject was clearly moribund. Abstraction ruled: the disciples of Emmy Noether, a student of Gordan, led the fight against the discredited computational empire, perhaps as a reaction to Noether's original, onerous thesis topic that involved computing the invariants for a quartic form in three variables.

Although the classical heritage had vanished from the scene by mid-century, all was not quiet. The profoundly influential, yet often frustratingly difficult, book by Weyl, [231], places the classical theory within a much more general framework; polynomials now become particular types of tensorial objects, while, motivated by simultaneous developments in algebra and physics, the action of linear or linear fractional transformations is now extended to the vast realm of group representations. Attempts to reconcile both the classical heritage and Weyl's viewpoint with modern algebra and geometry have served to inspire a new generation of invariant theorists. Among the most influential has been Mumford's far-reaching development of the incisive methods of Hilbert, leading to the deep but abstract geometrical invariant theory, [156]. New directions, inspired by recent developments in representation theory and physics, appear in the recent work of Howe, [112]. Particular mention must be made of Rota and his disciples, [94, 135], whose efforts to place the less than rigorous classical theory, particularly the symbolic method, on a firm theoretical foundation have had significant influence. The comprehensive text of Gurevich, [97], is a particularly useful source, which helped inspire a vigorous, new Russian school of invariant theorists, led by Popov, [181], and Vinberg, [226], who have pushed the theory into fertile new areas.

Of course, one cannot fail to mention the rise of modern computer algebra. Even the masters of the last century became stymied by the sheer complexity of the algebraic formulas and manipulations that the subject breeds. The theory of Gröbner bases, cf. [54], has breathed new life into the computational aspect of the subject. Sturmfels' elegant book, [204], gives an excellent survey of current work in this direction and is particularly recommended to the student wishing to continue beyond the material covered here. The influence of classical invariant theory can be felt throughout mathematics and extends to significant physical applications, ranging from algebra and number theory, [79], through combinatorics, [201], Riemannian geometry, [149, 150, 229], algebraic topology, [196], and ordinary differential equations, [235, 195]. Applications include continuum mechanics, [197], dynamical systems, [195], engineering systems and control theory, [213], atomic physics, [189], and even computer vision and image processing, [**157**]. This text should prove to be useful to students in all of these areas and many more.

Acknowledgments

Many people deserve thanks for helping inspire my interest in this subject and desire to put pen to paper (or, more accurately, finger to keyboard). They include John Ball, whose questions in nonlinear elasticity directly motivated my initial forays; Gian–Carlo Rota, whose wonderful lectures opened my vistas; and Bernd Sturmfels, who patiently introduced me to modern computational tools. I would particularly like to thank my student Irina Berchenko for her careful proofreading of the manuscript. Most of all, I must express my heartfelt gratitude to my wonderful wife, Cheri Shakiban, who directly collaborated with me on several invariant theoretic papers, [**171, 172**], which form the basis of significant parts of this text. I am incredibly fortunate that she has been such a major part of my life.

I have dedicated this book to my parents. My father, Frank W.J. Olver, is a great applied analyst and certainly played a direct, inspirational role in my choice of career. I wish my mother, Grace E. Olver, were still alive to see the further fruits of her love and care. They both set me on those important first steps in mathematics, and for this I am eternally grateful.

Minneapolis
May 1998

Chapter 1

Prelude —
Quadratic Polynomials and
Quadratic Forms

Classical invariant theory is the study of the intrinsic properties of polynomials. By "intrinsic", we mean those properties which are unaffected by a change of variables and are hence purely geometric, untied to the explicit coordinate system in use at the time. Thus, properties such as factorizability and multiplicities of roots, as well as their geometrical configurations, are intrinsic, whereas the explicit values of the roots and the particular coefficients of the polynomial are not. The study of invariants is closely tied to the problem of equivalence — when can one polynomial be transformed into another by a suitable change of coordinates — and the associated canonical form problem — to find a system of coordinates in which the polynomial takes on a particularly simple form. The solution to these intimately related problems, and much more, are governed by the invariants, and so the first goal of classical invariant theory is to determine the fundamental invariants. With a sufficient number of invariants in hand, one can effectively solve the equivalence, and canonical form problems, and, at least in principle, completely characterize the underlying geometry of a given polynomial.

All of these issues are already apparent in the very simplest example — that of a quadratic polynomial in a single variable. This case served as the original catalyst for Boole and Cayley's pioneering work in the subject, [**24**, **36**], and can be effectively used as a simple (i.e., just high school algebra is required) concrete example that will motivate our study of the subject. We shall devote this introductory chapter to the elementary theory of quadratic polynomials in a single variable, together with homogeneous quadratic forms in two variables. Readers who are unimpressed with such relative trivialities are advised to proceed directly to the true beginning of our text in Chapter 2.

Quadratic Polynomials

Consider a quadratic polynomial in a single variable p:

$$Q(p) = ap^2 + 2bp + c. \tag{1.1}$$

Before addressing the question of what constitutes an invariant in this context, we begin our analysis with the elementary problem of determining a canonical form for the polynomial Q. In other words, we are trying to make Q as simple as possible by use of a suitable change of variable. As long as $a \neq 0$, the two roots of Q are, of course, given by the justly famous *quadratic formula*

$$p_+ = \frac{-b + \sqrt{-\Delta}}{a}, \qquad p_- = \frac{-b - \sqrt{-\Delta}}{a}, \tag{1.2}$$

where

$$\Delta = ac - b^2 \tag{1.3}$$

is the familiar *discriminant*[†] of Q. The existence of the two roots implies that we can factor

$$Q(p) = a(p - p_+)(p - p_-) \tag{1.4}$$

into two linear, possibly complex-valued, factors.

At this point, we need to be a bit more specific as to whether we are dealing with real or complex polynomials. Let us first concentrate on the slightly simpler complex version. The most obvious changes of variables preserving the class of quadratic polynomials are the affine transformations

$$\bar{p} = \alpha p + \beta, \tag{1.5}$$

for complex constants $\alpha \neq 0$ and β. Here α represents a (complex) scaling transformation,[‡] and β a complex translation. The transformation (1.5) maps the original quadratic polynomial $Q(p)$ to a new quadratic polynomial $\bar{Q}(\bar{p})$, which is constructed so that

$$\bar{Q}(\bar{p}) = \bar{Q}(\alpha p + \beta) = Q(p). \tag{1.6}$$

In particular, if p_0 is a root of $Q(p)$, then $\bar{p}_0 = \alpha p_0 + \beta$ will be a root of $\bar{Q}(\bar{p})$. For example, if $Q(p) = p^2 - 1$, and we apply the transformation

---

[†] The sign chosen for the discriminant is in accordance with later generalizations.

[‡] If we write $\alpha = re^{i\theta}$, then the modulus r will act by scaling, whereas the exponential $e^{i\theta}$ will induce a rotation in the complex p-plane; see p. 46.

$\bar{p} = 2p - 1$, then $\overline{Q}(\bar{p}) = \frac{1}{4}\bar{p}^2 + \frac{1}{2}\bar{p} - \frac{3}{4}$. The roots $p_\pm = \pm 1$ of Q are mapped to the roots $\bar{p}_+ = 1$ and $\bar{p}_- = -3$ of $\overline{Q}$.

For a general complex quadratic polynomial, there are only two cases to consider. If its discriminant is nonzero, $\Delta \neq 0$, then the roots of Q are distinct. We can translate one root, say, p_-, to be zero and then scale so that the second root takes the value 1. Thus, by a suitable choice of α and β we can arrange that $\overline{Q}$ has its roots at 0 and 1. Consequently, under complex affine transformations (1.5), every quadratic polynomial with distinct roots can be placed in the canonical form $\overline{Q}(\bar{p}) = k(\bar{p}^2 - \bar{p})$ for some $k \in \mathbb{C}$.

Exercise 1.1. Find the explicit formulas for α, β that will reduce a quadratic polynomial Q to its canonical form. Is the residual coefficient k uniquely determined? Determine the formula(e) for k in terms of the original coefficients of Q.

Exercise 1.2. An alternative canonical form for such quadratics is $\widetilde{Q}(\tilde{p}) = \tilde{k}(\tilde{p}^2 + 1)$. Do the same exercise for this canonical form, and describe what is happening to the roots of Q.

On the other hand, if the discriminant of Q vanishes, so $ac = b^2$, then Q has a single double root p_0 and so factors as a perfect square: $Q(p) = a(p - p_0)^2$. Clearly this property is intrinsic — it cannot be altered by any change of coordinates. We can translate the double root to the origin, reducing Q to a multiple of the polynomial $\tilde{p}^2$, and then scale the coordinate $\tilde{p}$ to reduce the multiple to unity, leading to a canonical form, $\overline{Q} = \bar{p}^2$, for a quadratic polynomial with a double root.

We are not quite finished, since we began by assuming that the leading coefficient $a \neq 0$. If $a = 0$, but $b \neq 0$, then Q reduces to a linear polynomial with a single root, $p_0 = -c/b$. We can, as in the preceding case, translate this root to 0 and then scale, producing the canonical form $\overline{Q} = \bar{p}$ in this case. If $b = 0$ also, then Q is a constant, and, from the viewpoint of affine transformations (1.5), there is nothing that can be done. Thus, we have constructed a complete list of canonical forms for quadratic polynomials, under complex affine changes of coordinates. Note particularly that the discriminant Δ and the leading coefficient a play distinguished roles in the classification.

Exercise 1.3. Suppose Q and $\overline{Q}$ are related by an affine change of variables (1.5). Determine how their discriminants and leading coefficients are related.

Affine Canonical Forms for Complex Quadratic Polynomials

I.	$k(p^2 + 1)$	$\Delta \neq 0$, $a \neq 0$	distinct roots
II.	p^2	$\Delta = 0$, $a \neq 0$	double root
III.	p	$a = 0$, $b \neq 0$	linear
IV.	c	$a = b = 0$	constant

The case of real polynomials under real affine changes of coordinates is similar, but there are a few more cases to consider. First, note that the roots (1.2) of a real quadratic polynomial are either both real or form a complex conjugate pair, depending on the sign of the discriminant. If Q has complex conjugate roots, meaning that its discriminant is positive, then it can never be mapped, under a real change of variables, to a quadratic polynomial with real roots, and so our complex canonical form is not as universally valid in this case. However, if the two roots are $p_\pm = r \pm is$, then a translation by $\beta = -r$ will move them onto the imaginary axis; this may be followed by a scaling to place them at $\pm i$. Thus, the canonical form in this case is $k(p^2 + 1)$. On the other hand, if the discriminant is negative, then Q has two distinct real roots, which can be moved to ± 1, leading to the alternative canonical form $k(p^2 - 1)$. The remaining cases are as in the complex version, since a double root of a real quadratic polynomial is necessarily real. We therefore deduce the corresponding table of real canonical forms.

Affine Canonical Forms for Real Quadratic Polynomials

Ia.	$k(p^2 + 1)$	$\Delta > 0$, $a \neq 0$	complex conjugate roots
Ib.	$k(p^2 - 1)$	$\Delta < 0$, $a \neq 0$	distinct real roots
II.	p^2	$\Delta = 0$, $a \neq 0$	double root
III.	p	$a = 0$, $b \neq 0$	single root
IV.	c	$a = b = 0$	constant

Exercise 1.4. Determine the possible canonical forms for a complex cubic polynomial $Q(p) = ap^3 + bp^2 + cp + d$ under affine changes of coordinates. *Hint*: What are the possible root configurations?

Quadratic Forms and Projective Transformations

While affine changes of coordinates are immediately evident, they do not form the most general class that preserves the space of polynomials. In order to motivate a further extension, we begin by explaining the connection between homogeneous and inhomogeneous polynomials. Instead of the inhomogeneous polynomial (1.1) in a single variable, we consider the homogeneous quadratic polynomial

$$Q(x, y) = ax^2 + 2bxy + cy^2, \tag{1.7}$$

in two variables x, y, known classically as a *quadratic form*. Clearly we can recover the inhomogeneous quadratic polynomial $Q(p)$ from the associated quadratic form $Q(x, y)$ by setting $p = x$ and $y = 1$, so that $Q(p) = Q(p, 1)$. On the other hand, the homogeneous version (1.7) can be directly constructed from $Q(p)$ according to the basic formula

$$Q(x, y) = y^2 \, Q \left(\frac{x}{y} \right). \tag{1.8}$$

An affine change of coordinates (1.5) will induce a linear transformation mapping the quadratic form $Q(x, y)$ associated with $Q(p)$ to the quadratic form $\overline{Q}(\bar{x}, \bar{y})$ associated with $\overline{Q}(\bar{p})$, as defined in (1.6). Clearly, the upper triangular linear transformation

$$\bar{x} = \alpha x + \beta y, \qquad \bar{y} = y, \qquad \alpha \neq 0, \tag{1.9}$$

will have the desired effect on the quadratic forms:

$$\overline{Q}(\bar{x}, \bar{y}) = \overline{Q}(\alpha x + \beta y, y) = Q(x, y).$$

We conclude that the theory of inhomogeneous quadratic polynomials under affine coordinate changes is isomorphic to the theory of quadratic forms under linear transformations of the form (1.9).

Now, a crucial observation is that the class of quadratic forms is preserved under a much wider collection of transformation rules. Namely, *any* invertible linear change of variables

$$\bar{x} = \alpha x + \beta y, \qquad \bar{y} = \gamma x + \delta y, \qquad \alpha \delta - \beta \gamma \neq 0, \tag{1.10}$$

will map a homogeneous polynomial in x and y to a homogeneous polynomial in $\bar{x}$ and $\bar{y}$ according to

$$\overline{Q}(\bar{x}, \bar{y}) = \overline{Q}(\alpha x + \beta y, \gamma x + \delta y) = Q(x, y). \tag{1.11}$$

The coefficients of the transformed polynomial

$$\overline{Q}(\bar{x}, \bar{y}) = \bar{a} \, \bar{x}^2 + 2\bar{b} \, \bar{x}\bar{y} + \bar{c} \, \bar{y}^2$$

are constructed from those of the original polynomial (1.7) according to the explicit formulae

$$a = \alpha^2\,\bar{a} + 2\alpha\gamma\,\bar{b} + \gamma^2\,\bar{c},$$
$$b = \alpha\beta\,\bar{a} + (\alpha\delta + \beta\gamma)\,\bar{b} + \gamma\delta\,\bar{c}, \qquad (1.12)$$
$$c = \beta^2\,\bar{a} + 2\beta\delta\,\bar{b} + \delta^2\,\bar{c}.$$

Remarkably, the discriminant of the transformed polynomial is directly related to that of the original quadratic form — a straightforward computation verifies that they agree up to the square of the determinant of the coefficient matrix $A = \left(\begin{smallmatrix}\alpha & \beta\\ \gamma & \delta\end{smallmatrix}\right)$ for the linear transformation (1.10):

$$\Delta = ac - b^2 = (\alpha\delta - \beta\gamma)^2\,(\bar{a}\bar{c} - \bar{b}^2) = (\alpha\delta - \beta\gamma)^2\,\overline{\Delta}. \qquad (1.13)$$

The transformation rule (1.13) expresses the underlying invariance of the discriminant of a quadratic polynomial and provides the simplest example of an invariant (in the sense of classical invariant theory).

Remark: A linear transformation (1.10) is called *unimodular* if it has unit determinant $\alpha\delta - \beta\gamma = 1$ and hence preserves planar areas. For the restricted class of unimodular transformations, the discriminant is a bona fide invariant: $\overline{\Delta} = \Delta$.

What is the effect of a general linear transformation on the original inhomogeneous polynomial? For this purpose, it helps to refer back to the formula (1.8) relating the inhomogeneous polynomial and its homogeneous counterpart. Specifically, the inhomogeneous or projective variable p is identified with the ratio of the homogeneous variables, so $p = x/y$. Therefore, the effect of the linear transformation (1.10) is to transform the projective variable p according to the *linear fractional* or *Möbius* or *projective transformation*

$$\bar{p} = \frac{\alpha p + \beta}{\gamma p + \delta}, \qquad \alpha\delta - \beta\gamma \neq 0. \qquad (1.14)$$

Thus, each invertible 2×2 coefficient matrix $A = \left(\begin{smallmatrix}\alpha & \beta\\ \gamma & \delta\end{smallmatrix}\right)$ induces an invertible transformation mapping p to $\bar{p}$, which is defined everywhere except, when $\gamma \neq 0$, at the singular point $p = -\delta/\gamma$. Such transformations lie at the heart of projective geometry and are of fundamental importance, not just in invariant theory, but in a wide range of classical and modern disciplines, including complex analysis and geometry, [8], number theory, [79], and hyperbolic geometry, [20]. Indeed, this rather simple construction has led to some of the most profound consequences in all of mathematics.

Exercise 1.5. Prove directly that the composition of two linear fractional transformations is again a linear fractional transformation, whose coefficients are obtained by multiplying the associated 2×2 coefficient matrices. In particular, the inverse of a linear fractional transformation is the linear fractional transformation determined by the inverse coefficient matrix.

Exercise 1.6. Show that two coefficient matrices A and $\widetilde{A}$ determine the same linear fractional transformation (1.14) if and only if they are scalar multiples of each other: $A = \lambda \widetilde{A}$. Thus, in the complex case, any linear fractional transformation (1.14) can be implemented by a unimodular coefficient matrix: $\det A = 1$. What is the unimodular linear transformation associated with the affine transformation (1.5)? Is this result valid in the real case?

How should a general linear fractional transformation act on an inhomogeneous quadratic polynomial (1.1)? We want to maintain the transformation rules (1.12) on the coefficients, so that the action will be the inhomogeneous counterpart to the linear action (1.11) on homogeneous quadratic forms. This requires that the quadratic polynomials $Q(p)$ and $\overline{Q}(\bar{p})$ are related according to the basic formula

$$Q(p) = (\gamma p + \delta)^2 \, \overline{Q}(\bar{p}) = (\gamma p + \delta)^2 \, \overline{Q} \left(\frac{\alpha p + \beta}{\gamma p + \delta} \right). \tag{1.15}$$

Note that the additional factor $(\gamma p + \delta)^2$, known as the quadratic *multiplier*, is used to clear denominators so that the linear fractional transformation (1.14) will still map quadratic forms to quadratic forms. The reader might enjoy verifying that the transformation rules (1.15) does lead to exactly the same formulae (1.12) for the coefficients, and hence the discriminant continues to satisfy the basic invariance criterion (1.13). Note that, even though two coefficient matrices which are scalar multiples of each other determine the *same* linear fractional transformation (1.14), their action on quadratic polynomials (1.15) is *different* (unless $A = \pm \widetilde{A}$), owing to the effect of the multiplier.

Exercise 1.7. Show that the inversion $\bar{p} = 1/(p + 1)$ maps the quadratic polynomial $Q(p) = p^2 - 1$ to the linear polynomial $\overline{Q}(\bar{p}) = -2\bar{p} + 1$. Thus projective transformations do not necessarily preserve the degree of a polynomial. Given a linear fractional transformation (1.14), determine which quadratic polynomials $Q(p)$ are mapped to linear polynomials. Which are mapped to constant polynomials?

Let us return to the canonical form problem for quadratic polynomials, now rearmed with the more general projective transformations. Clearly, by suitably combining the transformations and appealing to Exercise 1.5, we can begin by placing the quadratic polynomial in canonical affine form. Consider first the complex canonical form $Q = k(p^2 + 1)$. If we scale according to the coefficient matrix $A = \lambda \mathbb{1}$, where $\mathbb{1}$ is the 2×2 identity matrix and $\lambda^2 = k$, then we can normalize $Q \mapsto p^2 + 1$. Furthermore, the transformation $\bar{p} = (p - i)/(p + i)$ will map $p^2 + 1$ to the linear polynomial $\bar{p}$. Therefore, if $\Delta \neq 0$, and so $Q(p)$ either has two distinct roots or is a nonzero linear polynomial, then there is just one canonical form, namely $Q(p) = p$. On the other hand, if we take the affine canonical form $Q = p^2$, we can apply the inversion $\bar{p} = 1/p$ to map it to the constant polynomial $\bar{Q} = 1$; further, any other constant (nonzero) polynomial can, by applying a diagonal scaling matrix, be mapped to the constant 1. Thus, for complex quadratic polynomials under general linear fractional transformations, there are only three canonical forms: the first is p, which occurs when $\Delta \neq 0$; the second is 1, which occurs when $\Delta = 0$ but Q is not identically 0; and the last is the most trivial case, namely $Q \equiv 0$.

Canonical Forms for Complex Quadratic Polynomials

I.	p	$\Delta \neq 0$	distinct roots
II.	1	$\Delta = 0,\ Q \not\equiv 0$	double root
III.	0	$Q \equiv 0$	

Thus, under projective transformations, every complex quadratic polynomial is equivalent to a linear or constant polynomial. Since the action of linear fractional transformations on inhomogeneous quadratic polynomials mirrors that of linear transformations on homogeneous quadratic forms, each of our canonical forms has a homogeneous counterpart. We conclude that, under complex linear transformations, there are also three different canonical quadratic forms: first, xy, or, alternatively, $x^2 + y^2$; second, x^2; and, third, the trivial zero form 0.

In the real case, note that the transformation rules (1.3) for the discriminant imply that its sign is invariant: if $\Delta > 0$, say, then $\bar{\Delta} > 0$ also. Of course, this just means that one cannot map real roots to complex roots by a real projective transformation. Moreover, the sign of Q itself is also invariant; one cannot map a positive definite quadratic polynomial to an indefinite or negative definite one. Consequently, there are three different canonical forms with nonvanishing discriminant. The sign of Q also affects the classification of quadratics with vanishing discriminant.

Canonical Forms for Real Quadratic Polynomials

Ia.	$p^2 + 1$	$\Delta > 0,\ Q \geq 0$	complex roots
Ib.	$-p^2 - 1$	$\Delta > 0,\ Q \leq 0$	complex roots
Ic.	p	$\Delta < 0$	distinct real roots
IIa.	1	$\Delta = 0,\ Q \geq 0$	double root
IIb.	-1	$\Delta = 0,\ Q \leq 0$	double root
III.	0	$Q \equiv 0$	

The corresponding homogeneous canonical forms are the positive definite, $x^2 + y^2$, negative definite, $-x^2 - y^2$, and indefinite, which can be taken as either xy or $x^2 - y^2$, all of which were complex-equivalent, followed by the degenerate cases x^2, $-x^2$, and 0.

Suppose we restrict to area-preserving transformations, with unimodular coefficient matrix: $\det A = 1$. In this case, the discriminant is strictly invariant, and hence we can no longer rescale to normalize it to be ± 1. Retracing the preceding arguments, we see that the only effect is to introduce an extra scaling factor into the list of canonical forms. Thus, for complex-valued quadratic polynomials under area-preserving changes of variables, the canonical forms having nonzero discriminant become a one-parameter family of linear forms kp. Note that the inversion $\bar{p} = -1/p$ will map kp to $-k\bar{p}$, both of which have discriminant $\Delta = k^2$, but otherwise one cannot transform between two different linear canonical forms. Therefore, a complete list of canonical forms for complex quadratic polynomials under unimodular linear fractional transformations consists of the linear forms kp, along with the constant forms 1 and 0. In the real case, one similarly finds two families of canonical forms, $k(p^2 + 1)$ and kp, which are distinguished by the sign of the discriminant. In the degenerate cases where $\Delta = 0$, the list of canonical forms remains the same as before.

Remark: A generic unimodular linear fractional transformation depends on three free parameters: α, β, and γ. Further, a quadratic polynomial (1.1) has three coefficients. Thus, one might expect that one could normalize all three coefficients via a suitable choice of the three parameters in the linear fractional transformation. The invariance of the discriminant proves that this naïve parameter count can be misleading. (See Chapter 8 for a more sophisticated and accurate version, which is based on the orbit dimensions.)

Exercise 1.8. Determine the canonical forms for complex-valued quadratic polynomials under the class of real linear (or linear fractional) transformations. In other words, the coefficients a, b, c in (1.7) or (1.1) are allowed to be complex, but the transformations (1.10) or (1.14) are restricted so that $\alpha, \beta, \gamma, \delta$ are all real.

This concludes our brief presentation of the admittedly elementary theory of quadratic polynomials in one complex or one real variable. Extensions to multi-dimensional quadratic forms are certainly of interest, and we shall briefly return to this topic in Chapters 3 and 10. However, our more immediate interest is in extending these basic considerations to higher degree polynomials in a single variable and/or homogeneous polynomials in two variables. In the classical literature, these are known as "binary forms". Their invariants, geometry, and canonical forms, under projective and/or linear transformations, constitute the heart and soul of the classical theory.

Chapter 2

Basic Invariant Theory for Binary Forms

Using the previous chapter as our motivational springboard, let us now dive into our chosen subject. Most of the classical literature, and indeed most of the present text, is devoted to the simplest case — that of a binary form or homogeneous polynomial in two variables, along with the inhomogeneous univariate counterpart. In this chapter, we shall introduce many of the fundamental concepts in the invariant theory of binary forms. The ideas will be illustrated by the next two most important cases — that of cubic and quartic polynomials. In each case, we shall exhibit a complete system of invariants, as well as a complete list of canonical forms. These examples will serve to motivate the general definitions of invariants and covariants. The emphasis here is on important particular examples, such as Hessians, resultants, and discriminants, and their role in the classification and geometry of binary forms. These initial constructions bring the basic problem of classifying the invariants into focus, leading to the fundamental Basis Theorem of Hilbert, whose proof will appear in Chapter 9. The chapter concludes with a brief discussion of the algebraic relationships, known as "syzygies", that exist among the fundamental invariants and covariants.

Binary Forms

In the classical literature, homogeneous polynomials are called *forms*.[†] The adjectives "binary", "ternary", and so on refer to the number of variables that the form depends on. The most important case, and the one we shall primarily concentrate on, is that of a *binary form*

$$Q(\mathbf{x}) = Q(x, y) = \sum_{i=0}^{n} \binom{n}{i} a_i \, x^i y^{n-i}, \tag{2.1}$$

[†] The classical term "form" (which replaced Cayley's older "quantic", [40]) as used here should not be confused with the modern term "differential form". In this book, all forms are symmetric (and hence polynomials) — as opposed to the anti-symmetric forms of importance in geometry and topology, [25, 168].

which is a homogeneous polynomial function of two variables $\mathbf{x} = (x, y)$.
(The binomial coefficients $\binom{n}{i}$ are introduced for later convenience.)
We shall consider both real and complex[†] forms, as the methods apply
equally well to both. The number $n \in \mathbb{N}$ is the *degree* of the form, and we
note that Q satisfies the basic homogeneity equation $Q(\lambda \mathbf{x}) = \lambda^n Q(\mathbf{x})$.

As with the correspondence between quadratic polynomials and
quadratic forms, each homogeneous polynomial (2.1) will correspond
to an inhomogeneous polynomial

$$Q(p) \equiv Q(p, 1) = \sum_{i=0}^{n} \binom{n}{i} a_i p^i, \qquad (2.2)$$

depending on a single scalar variable p. At the risk of initial confusion, we shall use the same symbol Q for both the homogeneous form
(2.1) as well as its inhomogeneous counterpart (2.2). The reader might
wish to insert some extra notation, such as $\widetilde{Q}(p)$, to distinguish the
inhomogeneous version (2.2), because two different homogeneous forms
might, ostensibly, reduce to the "same" inhomogeneous counterpart. For
example, the linear form $Q_1(x, y) = x + 2y$ has inhomogeneous version $\widetilde{Q}_1(p) = p + 2$; the quadratic form $Q_2(x, y) = xy + 2y^2$ also has
$\widetilde{Q}_2(p) = p + 2$. However, this identification is, in fact, illusory; the former is a linear (rather affine) polynomial, whereas the latter should be
regarded not as a linear polynomial, but rather as a degenerate quadratic polynomial! The distinction is, at the outset, certainly not evident,
but it will become so once the transformation rules are brought into
play. In point of fact, the use of a notation like $\widetilde{Q}$ tends, I believe, to
be *more* confusing than our agreement to use the same notation for homogeneous and inhomogeneous polynomials. In the preceding example,
then, we would have $Q_1(p) = p + 2$, and $Q_2(p) = p + 2$, but $Q_1 \neq Q_2$
because they come from different homogeneous representatives! Any
reader who is willing to persevere should soon recover from this initial
confusion.

Remark: Actually, the difficulty we are experiencing at this junc-
ture is reflective of the fact that the inhomogeneous representative of a
homogeneous function is not really a function at all, but, rather, a section of a "line bundle" over a one-dimensional base space, cf. [**25, 161**].
Thus, the fact that $Q_1 \neq Q_2$ even though they both have the same co-

---

ordinate formula, is because Q_1 is a section of the "linear line bundle", whereas Q_2 is a section of the "quadratic line bundle", and we just happen to have chosen the underlying coordinate p so that they have the same formula. But to keep the exposition reasonably elementary, I have chosen not to adopt this more advanced geometric framework.

Given an inhomogeneous polynomial $Q(p)$, we can recover its homogeneous form $Q(x, y)$ via the simple rule

$$Q(x, y) = y^n \, Q\left(\frac{x}{y}\right) , \tag{2.3}$$

provided we specify its degree n in advance. Formula (2.3) proves that there is a one-to-one correspondence $Q(x, y) \Longleftrightarrow (Q(p), n)$ between homogeneous forms and inhomogeneous polynomials once we append the latter's degree. The degree n of the form might be larger than the naïve degree of $Q(p)$, meaning the degree of its leading term. Note that the naïve degree of $Q(p)$ will be strictly less than the degree of (2.2) if and only if its leading coefficient vanishes: $a_n = 0$. In classical invariant theory, the naïve degree is more or less meaningless since it can be changed by a suitable transformation, whereas the true degree is intrinsic and extremely important.

Definition 2.1. The *degree* of a (nonzero) homogeneous form $Q(x, y)$ is the degree of any of its terms. The *degree* of an inhomogeneous polynomial $Q(p)$ is the degree of its homogeneous representative. Two inhomogeneous polynomials are considered to be equal if and only if they have the same coordinate formula *and* the same degree.

Transformation Rules

Classical invariant theory is concerned with the intrinsic geometric properties of forms, meaning those properties which do not depend on the introduction of a particular coordinate system. In the case of homogeneous forms (2.1), we are naturally led to consider the effect of invertible linear changes of variables

$$\bar{x} = \alpha x + \beta y, \qquad \bar{y} = \gamma x + \delta y, \qquad \alpha\delta - \beta\gamma \neq 0, \tag{2.4}$$

in which the coefficient matrix $A = \begin{pmatrix} \alpha & \beta \\ \gamma & \delta \end{pmatrix}$ is nonsingular, and either real or complex, depending on the type of form under consideration. Note that (2.4) defines (essentially) the most general class of transformations on a two-dimensional space which preserve the class of homogeneous polynomials of a fixed degree.[†] Under such a linear transformation, the polynomial $Q(x, y)$ is mapped to a new polynomial $\overline{Q}(\bar{x}, \bar{y})$, defined so that

$$\overline{Q}(\bar{x}, \bar{y}) = \overline{Q}(\alpha x + \beta y, \gamma x + \delta y) = Q(x, y). \tag{2.5}$$

Thus, the matrix A induces a transformation on the coefficients a_i of Q, mapping them into the corresponding coefficients $\bar{a}_i$ of $\overline{Q}$. It is not difficult to determine precise formulae for the coefficients of the transformed polynomial.

Theorem 2.2. *Let $Q(x, y)$ and $\overline{Q}(\bar{x}, \bar{y})$ be two binary forms related by the transformation rule (2.5). Then their coefficients are related by the explicit formulae*

$$a_i = \sum_{k=0}^{n} \bar{a}_k \left\{ \sum_{j=\max\{0, i+k-n\}}^{\min\{i,k\}} \binom{i}{j} \binom{n-i}{k-j} \alpha^j \beta^{k-j} \gamma^{i-j} \delta^{n+j-i-k} \right\},$$
$$i = 0, \dots, n. \tag{2.6}$$

Note that (2.6) reduces to the quadratic transformation formulae (1.12) when $n = 2$. Theorem 2.2 is a straightforward consequence of the Binomial Theorem. In essence, classical invariant theory for binary forms can be regarded as the analysis of the consequences of these specific transformation rules. However, we shall make surprisingly little use of the complicated explicit formulae (2.6); all of the major techniques can be developed without any direct reference to them.

The induced action of a linear transformation (2.4) on the projective coordinate $p = x/y$ is, as in the case of quadratic forms, governed by linear fractional transformations

$$\bar{p} = \frac{\alpha p + \beta}{\gamma p + \delta}, \qquad \alpha \delta - \beta \gamma \neq 0. \tag{2.7}$$

[†] However, specific polynomials may admit more general types of transformations which preserve their underlying form. Also, any invertible homogeneous polynomial transformation will map a homogeneous polynomial to another homogeneous polynomial, albeit of a different degree. It is outside the scope of this book to consider this more general class, which includes the Tschirnhaus transformations, [42], [58; p. 210], that are classically used to reduce higher degree polynomials to canonical form. Even the classification of invertible polynomial transformations remains rather rudimentary. For instance, the classical "Jacobian conjecture" — a polynomial transformation with constant Jacobian determinant is invertible — remains unsolved, [18].

The transformation rule for inhomogeneous polynomials is a simple consequence of the basic correspondence (2.3).

Proposition 2.3. *Let $Q(x,y)$ and $\overline{Q}(\bar{x},\bar{y})$ be homogeneous polynomials of the same degree n which are related by a linear change of variables according to (2.5). Then the associated inhomogeneous polynomials $Q(p)$ and $\overline{Q}(\bar{p})$ are related by the basic linear fractional transformation rule of degree n:*

$$Q(p) = (\gamma p + \delta)^n \, \overline{Q}(\bar{p}) = (\gamma p + \delta)^n \, \overline{Q}\left(\frac{\alpha p + \beta}{\gamma p + \delta}\right). \qquad (2.8)$$

As before, the role of the nth order *multiplier* $(\gamma p + \delta)^n$ is to clear denominators so that the linear fractional transformation (2.7) will map polynomials to polynomials of the same degree. It is easy to see that the coefficients a_i of the inhomogeneous form $Q(p)$ are subjected to the *same* transformation rules (2.6) under the linear fractional transformation rule (2.8) as those of the homogeneous representative $Q(x,y)$. Thus, the degree of an inhomogeneous polynomial is not necessarily specified by its local coordinate formula (2.2) but rather is distinguished by its behavior under linear fractional changes of coordinates.

The Geometry of Projective Space

Before proceeding further, it will help if we review elementary projective geometry, which underlies the correspondence between homogeneous forms and their inhomogeneous counterparts. Projective geometry dates back (at least) to the Renaissance, when European artists developed perspective representations of scenes. The mathematical foundations begin with the work of Desargues and Pascal, and were brought to maturity by Poncelet; see [26] for historical details. This subject was one of the mainstays of classical mathematics, while recent advances in image processing and computer vision, cf. [157], have underscored its continued relevance.

Definition 2.4. Given a vector space V, the associated *projective space* $\mathbb{P}(V)$ is defined as the set of all one-dimensional subspaces of V, i.e., the set of all lines through the origin in V.

If V is finite-dimensional, then its projective space forms a "manifold"[†] whose dimension is one less than that of V itself. The simplest

---

[†] See Chapter 8 for the precise definition.

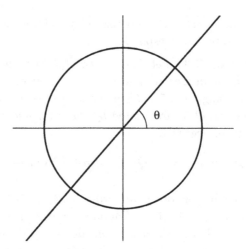

Figure 1.   The Real Projective Line.

example is the real projective line, $\mathbb{RP}^1 = \mathbb{P}(\mathbb{R}^2)$, which is the projective space associated with the real plane, i.e., the space of lines through the origin in $\mathbb{R}^2$. Each line intersects the unit circle $S^1 \subset \mathbb{R}^2$ twice — see Figure 1 — and thus we can identify $\mathbb{RP}^1$ with the circle obtained by identifying opposite (antipodal) points on S^1. (To see that this identification does, in fact, produce a circle, we note that the map $\theta \mapsto 2\theta$ from S^1 to itself will identify the antipodal points in a unique manner.) Thus, the angle $0 \le \theta < \pi$ that each line makes with the horizontal can be used to coordinatize the real projective line.

Classically, one views the Cartesian coordinates on $\mathbb{R}^2$ as defining *homogeneous coordinates* on $\mathbb{RP}^1$ and employs a square bracket to indicate this fact. Thus, a nonzero coordinate pair $0 \ne (x,y) \in \mathbb{R}^2$ defines the homogeneous coordinate $[x,y]$ of the line passing through it. Homogeneous coordinates are defined only up to scalar multiple, so that $[\lambda x, \lambda y] = [x,y]$ for any $\lambda \ne 0$. If a line is not horizontal, then it intersects the line $y = 1$ at a unique point $[p,1]$. Therefore, instead of the angle θ the line makes with the horizontal, we may adopt the horizontal component $p = \cot\theta$ of its intersection with the line $y = 1$ as our preferred coordinate; see Figure 2. If $[x,y]$ is any other homogeneous coordinate for the given line, then its canonical representative $[p,1] = [x/y,1]$ is obtained by multiplying by the scalar $\lambda = 1/y$. Thus, the open subset of $\mathbb{RP}^1$ consisting of the non-horizontal lines can be identified with $\mathbb{R}$ itself, with $p = x/y$ providing the *projective coordinate*. In

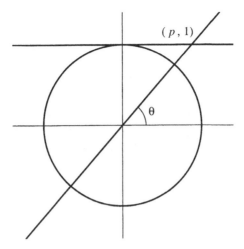

Figure 2.   The Projective Coordinate.

this way, we can regard the real projective line as the "completion" of the ordinary real line by adjoining a single point at infinity, which corresponds to the horizontal line through the origin. Alternatively, we can omit the vertical line and use the canonical coordinate $[1, q]$, where $q = y/x = \tan\theta$, to represent a different open subset of $\mathbb{RP}^1$. The change of coordinates from the non-horizontal to the non-vertical cases is the inversion $q = 1/p$.

A linear transformation (1.10) on $\mathbb{R}^2$ will induce a linear fractional transformation (2.7) on $\mathbb{RP}^1$. This is because the line with homogeneous coordinates $[p, 1]$ is mapped to the line with homogeneous coordinates $[\alpha p + \beta, \gamma p + \delta]$, whose canonical representative is (assuming $\gamma p + \delta \neq 0$) given by $[(\alpha p + \beta)/(\gamma p + \delta), 1]$. Moreover, the projective transformation (2.7) remains globally defined on $\mathbb{RP}^1$ — the point $p = -\delta/\gamma$ is mapped to the point at ∞ (indicating that the line through $(-\delta, \gamma)$ is mapped to the horizontal line), whereas the point at ∞ is mapped to the point α/γ. (If $\gamma = 0$, then the point at ∞ stays there, since such maps fix the horizontal line.) Note that the scalings $(x, y) \mapsto (\lambda x, \lambda y)$, corresponding to scalar multiples of the identity matrix, have trivial action on $\mathbb{RP}^1$ since they fix every line that passes through the origin.

Similar considerations apply to the complex projective line $\mathbb{CP}^1$, which is the projective space $\mathbb{P}(\mathbb{C}^2)$ corresponding to the two-dimensional complex linear space $\mathbb{C}^2$. One can identify $\mathbb{CP}^1$ with the usual Riemann sphere S^2 of complex analysis, [8], which can be viewed as the

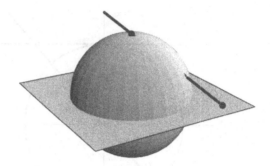

Figure 3.   Stereographic Projection.

completion of the complex line[†] $\mathbb{C}$ by adjoining a point at ∞. Explicitly, given $0 \neq (z, w) \in \mathbb{C}^2$, we define the real variables ξ, η, ζ by

$$\xi + i\eta = \frac{2\,z\,\overline{w}}{|z|^2 + |w|^2}, \qquad \zeta = \frac{|z|^2 - |w|^2}{|z|^2 + |w|^2}. \tag{2.9}$$

The reader can check that the point (ξ, η, ζ) necessarily lies on the unit sphere $S^2 \subset \mathbb{R}^3$, that is, $\xi^2 + \eta^2 + \zeta^2 = 1$. Moreover, two points (z, w) and (z', w') in $\mathbb{C}^2$ map to the same point $(\xi, \eta, \zeta) \in S^2$ if and only if they are complex scalar multiples of each other, so that $z' = \lambda z$, $w' = \lambda w$, for some $0 \neq \lambda \in \mathbb{C}$.

Exercise 2.5.   The usual method for mapping the unit sphere $\xi^2 + \eta^2 + \zeta^2 = 1$ to the x, y coordinate plane is by stereographic projection from the north pole $(0, 0, 1)$. Geometrically, the stereographic image of a point $(\xi, \eta, \zeta) \in S^2$ which is *not* the north pole is the point $(u, v, 0)$ obtained by intersecting the line connecting (ξ, η, ζ) to the north pole with the plane — see Figure 3. Show that the stereographic image of a point is defined by the formulae

$$u = \frac{\xi}{1 - \zeta}, \qquad v = \frac{\eta}{1 - \zeta}. \tag{2.10}$$

Prove further that the map from $\mathbb{C}^2 \setminus \{0\}$ to $\mathbb{C} \simeq \mathbb{R}^2$ obtained by first

---

[†] Here is a potential source of confusion. In elementary mathematics, one speaks of the "complex plane" since the set of complex numbers $\mathbb{C}$ is visualized as a two-dimensional plane (and, indeed, is a *real* two-dimensional vector space); however, as a complex vector space, $\mathbb{C}$ is one-dimensional and will therefore be referred to as the "complex line", with $\mathbb{C}^2$ being the "genuine complex plane". To minimize misunderstanding, we shall try to avoid using the term "complex plane" in this book.

mapping to the Riemann sphere via (2.9) and then applying stereographic projection (2.10) is the same as the projective coordinate map $u + iv = z/w$, for $(z, w) \in \mathbb{C}^2$ with $w \neq 0$.

All of our constructions for the real projective line have their complex counterparts, which are found just by letting all quantities assume arbitrary complex values. Indeed, we shall usually use x and y rather than z and w to denote complex coordinates on $\mathbb{C}^2$, so that the real and complex algebraic formulae are identical. The subset of complex lines not parallel to the x-axis can be given the homogeneous coordinate $[p, 1]$, where $p = x/y$ is the projective coordinate, and can thus be identified with the ordinary complex line $\mathbb{C}$ with coordinate p. Complex linear changes of variables on $\mathbb{C}^2$, as in (2.4), induce complex linear fractional transformations on $\mathbb{CP}^1$, as in (2.7).

Homogeneous Functions and Forms

The correspondence between a vector space and its associated projective space induces a correspondence between homogeneous functions and their inhomogeneous counterparts, generalizing the correspondence between quadratic forms and ordinary quadratic polynomials. Let us illustrate this correspondence in the particular case of the real projective line. Let $\pi\colon \mathbb{R}^2 \setminus \{0\} \to \mathbb{RP}^1$ denote the map that takes a point in $\mathbb{R}^2$ to the line that connects it to the origin. In terms of our projective coordinates, $p = \pi(x, y) = x/y$ for points (x, y) not on the horizontal axis. Any real-valued function[†] $F\colon \mathbb{RP}^1 \to \mathbb{R}$ on the projective line induces a function $Q\colon \mathbb{R}^2 \to \mathbb{R}$, which is given by composition: $Q = F \circ \pi$. In other words, given $F(p)$, the corresponding homogeneous function is

$$Q(x, y) = F\left(\frac{x}{y}\right). \tag{2.11}$$

Clearly, though, formula (2.11) does not reproduce the correspondence (2.3) between homogeneous and inhomogeneous polynomials. For instance, if $F(p) = p^2 + 1$, then $Q(x, y) = y^{-2}x^2 + 1$, which is not even defined on the x-axis. The key point is that (2.11) defines a function Q which is homogeneous of degree zero, $Q(\lambda \mathbf{x}) = Q(\mathbf{x})$, and hence can never (unless Q is constant) define a homogeneous polynomial on $\mathbb{R}^2$.

[†] The mapping notation $F\colon X \to Y$ does not necessarily imply that F is defined everywhere on X.

Definition 2.6. A function $Q: \mathbb{R}^2 \to \mathbb{R}$ is called *homogeneous* of *degree* n if it satisfies the basic homogeneity equation

$$Q(\lambda \mathbf{x}) = \lambda^n \, Q(\mathbf{x}), \qquad \text{for all} \ \ \mathbf{x} \in \mathbb{R}^2. \tag{2.12}$$

In particular, homogeneous polynomials of degree n are homogeneous functions, of positive integral degree n; on the other hand, not every homogeneous function, even those of positive integral degree, is a polynomial. The following simple characterization of homogeneous functions is attributed to Euler.

Theorem 2.7. *A differentiable function Q is homogeneous of degree n if and only if it satisfies Euler's formula, which is the first order partial differential equation*

$$x \, \frac{\partial Q}{\partial x} + y \, \frac{\partial Q}{\partial y} = nQ. \tag{2.13}$$

Proof: Equation (2.13) follows directly from (2.12) by differentiating with respect to λ:

$$\frac{\partial}{\partial \lambda} \, Q(\lambda x, \lambda y) = x \, \frac{\partial Q}{\partial x}(\lambda x, \lambda y) + y \, \frac{\partial Q}{\partial y}(\lambda x, \lambda y) = n\lambda^{n-1} Q(x, y). \tag{2.14}$$

Setting $\lambda = 1$ yields (2.13). Conversely, if Q satisfies (2.13), then the first equality in (2.14) implies that

$$\frac{\partial}{\partial \lambda} \, Q(\lambda x, \lambda y) = \frac{n}{\lambda} \, Q(\lambda x, \lambda y). \tag{2.15}$$

Fixing x and y, we regard (2.15) as an ordinary differential equation for the function $h(\lambda) = Q(\lambda x, \lambda y)$, namely, $dh/d\lambda = (n/\lambda) \, h$. This equation can be readily integrated; the resulting solution $h(\lambda) = \lambda^n h(1) = \lambda^n Q(x, y)$ recovers the homogeneity condition (2.12). *Q.E.D.*

Exercise 2.8. Show that the function

$$Q(x, y) = \begin{cases} x^2 \exp(x^2/y^2), & y \neq 0, \\ 0, & y = 0, \end{cases}$$

is a smooth (meaning infinitely differentiable), globally defined homogeneous function of degree 2. Are there any analytic, globally defined homogeneous functions other than homogeneous polynomials?

A simple modification of the direct formula (2.11) will allow us to construct homogeneous functions of arbitrary degree from functions on the projective space. First, we remark that the product of a homogeneous function of degree m with a homogeneous function of degree n

is also a homogeneous function of degree $m + n$. Therefore, if $Q_0(\mathbf{x})$ is a particular nonzero homogeneous function of degree n, then any other homogeneous function of degree n can be written as a product $Q(\mathbf{x}) = Q_0(\mathbf{x})R(\mathbf{x})$, where $R(\mathbf{x})$ is an arbitrary homogeneous function of degree 0. In particular, choosing $Q_0(x, y) = y^n$ allows us to conclude the general version of the correspondence (2.3).

Proposition 2.9. *Every homogeneous function $Q(x, y)$ of degree n can be written in the form $Q(x, y) = y^n F(x/y)$, where $F(p)$ is an arbitrary function on $\mathbb{RP}^1$.*

As with polynomials, the feature that distinguishes the different homogeneous representatives of a given inhomogeneous function is how they behave under changes of variables. Consequently, we cannot speak of a function $F: \mathbb{RP}^1 \to \mathbb{R}$ on projective space *in vacuo*, since (a) it does not tell us what degree its homogeneous representative should be, and (b) it does not tell us how it behaves under changes of variables. Only when we specialize to homogeneous functions of a fixed degree are the correspondences and transformation rules unambiguous.

Roots

As we saw in Chapter 1, the roots of quadratic polynomials play a critical role in their classification. We expect the geometrical configurations of the roots of more general polynomials to play a similar role in their classification and the structure of their invariants. We begin with the basic definition.

Definition 2.10. Let $Q(p)$ be a function defined on the projective line. A *root* of Q is a point p_0 where Q vanishes: $Q(p_0) = 0$.

The key remark is that the concept of a root is independent of the coordinate system used to characterize the inhomogeneous function. Indeed, referring to the basic transformation rule (2.8), we see that, provided $\gamma p_0 + \delta \neq 0$, then $Q(p_0) = 0$ if and only if $\overline{Q}(\bar{p}_0) = 0$, where $\bar{p}_0 = (\alpha p_0 + \beta)/(\gamma p_0 + \delta)$ is the transformed root. On the other hand, if $\gamma p_0 + \delta = 0$, then the transformed root is at $\bar{p}_0 = \infty$, and so the coordinate formula breaks down. Nevertheless, the root still persists, and one says that the transformed function $\overline{Q}(\bar{p})$ has a root at ∞.

Each root p_0 of the inhomogeneous representative will correspond to an entire line of solutions to the homogeneous equation $Q(\mathbf{x}) = 0$. Indeed, (2.12) implies that if $\mathbf{x}_0 = (x_0, y_0)$ is a solution, so is any nonzero scalar multiple $\lambda\mathbf{x}_0$, $\lambda \neq 0$. We will not distinguish between such solutions, since they all determine the same point in the projective space.

Definition 2.11. Let $Q(\mathbf{x})$ be a homogeneous function. By a *homogenized root* of Q we mean a line $\{\lambda\mathbf{x}_0\}$, $\mathbf{x}_0 \neq 0$, through the origin where (except possibly at the origin itself) Q vanishes.

Each homogenized root $\mathbf{x}_0 = [x_0, y_0]$ with $y_0 \neq 0$ corresponds to a root $p_0 = x_0/y_0$ of $Q(p)$. If $[x_0, 0]$ is a homogenized root, then it corresponds to the "infinite" root ∞ of the inhomogeneous form $Q(p)$. For instance, $Q(x, y) = xy + 2y^2$ has two homogenized roots: the lines through $(-2, 1)$ and $(1, 0)$; its inhomogeneous representative $Q(p) = p + 2$, which has degree 2, has roots at $p = -2$ and at $p = \infty$.

In the case of complex-valued polynomials, the Fundamental Theorem of Algebra tells us precisely how many roots there are. The key result is that every complex polynomial has at least one root.

Lemma 2.12. *Let* $Q(p)$ *be a nonconstant complex polynomial. Then there exists a point* $p_0 \in \mathbb{C}$ *such that* $Q(p_0) = 0$.

Many different proofs of this seminal result exist, and we refer the interested reader to [**71**], [**222**; Chapter 11] for details. Once we establish the existence of at least one complex root, then the polynomial admits a linear factor, and so the complete factorization of any complex polynomial follows by a straightforward induction.

Theorem 2.13. *Let*

$$Q(p) = c_m p^m + c_{m-1} p^{m-1} + \cdots + c_1 p + c_0 \qquad (2.16)$$

be a polynomial with nonzero leading coefficient, $c_m \neq 0$. *Then* Q *can be uniquely factored into a product of linear polynomials:*

$$Q(p) = c_m \prod_{\nu=1}^{m} (p - p_\nu), \qquad (2.17)$$

where $p_1, \ldots, p_m$ *are the finite complex roots of* Q.

Exercise 2.14. Prove that any real polynomial can be factored, over the reals, into a product of linear and quadratic factors. *Hint*: Use the fact that the complex roots of a real polynomial appear in complex conjugate pairs.

Definition 2.15. A root p_0 of a polynomial $Q(p)$ is said to have *multiplicity* k if we can write $Q(p) = (p - p_0)^k R(p)$, where $R(p)$ is a polynomial with $R(p_0) \neq 0$. In other words, p_0 is a root of of multiplicity

k if and only if the linear factor $p - p_0$ appears precisely k times in the factorization (2.17).

Exercise 2.16. Prove that p_0 is a root of $Q(p)$ of multiplicity k if and only if Q and its first $k - 1$ derivatives vanish there: $Q(p_0) = Q'(p_0) = \cdots = Q^{(k-1)}(p_0) = 0$, but $Q^{(k)}(p_0) \neq 0$.

So far, we have been a bit cavalier with our presentation, since we have been ignoring the "true" degree of the polynomial $Q(p)$, meaning the degree of its homogeneous representative $Q(x, y)$, in lieu of its naïve degree, as determined by the degree of its leading term. Since the naïve degree of a polynomial can change under projective transformations, we need to be a little more careful. The key remark is that not only roots, but also their multiplicities, are preserved under linear fractional transformations (2.7). Indeed, substituting (2.17) in the transformation rule (2.8), we deduce that if p_0 is a root of multiplicity k and $\gamma p_0 + \delta \neq 0$, so that p_0 is not mapped to ∞, then $\bar{p}_0 = (\alpha p_0 + \beta)/(\gamma p_0 + \delta)$ will be a root of multiplicity k also. On the other hand, if $\gamma p_0 + \delta = 0$, then p_0 will map to an infinite root $\bar{p}_0 = \infty$ of the same multiplicity as p_0, in accordance with the following definition.

Definition 2.17. Let $Q(p)$ be an inhomogeneous binary form of degree n. The point $p_0 = \infty$ is said to be a *root* with *multiplicity* k if and only if the point $\bar{p}_0 = 0$ is a root of multiplicity k for the inverted polynomial $\bar{Q}(\bar{p}) = \bar{p}^n Q(1/\bar{p})$.

Exercise 2.18. Prove that an inhomogeneous binary form (2.2) has an infinite root of multiplicity k if and only if its leading k coefficients vanish: $a_n = a_{n-1} = \cdots = a_{n-k+1} = 0$. Thus the naïve degree of an inhomogeneous polynomial is strictly less than its degree if and only if the polynomial has an infinite root.

Inclusion of the roots at infinity completes the projective version of the Fundamental Theorem of Algebra.

Theorem 2.19. *An inhomogeneous polynomial $Q(p)$ of degree n has precisely n complex roots, counting multiplicities and roots at ∞. Moreover, the linear fractional transformation (2.7) maps each root of $Q(p)$ to a root having the same multiplicity of the transformed polynomial $\bar{Q}(\bar{p})$, as given by (2.8).*

For example, if $Q(p) = p^2 - 3p$ has degree 2, then it has two roots, namely 0 and 3. Under the inversion $p = 1/\bar{p}$, the transformed polynomial is $\bar{Q}(\bar{p}) = -3\bar{p} + 1$, which has corresponding roots $\infty = \frac{1}{0}$ and $\frac{1}{3}$.

Note that any polynomial can be readily transformed to one that does not have ∞ as a root and so is genuinely of degree n.

On the homogeneous level, the factorization of an inhomogeneous polynomial (2.17) of degree n translates into a complete factorization of its homogeneous counterpart into n linear factors. We will, for later convenience, always choose the points representing the homogenized roots so that the factorization takes the *normal form*

$$Q(x,y) = \prod_{\nu=1}^{n} (y_\nu x - x_\nu y).$$ \hfill (2.18)

The normal factorization (2.18) requires that $\prod y_\nu = a_n$ be the leading coefficient of $Q(x,y)$, which can clearly be arranged by rescaling any one of the root representatives; leaving all the roots in general position gives a factorization of the same form (2.18), but with an additional nonzero coefficient in front of the product. If the horizontal line through $(1,0)$ is a root of $Q(x,y)$ having multiplicity k, so that the factor y^k appears in (2.18), then ∞ will be a root of multiplicity k of $Q(p)$, which must therefore satisfy the conditions of Exercise 2.18. Under the projective reduction (2.2) the "infinite" factors y^k all reduce to the constant 1, which accounts for the missing factors in (2.17).

Invariants and Covariants

We have now arrived at the key object of study in classical invariant theory — the concept of an invariant. Our motivational example is the discriminant (1.3) of a quadratic polynomial. The crucial property which we shall generalize is how it behaves under general linear (or, in the projective version, linear fractional) transformations. According to (1.13), the discriminant is not, strictly speaking, invariant, but rather is multiplied by a suitable power of the determinant of the matrix governing the transformation rule. This justifies the general definition.

Definition 2.20. An *invariant* of a binary form $Q(x,y)$ of degree n is a function $I(\mathbf{a}) = I(a_0, \ldots, a_n)$, depending on its coefficients $\mathbf{a} = (a_0, \ldots, a_n)$, which, up to a determinantal factor, does not change under the general linear transformation:

$$I(\mathbf{a}) = (\alpha\delta - \beta\gamma)^k\, I(\bar{\mathbf{a}}).$$ \hfill (2.19)

Here $\bar{\mathbf{a}} = (\bar{a}_0, \ldots, \bar{a}_n)$ are the coefficients of the transformed polynomial (2.5), given explicitly in (2.6). The determinantal power $k = \text{wt}\, I$ is called the *weight* of the invariant.

For example, according to (1.13), the discriminant of a binary quadratic is an invariant of weight 2. In fact, it is not difficult to prove that the discriminant is the only independent invariant of a quadratic polynomial, meaning that the other invariant is a power Δ^m of the discriminant, which has weight $k = 2m$.

While invariants are of fundamental importance in the geometry of binary forms, by themselves they do not paint the entire picture. Indeed, only when the discriminant of a quadratic form is nonzero does it completely determine its equivalence class and hence its canonical form. According to the table on p. 8, there are two possible canonical forms for quadratic polynomials with vanishing discriminant, either $Q(p) \equiv 1$ if the form is not identically zero and hence has a double root, or $Q(p) \equiv 0$. A similar situation holds for forms of higher degree, particularly for those with vanishing invariants, in which more subtle algebraic information is required than can be provided by the invariants. Classically, it was recognized that one needs to also consider functions depending not only on the coefficients of the binary form, but also on the independent variables x and y. This leads one to the more general definition of a "covariant".

Definition 2.21. A *covariant* of *weight* k of a binary form Q of degree n is a function $J(\mathbf{a}, \mathbf{x}) = J(a_0, \ldots, a_n, x, y)$ depending both on the coefficients a_i and on the independent variables $\mathbf{x} = (x, y)$ which, up to a determinantal factor, is unchanged under linear transformations:

$$J(\mathbf{a}, \mathbf{x}) = (\alpha\delta - \beta\gamma)^k \, \bar{J}(\bar{\mathbf{a}}, \bar{\mathbf{x}}). \qquad (2.20)$$

Note that invariants are just covariants that do not explicitly depend on $\mathbf{x}$. If the weight of a covariant (or invariant) J is $k = 0$, we call J an *absolute covariant*. The simplest covariant is the form Q itself, which in view of (2.5) forms an absolute covariant. For a binary quadratic, this is essentially the only covariant. More specifically, every polynomial covariant of a binary quadratic is given by a power product $J = \Delta^m Q^l$ depending on the form and its discriminant; the weight of J is $2m$. All of the important invariants and covariants are polynomial functions of the coefficients and the variables x, y, from which more general combinations, such as rational covariants, can be readily constructed.

The Simplest Examples

Let us now discuss some particular examples which will serve to illustrate and motivate the general features of the theory. Since we have already exhausted the study of the binary quadratic, we now turn to cubic and quartic polynomials.

Example 2.22. Consider a binary cubic form

$$Q(\mathbf{x}) = a_3 x^3 + 3a_2 x^2 y + 3a_1 x y^2 + a_0 y^3. \tag{2.21}$$

It turns out that there is just one fundamental invariant

$$\Delta = a_0^2 a_3^2 - 6a_0 a_1 a_2 a_3 + 4a_0 a_2^3 - 3a_1^2 a_2^2 + 4a_1^3 a_3, \tag{2.22}$$

called the *discriminant* of the cubic Q. The direct proof that Δ is an invariant of weight 6 is a lengthy computation, but this will follow directly from the general theory presented below. The vanishing of Δ has an immediate geometric interpretation: $\Delta = 0$ if and only if Q has a double or a triple root; see Theorem 2.39.

The most important covariant of a cubic or, indeed, of any binary form is its *Hessian*

$$H = Q_{xx} Q_{yy} - Q_{xy}^2. \tag{2.23}$$

Here, and below, we shall often use subscripts to denote partial derivatives, so that

$$Q_x = \frac{\partial Q}{\partial x}, \qquad Q_y = \frac{\partial Q}{\partial y}, \qquad Q_{xx} = \frac{\partial^2 Q}{\partial x^2}, \qquad Q_{xy} = \frac{\partial^2 Q}{\partial x \partial y},$$

and so on. If Q has degree n, its Hessian will be a polynomial of degree $2n - 4$, whose coefficients depend quadratically on the coefficients of Q itself. Moreover, the Hessian forms a covariant of weight 2. The covariance of the Hessian will be shown in Chapter 5, although the interested reader might wish to try to prove this directly here. Note that the Hessian of a quadratic form is just 4 times its discriminant. If Q is a cubic, then its Hessian is a quadratic polynomial and is given explicitly by

$$\tfrac{1}{36} H = (a_1 a_3 - a_2^2) x^2 + (a_0 a_3 - a_1 a_2) xy + (a_0 a_2 - a_1^2) y^2. \tag{2.24}$$

The geometrical significance of the Hessian is contained in the following basic result; the general proof can be found on p. 91.

Proposition 2.23. *A binary form $Q(x, y)$ has vanishing Hessian, $H \equiv 0$, if and only if $Q(x, y) = (cx + dy)^n$ is the nth power of a linear form.*

In particular, the Hessian of a cubic is identically zero if and only if the cubic has a triple root. (The reader is invited to prove this directly from the explicit formula (2.24).) Since the quadratic Hessian (2.24) is a covariant, its discriminant, which is

$$\widetilde{\Delta} = \tfrac{1}{4}(a_0 a_3 - a_1 a_2)^2 - (a_1 a_3 - a_2^2)(a_0 a_2 - a_1^2), \qquad (2.25)$$

is also an invariant for the cubic. (This is a special case of the general technique of composing covariants to be discussed shortly.) Expanding and comparing with (2.22), we see that this invariant is just a multiple of the original cubic discriminant: $\widetilde{\Delta} = \tfrac{1}{4}\Delta$. Consequently, a cubic has a multiple root if and only if its Hessian has a multiple root.

If K, L are any two covariants of a binary form Q, their *Jacobian*

$$J = \frac{\partial(K, L)}{\partial(x, y)} = K_x L_y - K_y L_x \qquad (2.26)$$

is also a covariant. This result can be proved directly, but will again follow from more general considerations to be discussed in Chapter 5. If K has degree m and weight k, and L has degree l and weight j, then their Jacobian J has degree $m + l - 2$ and weight $k + j + 1$. For a binary cubic, it turns out that, besides the form Q and its Hessian, H, there is only one other independent covariant — the Jacobian of Q and H:

$$\begin{aligned}
T &= Q_x H_y - Q_y H_x \\
&= -Q_y Q_{yy} Q_{xxx} + (2Q_y Q_{xy} + Q_x Q_{yy}) Q_{xxy} - \\
&\quad - (Q_y Q_{xx} + 2Q_x Q_{xy}) Q_{xyy} + Q_x Q_{xx} Q_{yyy}.
\end{aligned} \qquad (2.27)$$

If Q is a binary cubic, then T is also a cubic polynomial whose coefficients have degree 3 in the coefficients of Q and forms a covariant of weight 3. A classical result, which we shall prove in Chapter 7, states that every polynomial invariant or covariant of a binary cubic can be written in terms of the covariants Q, H, T, and the invariant Δ.

Remark: Both the Hessian (2.23) and the Jacobian covariant (2.27) are homogeneous *differential polynomials* of the function Q, meaning that they can be expressed as polynomials in Q and its derivatives. In fact, *every* polynomial covariant and invariant of a binary form can be written as a homogeneous, constant coefficient differential polynomial. The First Fundamental Theorem 6.14 of classical invariant theory provides the explicit mechanism for constructing the differential polynomials that give rise to classical covariants and invariants.

Cubics can also be completely characterized by their covariants. Suppose first that the cubic has three distinct roots and so is characterized by the invariant condition $\Delta \neq 0$. In the complex case, we can place them anywhere we like in $\mathbb{CP}^1$ by a suitable linear fractional transformation, e.g., $-1, 1$, and ∞, resulting in the canonical form $p^2 - 1$; see Example 4.30. The cubic has a double root if and only if its discriminant vanishes, but its Hessian is not identically zero; placing the double root at ∞ and the simple root at 0 leads to the canonical form p. A (nonzero) cubic has a single triple root if and only if its Hessian vanishes; the canonical form can be taken either to be p^3, by placing the root at 0, or to be 1, with the root sent to ∞. The complete list of complex canonical forms is given by the following table.

Canonical Forms for Complex Binary Cubics

I.	$p^2 - 1$	$\Delta \neq 0$	simple roots
II.	p	$\Delta = 0, H \not\equiv 0$	double root
III.	1	$H \equiv 0, Q \not\equiv 0$	triple root
IV.	0	$Q \equiv 0$	

In the real case, the first canonical form splits into two real forms, distinguished by the sign of its discriminant, depending on whether the cubic has any complex roots. If so, they can be placed at $\pm i$ and ∞ by a real linear fractional transformation. The remaining cases are unchanged. In this manner, we complete the classification of real canonical forms.

Canonical Forms for Real Binary Cubics

Ia.	$p^2 + 1$	$\Delta > 0$	two complex roots
Ib.	$p^2 - 1$	$\Delta < 0$	three simple real roots
II.	p	$\Delta = 0, H \not\equiv 0$	double root
III.	1	$H \equiv 0, Q \not\equiv 0$	triple root
IV.	0	$Q \equiv 0$	

Example 2.24. Consider next the binary quartic

$$Q(\mathbf{x}) = a_4 x^4 + 4a_3 x^3 y + 6a_2 x^2 y^2 + 4a_1 x y^3 + a_0 y^4. \tag{2.28}$$

There are two fundamental invariants:

$$i = a_0 a_4 - 4a_1 a_3 + 3a_2^2, \qquad (2.29)$$

which is of weight 4, and

$$j = \det \begin{vmatrix} a_4 & a_3 & a_2 \\ a_3 & a_2 & a_1 \\ a_2 & a_1 & a_0 \end{vmatrix}, \qquad (2.30)$$

which is of weight 6. (Again, these remarks can be verified directly, but will follow more simply from the subsequent general theory.) The vanishing of the invariants i and/or j has geometric meaning: $i = j = 0$ if and only if Q has a triple or a quadruple root. Furthermore, if $i = 0$, $j \neq 0$, the roots form an "equi-anharmonic quadruplet", whereas $j = 0$ if and only if Q can be written as the sum of two fourth powers, $Q = (ap+b)^4 + (cp+d)^4$, and the roots form an "anharmonic quadruplet"; see Gurevich, [97; Exercise 25.7], for definitions and details. Note further that since i has weight 4 and j has weight 6, both i^3 and j^2 are relative invariants of weight 12. Therefore the ratio i^3/j^2 is an absolute invariant, and its value is fixed. Any linear combination of i^3 and j^2 is again a relative invariant of weight 12. The most important of these is the *discriminant* $\Delta = i^3 - 27j^2$, which vanishes if and only if the quartic has a multiple root; see below.

If Q is a quartic polynomial, then its Hessian (2.23) is also a quartic,

$$\tfrac{1}{144}H = (a_2 a_4 - a_3^2)x^4 + 2(a_1 a_4 - a_2 a_3)x^3 y + \qquad (2.31)$$
$$+ (a_0 a_4 + 2a_1 a_3 - 3a_2^2)x^2 y^2 + 2(a_0 a_3 - a_1 a_2)xy^3 + (a_0 a_2 - a_1^2)y^4,$$

and is a covariant of weight 2. By Proposition 2.23, $H \equiv 0$ if and only if Q has a single quadruple root. As with the cubic, the only other

---

Canonical Forms for Complex Binary Quartics

I.	$p^4 + \mu p^2 + 1$	$\mu \neq \pm 2,\ \Delta \neq 0$	simple roots
II.	$p^2 + 1$	$\Delta = 0,\ T \not\equiv 0$	one double root
III.	p^2	$\Delta = 0,\ T \equiv 0,\ i \neq 0$	two double roots
IV.	p	$i = j = 0,\ H \not\equiv 0$	triple root
V.	1	$H \equiv 0,\ Q \neq 0$	quadruple root
VI.	0	$Q \equiv 0$	

independent covariant is the Jacobian $T = Q_x H_y - Q_y H_x$ of Q and H. As we shall prove in Chapter 7, every polynomial invariant or covariant of a binary quartic can be written in terms of the invariants i, j and the covariants Q, H, T. One can now use the invariants and covariants to provide a complete classification of binary quartics.

Exercise 2.25. Determine the real classification of binary quartics; see also [**97**; Exercises 25.13, 25.14].

Degree, Order, and Weight

Since the linear transformations (2.4) induce linear maps on the coefficients of a binary form, if J is any polynomial covariant, its homogeneous summands are individually polynomial covariants. Therefore we can, without loss of generality, restrict our attention to homogeneous covariants. We shall now make this requirement more precise and look at some elementary consequences.

Definition 2.26. Let $J(\mathbf{a}, \mathbf{x})$ be a homogeneous polynomial covariant of a binary form. The *degree* of J is its degree in the independent variables $\mathbf{x}$. The *order* of J is its degree in the coefficients $\mathbf{a}$ of the form.

So far we have been considering the case of a binary form that has weight zero, meaning that there is no extra determinantal factor in its transformation rules (2.5). More generally, we can assign a nonzero weighting to the original binary form.

Definition 2.27. A binary form $Q(\mathbf{x})$ is said to have *weight* m if, under the action of GL(2), its coefficients are subject to the transformation rules induced by the change of variables formula

$$\begin{aligned} Q(x, y) &= (\alpha\delta - \beta\gamma)^m \, \overline{Q}(\alpha x + \beta y, \gamma x + \delta y) \\ &= (\alpha\delta - \beta\gamma)^m \, \overline{Q}(\bar{x}, \bar{y}). \end{aligned} \tag{2.32}$$

Since reweighting a binary form only introduces an additional determinantal factor, all the homogeneous invariants and covariants of a weight 0 binary form remain invariants and covariants of a weight m form, albeit with a suitably modified weighting.

Proposition 2.28. *If $J(\mathbf{a}, \mathbf{x})$ is a homogeneous covariant of weight k and order j for a binary form Q of weight 0, then J will be a covariant of weight $k + jm$ and order j when Q has weight m.*

In particular, the Hessian $H = Q_{xx}Q_{yy} - Q_{xy}^2$ of a weight m form will have the modified weight $2 + 2m$. An interesting example, which turns out to be important for the study of differential operators, cf. [**169, 235**], is the case of a quartic polynomial of weight -2, so that its transformation rule includes the reciprocal of the square of the determinant. In this case, the invariants i and j both have weight 0, i.e., they are absolute invariants.

The degree, order, and weight of a covariant are intimately related, which implies that any two of these uniquely determine the third.

Proposition 2.29. *Let $J(\mathbf{a}, \mathbf{x})$ be a homogeneous polynomial covariant of a binary form $Q(\mathbf{x})$. Then*

$$\deg J + 2 \operatorname{wt} J = (\deg Q + 2 \operatorname{wt} Q) \operatorname{ord} J. \tag{2.33}$$

Proof: Let $n = \deg Q$, $m = \operatorname{wt} Q$, $j = \operatorname{ord} J$, $k = \operatorname{wt} J$, $i = \deg J$. Then, by homogeneity, $J(\mu\mathbf{a}, \nu\mathbf{x}) = \mu^j \nu^i J(\mathbf{a}, \mathbf{x})$. On the other hand, consider the effect of a scaling transformation $\overline{\mathbf{x}} = \lambda\mathbf{x}$, which has determinant $\det A = \lambda^2$. According to (2.32), the coefficients of the transformed polynomial $\overline{Q}$ are given by $\overline{\mathbf{a}} = \lambda^{-n-2m}\mathbf{a}$. The covariance of J implies that

$$J(\mathbf{a}, \mathbf{x}) = \lambda^{2k} J(\overline{\mathbf{a}}, \overline{\mathbf{x}}) = \lambda^{2k} J(\lambda^{-n-2m}\mathbf{a}, \lambda\mathbf{x}) = \lambda^{i+2k-(n+2m)j} J(\mathbf{a}, \mathbf{x}).$$

Consequently, the final exponent of λ in this equation must vanish, which suffices to prove (2.33). *Q.E.D.*

Exercise 2.30. Prove that a binary form of even degree has no nonzero polynomial covariants of odd degree. Prove that every nonzero polynomial covariant of a binary form of odd degree is either of even order and even degree or of odd order and odd degree.

Exercise 2.31. Let Q be a binary form of degree n and weight 0, with coefficients $\mathbf{a} = (a_0, \ldots, a_n)$. Suppose $I = \sum c_M \mathbf{a}^M$ is an invariant of weight k and order j with constituent monomials $\mathbf{a}^M = (a_0)^{m_0}(a_1)^{m_1}\cdots(a_n)^{m_n}$. Prove that

$$j = m_0 + m_1 + m_2 + \cdots + m_n,$$
$$k = \tfrac{1}{2}nj = m_1 + 2m_2 + 3m_3 + \cdots + nm_n \tag{2.34}$$
$$= nm_0 + (n-1)m_1 + (n-2)m_2 + \cdots + m_{n-1}.$$

Next, write $n = 2l$ or $n = 2l + 1$ depending on whether Q is of even or odd degree. Divide the coefficients into two subsets $\mathbf{a}_- = (a_0, \ldots, a_l)$ and $\mathbf{a}_+ = (a_{n-l}, \ldots, a_n)$. (Note that a_l appears in both subsets when

$n = 2l$ is even.) Prove that every term a^M in the invariant I must contain at least one factor from $\mathbf{a}_-$ and at least one factor from $\mathbf{a}_+$. In other words, no term in an invariant can depend solely on the coefficients $\mathbf{a}_-$ or solely on the coefficients $\mathbf{a}_+$. Is a similar result true for covariants?

Construction of Covariants

A wide variety of useful techniques for constructing covariants of binary forms have been proposed, including algebraic methods, symbolic methods, methods using differential polynomials and/or differential invariants, infinitesimal methods, methods based on the roots of the polynomials, and representation-theoretic methods. We begin by looking at the simplest algebraic methods that can be used to construct covariants.

The most trivial method is to multiply covariants. If J is a covariant of weight j and K has weight k, then the product $J \cdot K$ is a covariant of weight $j + k$. Therefore we can take general products of (powers of) covariants to straightforwardly construct other covariants, trivially related to the original covariants. However, these are typically not of great interest as they provide essentially the same information as their constituents. It is also possible to add covariants, but *only if they have the same weight*. Thus, $J+K$ will be a covariant if and only if both J and K have equal weight j, in which case their sum (or any other constant coefficient linear combination thereof) also has weight j. For example, if we begin with a binary quartic, having the standard (classical) weight 0, then its invariants i and j, cf. (2.29), (2.30), have respective weights 4 and 6, so $i+j$ is *not* an invariant since its components are multiplied by different determinantal powers. (However, its value is invariant if we only allow unimodular linear transformations.) The powers i^3 and j^2 have weight 12, and so any linear combination, including the discriminant $\Delta = i^3 - 27j^2$, is also an invariant of weight 12. On the other hand, if we give the original quartic weight -2, then, as remarked earlier, both i and j have weight 0, and so the sum $i+j$ is also an absolute invariant for this special weighting; indeed, so is any function $F(i,j)$.

A second method for constructing covariants, alluded to in our discussion of the binary cubic, is the method of composition. If $Q(\mathbf{x})$ is a binary form with coefficients $\mathbf{a}$, any polynomial covariant $J(\mathbf{a},\mathbf{x})$ can itself be considered as a binary form, whose weight is the weight of J. Let $\mathbf{b} = \varphi(\mathbf{a})$ denote the coefficients of J, which are certain polynomials in the coefficients $\mathbf{a}$ of Q. It is not hard to see that if $K(\mathbf{b},\mathbf{x})$

is a covariant depending on the coefficients of J, then the polynomial $\widetilde{K}(\mathbf{a}, \mathbf{x}) = K(\varphi(\mathbf{a}), \mathbf{x})$ obtained by replacing the coefficients of J by their formulae in terms of the coefficients of Q provides a covariant of the original form.

For example, if Q is a quartic polynomial (2.28), then its Hessian $H = H(Q)$ is itself a quartic polynomial, cf. (2.23). Thus, the i and j invariants of H, denoted $i \circ H = i(H(Q))$ and $j \circ H = j(H(Q))$, will in turn yield new invariants of Q. To compute these, we replace the coefficient a_i of $x^i y^{4-i}$ in (2.29) and (2.30) by the corresponding coefficients b_i of $x^i y^{4-i}$ in H itself, so, for instance, a_0 is replaced by $144(a_0 a_2 - a_1^2)$, and so on. However, if we already know that i and j are the only independent invariants of Q, it will not be surprising that we discover that these new invariants can be re-expressed in terms of i and j. For instance, $i \circ H = 1728 \, i^2$. In Chapter 6 we shall discover more efficient methods for determining such identities.

Exercise 2.32. Determine the general rule for the behavior of weights under composition of covariants.

Joint Covariants and Polarization

More generally, if we are given a system $Q_1(\mathbf{x}), \ldots, Q_l(\mathbf{x})$ of homogeneous polynomials, their common or correlated geometrical properties will be classified by their *joint invariants* and *covariants*. By definition, these are functions $J(\mathbf{a}^1, \ldots, \mathbf{a}^l, \mathbf{x})$ depending on all the coefficients $\mathbf{a}^\kappa = (\ldots a_i^\kappa \ldots)$ of the Q_κ, and, in the case of covariants, the variables $\mathbf{x} = (x, y)$, which, when all the forms are simultaneously subjected to a linear transformation, satisfy the same basic transformation rule (2.20). The determinantal power k is, as before, the weight of the joint covariant. The forms themselves may be of varying weights, the most common case occurring when they all have weight 0. We shall say that the joint covariant has *order* $i = (i_1, \ldots, i_l)$ if it is a homogeneous function of degree i_κ in the coefficients $\mathbf{a}^\kappa$ of Q_κ. The most important joint covariants typically arise as differential polynomials $J = J[Q_1, \ldots, Q_l]$ depending on the forms and their derivatives.

For example, if $Q(x, y) = ax + by$, $R(x, y) = cx + dy$ are linear forms of weight 0, their determinant $ad - bc$ is a bilinear, i.e., order $(1, 1)$, joint invariant of weight 1. This is a special case of the general Jacobian covariant $J = Q_x R_y - Q_y R_x$ already considered in (2.26). Another important example is the bilinear (or polarized) version of the Hessian,

$$H[Q, R] = Q_{xx}R_{yy} - 2Q_{xy}R_{xy} + Q_{yy}R_{xx}; \qquad (2.35)$$

the Hessian itself, (2.23), is recovered by setting $Q = R$. If Q and R have weight 0, then $H[Q, R]$ has weight 2. As in the case of a single form, if $J = J[Q_1, \ldots, Q_l]$ is joint covariant of order $(i_1, \ldots, i_l)$ and weight k when each Q_α has weight 0, then J remains a joint covariant of weight $k + \sum i_\kappa m_\kappa$ when Q_κ has revised weight m_κ.

The connection between the Hessian and its polarized counterpart is a special case of a general procedure for relating joint covariants and ordinary covariants, first noted in Boole's original paper, [24]. In the simplest version, suppose $K(\mathbf{a}, \mathbf{x})$ is any polynomial depending on the coefficients $\mathbf{a} = (a_0, \ldots, a_n)$ of the degree n binary form Q. Define its *polarization* to be the joint polynomial

$$J(\mathbf{a}, \mathbf{b}, \mathbf{x}) = \sum_{i=0}^{n} b_i \frac{\partial K}{\partial a_i}(\mathbf{a}, \mathbf{x}), \qquad (2.36)$$

depending on the respective coefficients $\mathbf{a}$, $\mathbf{b}$ of two binary forms Q, R of *the same degree*. If $K = K[Q]$ is a differential polynomial in Q, then its polarization $J[Q, R]$ is obtained by formally applying the differentiation process $R\, \partial/\partial Q$ to J. The formal differential operator $\partial/\partial Q$ does not affect the x, y coordinates, or derivatives with respect to them. For example, if $K = QQ_yQ_{xxy}$, then

$$J = R\frac{\partial K}{\partial Q} = RQ_yQ_{xxy} + QR_yQ_{xxy} + QQ_yR_{xxy},$$

while the polarization of the Hessian (2.23) is precisely (2.35).

Given a joint function $J[Q, R]$ depending on two binary forms of the *same* degree, we define its *trace* to be the function $K[Q] = J[Q, Q]$ obtained by setting $Q = R$. If $J[Q, R]$ is a joint covariant, then its trace is an ordinary covariant. The trace operation is, in a sense, the inverse process to polarization. If $K[Q]$ has order k, and $J[Q, R] = R\, \partial K/\partial Q$ is its polarization, then Euler's formula (2.13) implies that the trace of J recovers the original function up to a multiple: $J[Q, Q] = k\, K[Q]$. For example, setting $Q = R$ in (2.35) gives twice the Hessian covariant. On the other hand, the trace of the Jacobian joint covariant is trivial, and so one cannot obtain it by polarizing an ordinary covariant.

Proposition 2.33. *If $K[Q]$ is a covariant of weight k and order l for the single binary form Q of degree n and weight m, then its polarization $J[Q, R] = R\, \partial K/\partial Q$ is a joint covariant of weight k and order*

$(l-1,1)$ for the pair of degree n, weight m forms Q, R. Conversely, if $J[Q, R]$ is a joint covariant of weight k and order (i, j) for two forms Q, R, of the same degree and weight, then its trace $K[Q] = J[Q, Q]$ is a weight k and order $i + j$ covariant for the single form Q.

Both polarization and trace can be readily generalized to joint covariants $J[Q_1, \ldots, Q_l]$ depending on several forms. If the Q_κ's all have the same degree and weight, then the trace $J[Q, \ldots, Q]$, which is obtained by equating all the forms, is a covariant of the single form Q of the given degree and weight. One can also take partial traces by equating only some of the forms. Conversely, if Q_α and Q_β have the same degree, then the general *polarization process*

$$\tilde{J}[Q_1, \ldots, Q_l] = Q_\beta \frac{\partial J}{\partial Q_\alpha} [Q_1, \ldots, Q_l] \qquad (2.37)$$

defines another covariant, whose order in Q_β has increased by one, and whose order in Q_α has decreased by one. One can iterate this procedure to provide joint covariants depending on more and more forms (all of the same degree). An important problem then is to find a minimal system of joint covariants, from which all others can be constructed by polarization and algebraic operations. See Weyl, [**231**; p. 251], and Chapter 8 for further results in this direction.

Exercise 2.34. Find the general formula, analogous to (2.33), for the weight of a joint covariant.

Resultants and Discriminants

A particularly important joint invariant of two polynomials is their resultant, which indicates the existence of common roots. Let

$$P(\mathbf{x}) = \tilde{a}_m x^m + \tilde{a}_{m-1} x^{m-1} y + \cdots + \tilde{a}_0 y^m,$$
$$Q(\mathbf{x}) = \tilde{b}_n x^n + \tilde{b}_{n-1} x^{n-1} y + \cdots + \tilde{b}_0 y^n, \qquad (2.38)$$

be homogeneous polynomials of respective degrees m and n. (The formulae are a bit easier to read if we omit our usual binomial coefficients.) If P and Q have a common nonconstant factor F, then we can write $P = F \cdot R$, $Q = F \cdot S$, and hence

$$S(\mathbf{x}) P(\mathbf{x}) = R(\mathbf{x}) Q(\mathbf{x}), \quad \text{where} \quad \begin{matrix} \deg R < \deg P, \\ \deg S < \deg Q, \end{matrix} \quad R, S \not\equiv 0. \quad (2.39)$$

The unique prime factorization of complex polynomials implies that
(2.39) is necessary and sufficient for the existence of a common factor. In fact, by multiplying both sides by a common nonzero factor, we
can assume that the degrees of R and S are precisely one less than those
of P and Q, and so

$$
\begin{aligned}
R(\mathbf{x}) &= \tilde{c}_{m-1}x^{m-1} + \tilde{c}_{m-2}x^{m-2}y + \cdots + \tilde{c}_0 y^{m-1}, \\
S(\mathbf{x}) &= \tilde{d}_{n-1}x^{n-1} + \tilde{d}_{n-2}x^{n-2}y + \cdots + \tilde{d}_0 y^{n-1}.
\end{aligned}
\tag{2.40}
$$

Substituting into (2.39) and equating the various coefficients of powers
of x, y, we deduce that the coefficients of R, S must satisfy the system
of $m + n$ linear equations

$$
\begin{aligned}
\tilde{a}_m \tilde{d}_{n-1} &= \tilde{b}_n \tilde{c}_{m-1}, \\
\tilde{a}_m \tilde{d}_{n-2} + \tilde{a}_{m-1}\tilde{d}_{n-1} &= \tilde{b}_n \tilde{c}_{m-2} + \tilde{b}_{n-1}\tilde{c}_{m-1}, \\
\tilde{a}_m \tilde{d}_{n-3} + \tilde{a}_{m-1}\tilde{d}_{n-2} + \tilde{a}_{m-2}\tilde{d}_{n-1} &= \tilde{b}_n \tilde{c}_{m-3} + \tilde{b}_{n-1}\tilde{c}_{m-2} + \tilde{b}_{n-2}\tilde{c}_{m-1}, \\
&\vdots \\
\tilde{a}_0 \tilde{d}_0 &= \tilde{b}_0 \tilde{c}_0.
\end{aligned}
\tag{2.41}
$$

If this homogeneous system of linear equations is to have a nonzero
solution, then its associated $(m + n) \times (m + n)$ coefficient matrix must
have zero determinant. The determinant[†] in question is

$$
\mathbf{R}[P,Q] = \det
\begin{vmatrix}
\tilde{a}_m & \tilde{a}_{m-1} & \cdots & & \cdots & \tilde{a}_0 & & & \\
 & \tilde{a}_m & \tilde{a}_{m-1} & \cdots & & \cdots & \tilde{a}_0 & & \\
 & & \ddots & & & & & \ddots & \\
 & & & \tilde{a}_m & \tilde{a}_{m-1} & \cdots & \cdots & & \tilde{a}_0 \\
\tilde{b}_n & \tilde{b}_{n-1} & \cdots & & \cdots & \tilde{b}_0 & & & \\
 & \tilde{b}_n & \tilde{b}_{n-1} & \cdots & & \cdots & \tilde{b}_0 & & \\
 & & \ddots & & & & & \ddots & \\
 & & & \tilde{b}_n & \tilde{b}_{n-1} & \cdots & \cdots & & \tilde{b}_0
\end{vmatrix},
\tag{2.42}
$$

in which there are n rows of $\tilde{a}$'s and m rows of $\tilde{b}$'s, and all blank spaces are
0. The resulting polynomial in the coefficients of P and Q is called their
resultant. For example, the resultant of a quadratic $P = ax^2 + 2bxy + cy^2$

[†] We have, in accordance with standard practice, [**222**; §5.8], transposed
the coefficient matrix of (2.41).

and a linear form $Q = dx + ey$ is the 3×3 determinant

$$\mathbf{R} = \det \begin{pmatrix} a & 2b & c \\ d & e & 0 \\ 0 & d & e \end{pmatrix} = ae^2 - 2bde + cd^2,$$

which vanishes if and only if $[-e, d]$ is a homogenized root of P.

Theorem 2.35. *The resultant of two polynomials vanishes if and only if they have a common nonconstant factor and hence have a common complex (possibly infinite) root.*

There is an alternative formula for the resultant in terms of the roots of the polynomials, which immediately proves its invariance under linear transformations.

Theorem 2.36. *Let P have homogenized roots $\mathbf{x}_1, \ldots, \mathbf{x}_m$ and Q homogenized roots $\widehat{\mathbf{x}}_1, \ldots, \widehat{\mathbf{x}}_n$, both of which are taken in normal factored form (2.18). Then the resultant of P and Q can be written as the product of the differences of the roots*

$$\mathbf{R}[P, Q] = \prod_{\alpha=1}^{m} \prod_{\beta=1}^{n} (x_\alpha \widehat{y}_\beta - y_\alpha \widehat{x}_\beta) = \prod_{\alpha=1}^{m} Q(\mathbf{x}_\alpha) = (-1)^{mn} \prod_{\beta=1}^{n} P(\widehat{\mathbf{x}}_\beta). \tag{2.43}$$

Proof: Let us regard $R = \mathbf{R}[P, Q]$ as a polynomial function of the roots $\mathbf{x}_\alpha$, $\widehat{\mathbf{x}}_\beta$ of P and Q. Since $R = 0$ whenever two roots coincide, $\mathbf{x}_\alpha = \widehat{\mathbf{x}}_\beta$, it must admit the linear polynomial $x_\alpha \widehat{y}_\beta - y_\alpha \widehat{x}_\beta$ as a factor. The degree of R in the roots equals the degree of the product of all these factors, and hence R is a constant multiple of the right-hand side in (2.43). Our assumption that P and Q are in normal factored form can be used to show that the constant must be 1. *Q.E.D.*

Corollary 2.37. *If P, Q have respective degrees m, n and weights j, k, then the resultant $\mathbf{R}[P, Q]$ is a joint invariant of weight $mn + mk + nj$.*

Exercise 2.38. The kth *subresultant* $R_k = \mathbf{R}_k[P, Q]$ of the polynomials P, Q is the $(m + n - 2k) \times (m + n - 2k)$ determinant obtained by deleting the first and last k rows and columns from the resultant determinant (2.42). Prove that P and Q have precisely k roots in common (counting multiplicities) if and only if their first k subresultants $R_0 = R, R_1, \ldots, R_{k-1}$ vanish, while $R_k \neq 0$; see also [**23**; p. 197].

The *discriminant* of a binary form $Q(\mathbf{x})$ of degree n is, up to a factor, just the resultant of Q and its derivative, namely,

$$\Delta[Q] = \frac{\mathbf{R}[Q, Q_x]}{n^n \, a_n} = \frac{\mathbf{R}[Q, Q_y]}{n^n \, a_0}. \qquad (2.44)$$

Note that we can identify $Q_x = \partial Q / \partial x$ with the derivative $Q'(p)$ of the inhomogeneous version $Q(p) = Q(p, 1)$. Theorem 2.35 implies that the discriminant will detect the presence of common roots of $Q(p)$ and $Q'(p)$. These are precisely the multiple roots of Q.

Theorem 2.39. *Let Q be written in the normal factored form (2.18) with roots $\mathbf{x}_1, \ldots, \mathbf{x}_n$. The discriminant of Q equals the product of the squares of the differences of the roots*

$$\Delta[Q] = \prod_{\alpha \neq \beta} (x_\alpha y_\beta - y_\alpha x_\beta)^2. \qquad (2.45)$$

The discriminant vanishes if and only if Q has a multiple root. Moreover, if Q has degree n and weight m, then its discriminant is an invariant of weight $(n-1)(n+2m)$.

Proof: We compute the derivative of the factored form directly:

$$Q(x, y) = \prod_{\alpha=1}^{n} (xy_\alpha - yx_\alpha), \qquad \text{so} \qquad \frac{\partial Q}{\partial x} = \sum_{\beta=1}^{n} \prod_{\alpha \neq \beta} (xy_\alpha - yx_\alpha).$$

Substituting into the final expression in (2.43) produces (2.45). *Q.E.D.*

For example, the discriminant of a binary quadratic (1.7) is

$$\Delta = \frac{1}{4a_2} \det \begin{vmatrix} a_2 & 2a_1 & a_0 \\ 2a_2 & 2a_1 & 0 \\ 0 & 2a_2 & 2a_1 \end{vmatrix} = a_0 a_2 - a_1^2,$$

which is $\frac{1}{4}$ times its Hessian. Similarly, the discriminant of a binary cubic is

$$\Delta = \frac{1}{27a_3} \det \begin{vmatrix} a_3 & 3a_2 & 3a_1 & a_0 & 0 \\ 0 & a_3 & 3a_2 & 3a_1 & a_0 \\ 3a_3 & 6a_2 & 3a_1 & 0 & 0 \\ 0 & 3a_3 & 6a_2 & 3a_1 & 0 \\ 0 & 0 & 3a_3 & 6a_2 & 3a_1 \end{vmatrix}.$$

Expanding the determinant, we find that this agrees with the previous formula (2.22).

Exercise 2.40. Prove that the discriminant of a quartic is equal to the particular combination $i^3 - 27j^2$ of the invariants (2.29), (2.30).

The Hilbert Basis Theorem

Since appropriately homogeneous polynomial combinations of covariants are also covariants, an important algebraic problem is to find a *minimal* list of "fundamental" polynomial covariants, known as a Hilbert basis, that generate all others. Knowledge of a Hilbert basis for a given system of forms allows one to straightforwardly describe all covariants, and hence (presumably) all the intrinsic geometric properties of such forms.

Definition 2.41. Suppose $Q_1, \ldots, Q_l$ are a collection of binary forms. A finite collection of invariants $I_1, \ldots, I_m$ forms a *Hilbert basis* if every other invariant can be written as a polynomial function of the basis invariants: $I = P(I_1, \ldots, I_m)$. Similarly, a finite collection of covariants $J_1, \ldots, J_k$ forms a *Hilbert basis* if every other covariant J can be written as a polynomial in the basis covariants: $J = P(J_1, \ldots, J_k)$.

For example, a Hilbert basis for the covariants of a binary quadratic consists of the form Q itself, and its discriminant, which is the only independent invariant. A binary cubic has 4 fundamental covariants, consisting of Q, the Hessian H, the Jacobian covariant T given in (2.27), and the discriminant Δ, which is the one invariant. A quartic has two invariants, i, j, and three covariants Q, H, T. These results, as well as those for the quintic and sextic, were known to Cayley, who then stated, [41], that binary forms of degree 7 or more do not have a finite Hilbert basis for their invariants. In 1868, Gordan, [86], succeeded in proving his finiteness theorem, which meant that Cayley was mistaken — *every* binary form admits a Hilbert basis. Gordan's method of proof is constructive, and so, at least in principle, one was now able to produce complete systems of invariants and covariants for general binary forms. However, Gordan's method has only been completely carried out for binary forms of degree at most 8, cf. [92; p. 132], [226]. The number of polynomial independent covariants rapidly increases with the order of the form, and the implementation of the method in higher degrees becomes infeasible (although modern computer algebra packages might come to the rescue). For example, quintic forms have 4 invariants and 23 covariants (including the invariants) in a complete Hilbert basis; while sextics have 5 invariants and 26 covariants.

Sylvester, [208, 209], produced tables of Hilbert bases for the covariants of binary forms of degree $n \leq 10$ and $n = 12$. However, as shown by Dixmier and Lazard, [60, 61], Sylvester's entry for the form of degree 7 is not correct — he misses several invariants, and so the higher order computations are rather suspect. (Indeed, Sylvester makes several remarks about the anomalous nature of the binary septic — the case that also led Cayley astray — but does not conclude that his calculations are incorrect.) Sylvester does get the invariants correct for a form of degree 8, as was re-proved by Shioda, [**194**], but I do not know whether Sylvester's list of covariants is correct. For historical interest, Sylvester's tables, with known corrections in parentheses, are as follows. The proliferation of invariants and covariants at the higher orders is striking. However, one should trust the listed number of covariants only up to degree 6 and invariants only up to degree 8.

degree	2	3	4	5	6	7	8	9	10	12
# invariants	1	1	2	4	5	26 (30)	9	89	104	109
# covariants	2	4	5	23	26	124 (130)	69	415	475	949

Following Gordan's triumph with binary forms, the focus shifted to polynomials in three or more homogeneous variables. (See Chapter 10 for the precise definitions of invariants and covariants in the multivariate context.) Progress was slow, until the mathematical world was stunned when David Hilbert, at age 26, [**105**], suddenly and unexpectedly proved the existence of a Hilbert basis for *any* number of forms in *any* number of variables. Hilbert's celebrated theorem is the following:

Theorem 2.42. *Any finite system of homogeneous polynomials admits a Hilbert basis for its invariants, as well as for its covariants.*

Hilbert's original proof of the Finiteness Theorem was existential, thereby provoking Gordan's famous (perhaps apocryphal) exclamation "Das ist Theologie und nicht Mathematik." In response to such criticisms, Hilbert published a second, more difficult constructive proof, [**106**], although this is less well known, and Hilbert has been unjustly saddled with the reputation of killing off constructive invariant theory.[†] As recently emphasized by Sturmfels, [**204**], Hilbert's second proof, combined with the modern theory of Gröbner bases, [**28, 54**], has the potential to be formed into a constructive algorithm for producing the Hilbert

---

[†] This and other invariant-theoretic apocrypha can mostly be traced to Weyl's incomplete and at times misleading historical remarks, [**231**; p. 27].

basis of a general system of forms. However, the actual implementation has yet to be completed, and there are counterclaims, [226], that one can "effectively" complete the classification only in the known cases.

Exercise 2.43. Let Q be a binary form. According to the general composition method, if J, K are any two covariants, their resultant $\mathbf{R}[J, K]$ and discriminants $\Delta[J]$ and $\Delta[K]$ will be invariants of Q. Consider the particular case when Q is a binary cubic. Since its discriminant $\Delta = \Delta[Q]$ is the only independent invariant, all such composed resultants and discriminants must be constant multiples of suitable powers of Δ. Prove the following formulae:

$$
\begin{aligned}
\Delta[Q] &= \Delta, & \Delta[H] &= -324\,\Delta, & \Delta[T] &= 2^8\,3^{12}\,\Delta^3, \\
\mathbf{R}[Q, H] &= 6^6\,\Delta^2, & \mathbf{R}[Q, T] &= -6^9\,\Delta^3, & \mathbf{R}[H, T] &= 2^{10}3^{12}\,\Delta^3.
\end{aligned}
\tag{2.46}
$$

(A computer algebra package might come in handy.) Discuss implications for the possible joint root configurations of the cubic and its covariants.

Syzygies

While polynomial independence of the fundamental covariants appearing in a Hilbert basis has received the lion's share of interest in the algebraic approach to invariant theory, applications to geometry do not typically require such detailed, elusive information. Indeed, if one relaxes the requirement of polynomial independence to either rational, algebraic, or, most generally, functional independence, then complete results are much easier to obtain.

Although the number of independent invariants in a Hilbert basis of a binary form increases rapidly with its degree, a simple dimension count based on the orbits shows that the number of functionally independent invariants (for a generic form) cannot exceed $n - 2$, where $n \geq 3$ is the degree of the form; see Chapter 9. In Chapter 7, we will construct an explicit rational basis consisting of n rationally independent covariants, having the property that any other invariant or covariant can be written as a rational function thereof. Even better, in Chapter 8, we will find that a complete solution to the equivalence and symmetry problems for binary forms can be based on merely two absolute rational covariants, which involve only three particular polynomial covariants. This has the remarkable implication that the complete geometry of any binary form is encapsulated in these two covariants and their functional dependencies!

The reason that one does not require so many rationally or functionally independent covariants is that there exist certain polynomial identities, known as *syzygies*, among the basis covariants. For example, the four covariants of a binary cubic are related by the single syzygy

$$T^2 = 2^4 3^6 \, \Delta Q^2 - H^3. \tag{2.47}$$

Therefore, if one is willing to forgo reliance on polynomial covariants, one needs to understand only three of the cubic covariants. Similarly, the covariants of a quartic are also related by a single syzygy

$$T^2 = -\tfrac{16}{9} H^3 + 2^{10} 3^2 \, i \, Q^2 H - 2^{14} 3^4 \, j \, Q^3. \tag{2.48}$$

Methods for deriving such identities will be discussed in detail later.

One important application of the syzygy (2.47) is the following rather pretty solution to a general cubic equation $Q(p) = 0$. Suppose first that the discriminant $\Delta \neq 0$, so the cubic has three simple roots. We factor the syzygy as

$$H^3 = (108\sqrt{\Delta}\, Q - T)\,(108\sqrt{\Delta}\, Q + T). \tag{2.49}$$

The two cubic factors $108\sqrt{\Delta}\, Q \pm T$ do not have a common linear factor, since if they did, then T and Q would have a common root, and hence their resultant $\mathbf{R}[Q, T]$ would vanish. But, according to (2.46), the resultant is a multiple of Δ^3, and we assumed that $\Delta \neq 0$. Moreover, again by (2.46), the discriminant of the Hessian does not vanish, and so we factor the quadratic Hessian into *distinct* linear factors: $H = L \cdot M$. Equation (2.49) implies that (perhaps by relabeling the linear factors)

$$108\sqrt{\Delta}\, Q - T = L^3, \qquad 108\sqrt{\Delta}\, Q + T = M^3,$$

and hence Q is expressed as a sum of two cubes. This expression can be directly factored:

$$Q = \frac{L^3 + M^3}{216\sqrt{\Delta}} = \frac{(L+M)(L+\varepsilon M)(L+\varepsilon^2 M)}{216\sqrt{\Delta}}, \tag{2.50}$$

where $\varepsilon = \sqrt[3]{1}$ is a primitive cube root of unity.

On the other hand, if $\Delta = 0$, but $H \not\equiv 0$, then (2.46) implies that the discriminant of the quadratic Hessian vanishes, and so $H = L^2$ is a perfect square. Moreover, since the resultant of Q and H vanishes, Q admits L as a factor; in fact, it is not hard to see that the linear form L provides the double root of $Q = L^2 \cdot M$, and hence the solution is straightforward. The final nonzero case is when $H \equiv 0$, in which case $Q = L^3$ is a perfect cube, and immediately solvable.

Exercise 2.44. Use the quartic syzygy (2.48) to solve the general quartic equation. *Hint*: You will need to apply the preceding solution to a cubic equation; see [**107**; Lecture XXII].

A modern result, due to Hochster and Roberts, [**111**], states that the ring of covariants of a (system of) binary forms has the structure of a *Cohen–Macaulay domain*. A precise definition of this more subtle algebraic concept would be out of place here; the interested reader can consult [**204, 205**] for details. An important consequence of this result is the existence of a *Hironaka decomposition* of the ring of covariants; this means that every covariant can be written, uniquely, in the form

$$I = \Phi_0(I_1, \ldots, I_n) + \sum_{\nu=1}^{k} \Phi_\nu(I_1, \ldots, I_n) \, J_\nu, \qquad (2.51)$$

where $I_1, \ldots, I_n$ are algebraically independent covariants, and $J_1, \ldots, J_k$ are additional covariants needed to complete the Hilbert basis. In fact, for a single binary form of degree n, the number of algebraically independent covariants equals the degree of the form. However, the number of auxiliary covariants is not known except in low order cases. For example, in the case of a cubic, in view of the syzygy (2.47), any covariant can be written as $C = \Phi(Q, H, \Delta) + \Psi(Q, H, \Delta) \, T$, while for a quartic, $C = \Phi(Q, H, i, j) + \Psi(Q, H, i, j) \, T$. A similar result holds for the rings of invariants, although this only becomes nontrivial for forms of degree 5 or more.

Chapter 3

Groups and Transformations

Following our preliminary foray into the basic ideas of invariant theory, it is now time to understand, in more detail, the mathematical foundations of our subject. Of course, one could continue to focus solely on invariant theory, but the full ramifications of our investigations would remain obscure without a proper appreciation for the underlying, modern mathematical theories, most of which can trace their genesis back to the problems of classical invariant theory itself. This chapter is devoted to a brief survey of the basic theory of transformation groups, starting with the properties of groups themselves. For our purposes, the most important examples are provided by simple actions on a linear space and their projective counterparts. Although our primary focus is on certain infinite, continuous groups, the present chapter will develop the general theory, which includes finite, discrete, infinite, and topological groups. More advanced methods that rely on the additional analytic structure of Lie groups will be postponed until Chapters 8 and 9.

Basic Group Theory

The theory of groups has its origins in the classical work of Lagrange, Abel, and Galois on the solubility of polynomials. (See, for example, [237, 239], for historical surveys of group theory.) These mathematical giants discovered that the symmetries of a geometric object (in their case, the object was the set of roots to a polynomial equation) admit a certain underlying structure, which is crystallized in the definition of a "group". A half century later, Felix Klein clarified the foundational role of groups in geometry, and his justly famous *Erlanger Programm*, [128], showed how each type of geometry (Euclidean geometry, affine geometry, projective geometry, etc.) is completely characterized by an underlying transformation group. Simultaneously, motivated by the study of partial differential equations, Sophus Lie introduced and developed the theory of continuous or Lie groups, [138], which are manifested as symmetry groups of the solutions to a differential equation. Groups are ubiquitous in mathematics and have an astounding variety of applications — to physics, to mechanics, to computer vision, to biology, and elsewhere. As quoted in [**239**], Alexandroff proclaims that "... the concepts of *number*, *set*, *function* and *group* are the four cornerstones on which the entire edifice of modern mathematics rests and to which any other mathematical concept reduces".

We begin our presentation with the fundamental definition.

Definition 3.1. A *group* is a set G admitting a binary multiplication operation, denoted $g \cdot h$ for group elements $g, h \in G$, which is subject to the following axioms:

(a) *Associativity*: $g \cdot (h \cdot k) = (g \cdot h) \cdot k$ for $g, h, k \in G$.

(b) *Identity*: The group contains a distinguished identity element, denoted e, satisfying $e \cdot g = g = g \cdot e$ for all $g \in G$.

(c) *Invertibility*: Each group element g has an inverse $g^{-1} \in G$ satisfying $g \cdot g^{-1} = g^{-1} \cdot g = e$.

Example 3.2. The simplest example of a group is the set $\mathbb{R}$ of real numbers, with addition being the group operation. The identity element is 0, and the inverse of x is its negative $-x$. Both the set of nonzero real numbers $\mathbb{R}^* = \mathbb{R} \setminus \{0\}$ and the set of positive reals $\mathbb{R}^+$ form groups when the group operation is given by multiplication. The identity is the number 1, and inversion means taking reciprocals. All three groups are commutative, so $g \cdot h = h \cdot g$ for all group elements g, h. In group theory, commutative groups are called *abelian* in honor of Abel.

Example 3.3. The set of all invertible $n \times n$ real matrices forms a group, known as the (real) general linear group, and denoted $\mathrm{GL}(n, \mathbb{R})$. The group operation is matrix multiplication, and the identity element is the identity matrix; matrix inversion defines the inverse. Except in the case $n = 1$, which corresponds to the multiplicative group $\mathbb{R}^*$, the general linear group $\mathrm{GL}(n, \mathbb{R})$ forms a non-abelian group. Analogously, the complex general linear group $\mathrm{GL}(n, \mathbb{C})$ consists of all invertible $n \times n$ complex matrices. We will, at times, employ the abbreviated notation $\mathrm{GL}(n)$ to mean either the real or complex general linear group — the precise version will either be irrelevant or clear from the context.

Exercise 3.4. Let G and H be groups. Show how their Cartesian product $G \times H$ can be naturally endowed with the structure of a group.

Definition 3.5. A subset $H \subset G$ of a group G forms a *subgroup* provided the group operations on G define a group structure on H.

Example 3.6. The set of integers $\mathbb{Z} \subset \mathbb{R}$ forms a subgroup of the additive group of real numbers. According to Exercise 3.4, the vector space $\mathbb{R}^n$ forms a group with matrix addition defining the group operation. The set $\mathbb{Z}^n$ of integral vectors forms a discrete subgroup.

Exercise 3.7. Suppose G is a group. Prove that $H \subset G$ is a subgroup if and only if it is closed under the group operations, meaning that if $h, k \in H$, then $h \cdot k \in H$ and $h^{-1} \in H$.

Example 3.8. One of the most important subgroups of the general linear group $\mathrm{GL}(n, \mathbb{R})$ is the *special linear group*

$$\mathrm{SL}(n, \mathbb{R}) = \{A \mid \det A = 1\}, \tag{3.1}$$

consisting of all unimodular (unit determinant) matrices. It forms a subgroup because (*a*) the determinant of the product of two matrices equals the product of their determinants, and hence the product of two unimodular matrices is unimodular, and (*b*) the determinant of the inverse of a matrix is the reciprocal of its determinant, so that the inverse of a unimodular matrix is unimodular. In general, subgroups of $\mathrm{GL}(n)$ are known as *matrix groups*.

Exercise 3.9. Prove that the set $\mathrm{SL}(n, \mathbb{Z})$ consisting of all unimodular $n \times n$ matrices having integer entries forms a subgroup of $\mathrm{GL}(n, \mathbb{R})$. On the other hand, show that the set $\mathrm{GL}(n, \mathbb{Z})$ of *all* $n \times n$ integer matrices is not a subgroup.

Exercise 3.10. Prove that the set of all nonzero real matrices of the form $\begin{pmatrix} \alpha & -\beta \\ \beta & \alpha \end{pmatrix}$ forms a subgroup of $\mathrm{GL}(2, \mathbb{R})$. Show that the group operation coincides with that of the multiplicative group $\mathbb{C}^* = \mathbb{C} \setminus \{0\}$ consisting of all nonzero complex numbers $\alpha + i\beta$. The *circle group*

$$S^1 = \{\, e^{i\theta} = \cos\theta + i\sin\theta \,\} \subset \mathbb{C}^* \tag{3.2}$$

containing all complex numbers of unit modulus forms a subgroup of $\mathbb{C}^*$. Its counterpart in $\mathrm{GL}(2, \mathbb{R})$ is the subgroup $\mathrm{SO}(2)$ consisting of all planar rotations $\begin{pmatrix} \cos\theta & -\sin\theta \\ \sin\theta & \cos\theta \end{pmatrix}$; see Example 3.35.

Exercise 3.11. Suppose $H \subset G$ is a subgroup. Let $g \in G$ be a fixed element. Prove that the set $g \cdot H \cdot g^{-1} = \{ghg^{-1} \mid h \in H\}$ is also a subgroup, called the *conjugate subgroup* to H under g.

Definition 3.12. A subgroup $H \subset G$ is called *normal* if it equals its conjugate subgroups, so $gHg^{-1} = H$ for all $g \in G$.

Example 3.13. The set $Z = \{\lambda \mathbb{1} \,|\, 0 \neq \lambda \in \mathbb{R}\}$ consisting of all nonzero scalar multiples of the identity matrix is a normal subgroup of $\mathrm{GL}(n, \mathbb{R})$. This example is a special case of the following basic result.

Exercise 3.14. Let G be a group. The *center* $Z \subset G$ is the subset consisting of all group elements which commute with every element in G. Thus, $z \in Z$ if and only if $z \cdot g = g \cdot z$ for all $g \in G$. In particular, $Z = G$ if and only if G is abelian. Prove that the center of a group is a subgroup and, in fact, a normal subgroup. Is every normal subgroup contained in the center?

Group Homomorphisms

In the foundations of group theory, the maps that respect the group operations play a distinguished role. These are the "morphisms" of the category of groups.

Definition 3.15. A map $\rho\colon G \to H$ between groups G and H is called a *group homomorphism* if it satisfies

$$\rho(g \cdot h) = \rho(g) \cdot \rho(h), \qquad \rho(e) = e, \qquad \rho(g^{-1}) = \rho(g)^{-1}, \qquad (3.3)$$

for all $g, h \in G$.

Exercise 3.16. Prove that the image $\rho(G) \subset H$ of a group homomorphism forms a subgroup of the target group H.

A group homomorphism $\rho\colon G \to H$ is called a *group isomorphism* if it is one-to-one and onto, in which case G and H are isomorphic (meaning identical) groups. More generally, if ρ is one-to-one, then it is called a group *monomorphism*, in which case its image $\rho(G) \subset H$ is a subgroup which is isomorphic to G itself.

For example, the map $\rho\colon \mathbb{C}^* \to \mathrm{GL}(2, \mathbb{R})$ that takes a complex number $\alpha + i\beta$ to the 2×2 matrix introduced in Exercise 3.10 defines a group monomorphism, which restricts to an isomorphism $\rho\colon S^1 \xrightarrow{\sim} \mathrm{SO}(2)$ between the circle group and the group of planar rotations.

Exercise 3.17. Show that any map $\rho\colon G \to H$ which satisfies the first two properties in (3.3) automatically preserves the inverse and hence defines a group homomorphism.

Example 3.18. The map $t \mapsto e^t$ defines a group isomorphism $\rho\colon \mathbb{R} \to \mathbb{R}^+$ from the additive group of real numbers to the multiplicative group of positive real numbers. The particular maps

$$\rho_1(t) = \begin{pmatrix} 1 & t \\ 0 & 1 \end{pmatrix}, \quad \rho_2(t) = \begin{pmatrix} e^t & 0 \\ 0 & e^{-t} \end{pmatrix}, \quad \rho_3(t) = \begin{pmatrix} \cos t & -\sin t \\ \sin t & \cos t \end{pmatrix},$$

define different monomorphisms from $\mathbb{R}$ to the general linear group GL(2). All of these examples are special cases of the general matrix exponential function $\rho_J(t) = e^{tJ}$, where J is a fixed $n \times n$ matrix. The fact that $\rho_J \colon \mathbb{R} \to \text{GL}(n)$ defines a group homomorphism is an immediate consequence of the usual properties of the matrix exponential:

$$e^{(t+s)J} = e^{tJ} \cdot e^{sJ}, \qquad e^{0J} = \mathbb{1}, \qquad e^{-tJ} = \left(e^{tJ}\right)^{-1}. \qquad (3.4)$$

Thus, the image $H_J = \{\exp tJ\}$ of ρ_J forms an abelian subgroup GL(n). These "one-parameter subgroups" play an extremely important role in the theory of Lie groups and will be discussed in detail in Chapter 9.

Exercise 3.19. Prove that the *contragredient* map $\rho_c \colon A \mapsto A^{-T}$ defines a group isomorphism $\rho_c \colon \text{GL}(n) \xrightarrow{\sim} \text{GL}(n)$.

If G is a group, and $H \subset G$ a subgroup, then the *quotient space* G/H is defined as the set of all left cosets $g \cdot H = \{g \cdot h \mid h \in H\}$ for each $g \in G$. In general, G/H does not carry any natural group structure. Let $\pi \colon G \to G/H$ be the natural projection that maps a group element g to its coset $g \cdot H$.

Proposition 3.20. *If $H \subset G$ is a normal subgroup, then the quotient space G/H can be naturally endowed with the structure of a group such that the projection $\pi \colon G \to G/H$ is a group homomorphism.*

Proof: To make π into a group homomorphism, we should define the group operations on G/H so that $\pi(g) \cdot \pi(\hat{g}) = \pi(g \cdot \hat{g})$ and $\pi(g^{-1}) = \pi(g)^{-1}$ for any $g, \hat{g} \in G$. The identity element in G/H will correspond to the identity coset $\pi(e) = H$. We need to show that the product is well defined. Two group elements g and g' will map to the same coset, $\pi(g) = \pi(g')$, if and only if $g' = gh$ for some $h \in H$. Suppose $\hat{g}$ and $\hat{g}'$ also map to the same coset, $\pi(\hat{g}) = \pi(\hat{g}')$ so $\hat{g}' = \hat{g}\hat{h}$ for some $\hat{h} \in H$. Then $g'\hat{g}' = g\hat{g}(\hat{g}^{-1}h\hat{g})\hat{h}$. Now, since H is a normal subgroup, $\hat{g}^{-1}h\hat{g} = \tilde{h}$ is also an element of H. Thus, $g'\hat{g}' = gh\hat{g}\hat{h} = g\hat{g}\tilde{h}\hat{h} = (g\hat{g})\bar{h}$, where $\bar{h} = \tilde{h}\hat{h} \in H$ also. This implies that $g \cdot \hat{g}$ and $g' \cdot \hat{g}'$ lie in the same coset, proving that the induced group multiplication on G/H is well defined. The fact that the group inversion on G/H is also well defined follows by a similar computation.   *Q.E.D.*

Exercise 3.21. Show that the set $2\pi\mathbb{Z} = \{0, \pm 2\pi, \pm 4\pi, \ldots\}$ of integer multiples of 2π forms a normal subgroup of $\mathbb{R}$; moreover the quotient group $\mathbb{R}/2\pi\mathbb{Z} \simeq S^1$ is isomorphic to the circle group. In a similar vein, show that $\mathbb{Z}_n = \mathbb{Z}/n\mathbb{Z}$ forms an abelian group containing n elements. The group operation in $\mathbb{Z}_n$ is "addition mod n".

We next describe a more refined version of the result in Exercise 3.16, which provides a group-theoretic counterpart to the fundamental theorem for linear transformations between vector spaces.

Theorem 3.22. *Let $\rho\colon G \to H$ be a group homomorphism. The image $\rho(G)$ forms a subgroup of H. The kernel $K = \{k \in G \mid \rho(k) = e\}$ forms a normal subgroup of G. Moreover, the quotient group G/K is naturally isomorphic to the image $\rho(G)$ under the induced group monomorphism $\widetilde{\rho}\colon G/K \to H$.*

Proof: First, to show K is normal, for any $k \in K$, $g \in G$, we have

$$\rho(gkg^{-1}) = \rho(g)\rho(k)\rho(g^{-1}) = \rho(g) \cdot e \cdot \rho(g)^{-1} = e,$$

and hence $gkg^{-1} \in K$. We define $\widetilde{\rho}$ as in the statement of the theorem, so that $\widetilde{\rho}(\pi(g)) = \rho(g)$ for $g \in G$. This is well defined since

$$\widetilde{\rho}(\pi(g \cdot k)) = \rho(g \cdot k) = \rho(g)\rho(k) = \rho(g)e = \widetilde{\rho}(\pi(g)).$$

Finally, the fact that $\widetilde{\rho}$ is a monomorphism follows immediately from the definition of K.   *Q.E.D.*

Corollary 3.23. *A group homomorphism $\rho\colon G \to H$ forms a monomorphism if and only if $\ker \rho = \{e\}$, that is, the only element of G mapped to the identity element of H is the identity $e \in G$.*

Example 3.24. The quotient group of the general linear group $\mathrm{GL}(n, \mathbb{R})$ by its center $Z = \{\lambda \mathbb{1}\}$ is known as the *projective linear group* and denoted by $\mathrm{PSL}(n, \mathbb{R}) = \mathrm{GL}(n, \mathbb{R})/\{\lambda \mathbb{1}\}$. This group plays the underlying role in the geometry of real projective space. If n is odd, then we can identify $\mathrm{PSL}(n, \mathbb{R}) \simeq \mathrm{SL}(n, \mathbb{R})$ with the special linear group. This follows from Theorem 3.22 if we use the group homomorphism $\rho(A) = (\det A)^{-1/n} A$, which maps $\mathrm{GL}(n, \mathbb{R})$ to $\mathrm{SL}(n, \mathbb{R})$. (We are using the fact that, for n odd, each real number has a unique real n^{th} root.) For n even, this is not quite correct; we can identify $\mathrm{PSL}(n, \mathbb{R}) \simeq \mathrm{SL}(n, \mathbb{R})/\{\pm \mathbb{1}\}$, so that $\mathrm{SL}(n, \mathbb{R})$ forms a two-fold covering of the projective group.

Exercise 3.25. Motivated by (1.12) (although the reader should note that the matrix has been transposed here), prove that the map

$$\rho \begin{pmatrix} \alpha & \beta \\ \gamma & \delta \end{pmatrix} = \begin{pmatrix} \alpha^2 & 2\alpha\beta & \beta^2 \\ \alpha\gamma & \alpha\delta + \beta\gamma & \beta\delta \\ \gamma^2 & 2\gamma\delta & \delta^2 \end{pmatrix} \tag{3.5}$$

forms a group homomorphism $\rho: \mathrm{GL}(2) \to \mathrm{GL}(3)$. Determine the kernel.

Exercise 3.26. For which matrices J does the matrix exponential $\rho_J(t) = e^{tJ}$ form a group monomorphism? (*Hint:* Look at the Jordan canonical form of J.) If ρ_J is not a monomorphism, then either $J = 0$ or the image $\rho_J(\mathbb{R})$ is isomorphic to the circle group S^1.

Transformation Groups

Although the abstract theory of groups is well worth developing for its own intrinsic beauty, the real power of the concept is only revealed when the group acts on some space. Indeed, in the last century, groups *per se* did not exist in the abstract, as is now standard, but always arose concretely through their action as groups of transformations. Such "transformation groups" all arise as subgroups of the following general example.

Example 3.27. Let X be any set. Let $\mathcal{G} = \mathcal{G}(X)$ denote the set of all one-to-one maps $\varphi: X \to X$. Then $\mathcal{G}$ forms a group in which the group operation is defined by composition of maps. The identity transformation $\mathbb{1}_X$ plays the role of the group identity element, and the inverse of a transformation is defined as the usual functional inverse. The basic properties of functional composition and inversion automatically imply that $\mathcal{G}$ satisfies the group axioms in Definition 3.1.

Example 3.28. Let X be a finite set with $n = \#X$ elements. Any invertible transformation $\varphi: X \to X$ induces a permutation of the elements of X, and hence we can identify $\mathcal{G}(X)$ with the group of permutations of n objects. Thus, $\mathcal{G}(X)$ forms a finite group having $n!$ elements, known as the *symmetric group* on n objects, and denoted $\mathbb{S}^n$. For example, if $X = \{1, 2, 3\}$ has three elements, labeled by the numbers from 1 to 3, then $\mathcal{G}(X) = \mathbb{S}^3$ consists of the $6 = 3!$ permutations

$$\mathbb{S}^3 = \Big\{ (123), (132), (213), (231), (312), (321) \Big\}, \tag{3.6}$$

where (ijk) denotes the permutation that maps 1 to i, 2 to j, 3 to k. Note that $\mathbb{S}^n$ is a non-abelian group when $n \geq 3$, while $\mathbb{S}^2 \simeq \mathbb{Z}_2$.

If the space X comes equipped with additional structure, then one might impose corresponding restrictions on the class of allowable maps, leading to important subgroups of the vast group $\mathcal{G}(X)$. The main requirement is that the relevant constraints on the maps must be preserved under composition and inversion.

Example 3.29. If $X = \mathbb{R}^n$, then the allowable transformations in $\mathcal{G}(X)$ are usually required to satisfy topological or differentiability constraints. For example, the subgroup $\mathcal{C}^0(X)$ consists of all continuous, invertible maps $\varphi \colon X \to X$. Here we are using the fact that the composition of two continuous functions is continuous, as is the inverse of a continuous one-to-one map. Similarly, $\mathcal{C}^k(X)$, where $0 \leq k \leq \infty$, is defined to be the group of all invertible continuously k times differentiable maps — *diffeomorphisms*. Even more restrictively, the analytic structure of X serves to define the subgroup $\mathcal{A}(X)$ of *analytic diffeomorphisms*. This example can be readily extended to the case when X is an analytic manifold, e.g., a surface, as defined in Chapter 8.

Definition 3.30. A *transformation group* acting on a space X is defined by a group homomorphism $\rho \colon G \to \mathcal{G}(X)$ mapping a given group G to the group of invertible maps on X.

In other words, each element $g \in G$ will induce an invertible map $\rho(g) \colon X \to X$. In order that this identification define a group homomorphism, we must require that ρ satisfy the basic properties

$$\rho(g \cdot h) = \rho(g) \circ \rho(h), \qquad \rho(e) = \mathbb{1}_X, \qquad \rho(g^{-1}) = \rho(g)^{-1}, \qquad (3.7)$$

for each $g, h \in G$. According to Exercise 3.7, any set $\widetilde{G} \subset \mathcal{G}(X)$ consisting of invertible maps $\varphi \colon X \to X$ which is closed under composition $\varphi \circ \psi$ and inversion φ^{-1} will form a subgroup of $\mathcal{G}(X)$ and hence determines a transformation group on X. In all the examples that will be considered in this book, the space X carries an analytic structure, and the transformation group $\rho \colon G \to \mathcal{A}(X)$ consists of analytic diffeomorphisms.

For a fixed group action, it is common to write $g \cdot x$ for the action of the group element $g \in G$ on the point $x \in X$, instead of the more cumbersome notation $\rho(g)(x)$. Thus, conditions (3.7) become

$$g \cdot (h \cdot x) = (g \cdot h) \cdot x, \qquad e \cdot x = x, \qquad g \cdot (g^{-1} \cdot x) = x, \qquad (3.8)$$

for all $g, h \in G$, $x \in X$. In the first condition, $h \cdot x = \rho(h)(x)$ denotes the action of h on x, whereas $g \cdot h$ denotes the group multiplication, which can be identified with composition $\rho(g) \circ \rho(h)$ between the corresponding transformations.

Example 3.31. Any group acts on itself by left multiplication. In other words, given G, we set $X = G$ also and let $\lambda \colon G \to \mathcal{G}(G)$ map the group element g to the left multiplication map $\lambda(g) \colon h \mapsto g \cdot h$. Note that this action is compatible with our notational convention (3.8). A closely related action is given by right multiplication $\rho(g) \colon h \mapsto h \cdot g^{-1}$ acting on $X = G$. The inverse is required so that ρ forms a group homomorphism: $\rho(g \cdot h) = \rho(g) \cdot \rho(h)$.

Exercise 3.32. Define the notions of homomorphism and isomorphism for transformation group actions. When are the left and right actions of a group on itself isomorphic?

Example 3.33. Let $X = V$ be a real vector space. Consider the group $\mathrm{GL}(V) \subset \mathcal{G}(V)$ consisting of all invertible linear transformations $T \colon V \to V$, i.e., one-to-one maps that satisfy $T(x + y) = T(x) + T(y)$, for $x, y \in V$, and $T(\lambda x) = \lambda T(x)$ for λ a scalar. If V is finite-dimensional, then we can introduce a basis $\{e_1, \ldots, e_n\}$ so as to identify $V \simeq \mathbb{R}^n$ and thereby identify each invertible linear transformation with its $n \times n$ matrix representative. In this manner, the abstract general linear group $\mathrm{GL}(n, \mathbb{R})$ introduced in Example 3.3 is naturally realized as the group of all invertible linear transformations on an n-dimensional vector space, where composition serves to define the group operation. Note that any matrix subgroup of $\mathrm{GL}(n, \mathbb{R})$ also acts on $\mathbb{R}^n$ via linear transformations. These are the linear group actions or representations, destined to play the pivotal role in classical invariant theory.

Example 3.34. A smooth transformation $\varphi \colon \mathbb{R}^n \to \mathbb{R}^n$ is called *orientation-preserving* if its Jacobian matrix has positive determinant everywhere. The set of invertible orientation-preserving transformations forms a subgroup of $\mathcal{C}^k(\mathbb{R}^n)$ for $k \geq 1$. In particular the group of linear, orientation-preserving transformations is the subgroup $\mathrm{GL}(n, \mathbb{R})^+ = \{\det A > 0\}$ consisting of $n \times n$ matrices with positive determinant. A transformation φ is called *volume-preserving* if $\mathrm{vol}\, \varphi(S) = \mathrm{vol}\, S$ for every subset $S \subset \mathbb{R}^n$, where vol denotes the ordinary Lebesgue measure. The special linear group $\mathrm{SL}(n, \mathbb{R})$, (3.1), consists of linear transformations that preserve both volume and orientation.

Example 3.35. We denote the usual Euclidean norm on $\mathbb{R}^n$ by $\| x \| = \sqrt{(x_1)^2 + \cdots + (x_n)^2}$. A transformation $\varphi \colon \mathbb{R}^n \to \mathbb{R}^n$ is called an *isometry* if $\| \varphi(x) - \varphi(y) \| = \| x - y \|$ for all $x, y \in \mathbb{R}^n$, i.e., it preserves (Euclidean) distances. The group of linear isometries forms a subgroup $\mathrm{O}(n) \subset \mathrm{GL}(n, \mathbb{R})$ of the general linear group, known as the *orthogonal* *group*; it can be characterized as $O(n) = \{A \in GL(n, \mathbb{R}) \mid A^T A = \mathbf{1}\}$. The orthogonal group contains both rotations and reflections, distinguished by the sign of their determinant. The *rotation* or *special orthogonal group* $SO(n) = \{A \in O(n) \mid \det A = +1\}$ consists of all orientation-preserving orthogonal transformations.

Example 3.36. Slightly generalizing Example 3.33, we recall that an *affine transformation* of the linear space $\mathbb{R}^n$ is a combination of a linear transformation and a translation, and hence has the general form $x \mapsto A\,x + a$, where $A \in GL(n, \mathbb{R})$ is an invertible matrix and $a \in \mathbb{R}^n$ a fixed vector. The composition of two affine transformations is also affine, as is the inverse. Therefore the set $A(n) = A(n, \mathbb{R})$ of all affine transformations forms a group — the *affine group*— which is parametrized by the pair $(A, a) \in GL(n, \mathbb{R}) \times \mathbb{R}^n$. Although as a set $A(n)$ can be identified with the Cartesian product of the groups $GL(n, \mathbb{R})$ and $\mathbb{R}^n$, it is *not* isomorphic to the Cartesian product group $GL(n, \mathbb{R}) \times \mathbb{R}^n$ because its group multiplication law, $(A, a) \cdot (B, b) = (AB, a + Ab)$, is *not* the same as the Cartesian product group action. This forms a particular case of a general construction known as the *semi-direct product*, cf. [**169**; p. 37], and often denoted by $A(n) = GL(n, \mathbb{R}) \ltimes \mathbb{R}^n$.

Exercise 3.37. Prove that the map $\rho(A, b) = \begin{pmatrix} A & b \\ 0 & 1 \end{pmatrix}$ defines a group monomorphism $\rho \colon A(n) \to GL(n+1, \mathbb{R})$, realizing the affine group as a matrix group in one higher dimension.

Exercise 3.38. An affine transformation is called *equi-affine* if it preserves volume. Show that the set of equi-affine transformations forms a subgroup $SA(n) \subset A(n)$ of the affine group, which is defined by the unimodularity constraint $\det A = 1$. The *Euclidean group* is defined as the subgroup of $A(n)$ whose linear part is orthogonal, so

$$E(n) = \{(R, a) \mid R \in O(n), a \in \mathbb{R}^n\} = O(n) \ltimes \mathbb{R}^n \subset A(n).$$

Prove that the Euclidean group is the group of affine isometries. In fact, it can be shown that every isometry is necessarily affine, cf. [**240**; §2.3], and thus the Euclidean group is the full isometry group of Euclidean space. As such, it plays the foundational role in Euclidean geometry. The *proper Euclidean group* $SE(n) = E(n) \cap SA(n) \simeq SO(n) \ltimes \mathbb{R}^n$ consists of the orientation-preserving Euclidean transformations.

Given a transformation group action defined by a homomorphism $\rho \colon G \to \mathcal{G}(X)$, the image $\rho(G)$ will form a subgroup of the group of all invertible maps of the space X. The kernel

$$K = \{g \in G \mid \rho(g) = \mathbb{1}_X\} = \{g \mid g \cdot x = x \text{ for all } x \in X\}$$

is known as the *global isotropy subgroup*. It forms a normal subgroup of the transformation group G and consists of all group elements which act completely trivially on the space. According to Theorem 3.22, we can identify $\rho(G)$ with the quotient group G/K.

Definition 3.39. A transformation group is said to act *effectively* or *faithfully* if it has trivial global isotropy subgroup: $K = \{e\}$.

The condition means that the only group element acting as the identity transformation on X is the identity element of G. A group acts effectively if and only if the map $\rho\colon G \to \mathcal{G}(X)$ is a monomorphism, which means that different group elements have different effects: $g \cdot x = h \cdot x$ for all $x \in X$ if and only if $g = h$. Theorem 3.22 shows that any non-effective transformation group G can, without any significant loss of information or generality, be replaced by the effectively acting quotient group $\widehat{G} = G/K$.

Proposition 3.40. *Suppose G is a transformation group acting on a space X, and let K denote the global isotropy subgroup. There is a well defined effective action of the quotient group $\widehat{G} = G/K$ on X, which "coincides" with that of G in the sense that two group elements g and $\tilde{g}$ have the same effect on X, so $g \cdot x = \tilde{g} \cdot x$ for all $x \in X$, if and only if they have the same image in $\widehat{G}$, so $\tilde{g} = g \cdot k$ for some $k \in K$.*

Example 3.41. The linear fractional transformations

$$A \cdot p = \frac{\alpha p + \beta}{\gamma p + \delta}, \qquad A = \begin{pmatrix} \alpha & \beta \\ \gamma & \delta \end{pmatrix} \in \mathrm{GL}(2), \qquad (3.9)$$

define an action of the general linear group $\mathrm{GL}(2, \mathbb{R})$ on the real projective line $\mathbb{RP}^1$. The action is *not* effective since multiples of the identity matrix act trivially. The global isotropy subgroup is $K = \{\lambda \mathbb{1}\}$, and the effectively acting quotient group given in Proposition 3.40 is the projective linear group $\mathrm{PSL}(2, \mathbb{R}) = \mathrm{GL}(2, \mathbb{R})/\{\lambda \mathbb{1}\}$ defined in Example 3.24. This example can be straightforwardly generalized to real and complex projective actions in higher dimensions.

Symmetry Groups, Invariant Sets, and Orbits

As remarked earlier, groups originally arose because they effectively crystallized the intuitive notion of symmetry and as such lie at the foundations of geometry, physics, art, and human perception. In general, a transformation group is a symmetry group of an object if its action leaves the object unchanged.

Definition 3.42. Let $Y \subset X$. A *symmetry* of Y is an invertible transformation $\varphi \colon X \to X$ that leaves Y fixed, so $\varphi(Y) = Y$.

The crucial observation, dating back to Lagrange and Galois, is that the collection of all symmetries of a subset $Y \subset X$ forms a subgroup $\mathcal{S}(Y) \subset \mathcal{G}(X)$ of the group of all invertible transformations on X, known as the *symmetry group* of Y. Indeed, the identity transformation $\mathbb{1}_X \in \mathcal{S}(Y)$ is clearly always a symmetry; further, if $\varphi(Y) = Y$ and $\psi(Y) = Y$, then $\psi \circ \varphi(Y) = \psi(\varphi(Y)) = Y$, and $\varphi^{-1}(Y) = Y$.

Remark: If we relax the original assumption and require only that $\varphi(Y) \subset Y$, then the inversion property does not hold and the set of such maps generally only forms a "semi-group". For example, if $Y = (-1, 1) \subset \mathbb{R}$, then the transformation $\varphi(x) = \frac{1}{2}x$ satisfies $\varphi(Y) \subset Y$, whereas $\varphi^{-1}(x) = 2x$ clearly does not.

Generally, one does not deal with the entire symmetry group of a given subset but rather admits only those symmetries satisfying suitable constraints — for example, continuous symmetries, analytic symmetries, linear symmetries, and projective symmetries. In other words, starting with a given transformation group $\rho \colon G \to \mathcal{G}(X)$, the *symmetry subgroup* of a subset $Y \subset X$ is $G_Y = \rho^{-1}(\mathcal{S}(Y)) = \{g \in G \mid g \cdot Y = Y\}$. A transformation group G acting on X is said to be a *symmetry group* of the subset Y if *every* group element is a symmetry, so that $G_Y = G$. In this case, Y is designated a *G–invariant subset* of X.

Example 3.43. Consider the planar equilateral triangle with vertices $\Delta = \left\{ (1,0), (-\frac{1}{2}, \frac{\sqrt{3}}{2}), (-\frac{1}{2}, -\frac{\sqrt{3}}{2}) \right\} \subset \mathbb{R}^2$. A linear transformation on $\mathbb{R}^2$ defines a symmetry of Δ if and only if it has one of the following six matrix representatives:

$$\begin{pmatrix} 1 & 0 \\ 0 & 1 \end{pmatrix}, \quad \begin{pmatrix} 1 & 0 \\ 0 & -1 \end{pmatrix}, \quad \begin{pmatrix} -\frac{1}{2} & \frac{\sqrt{3}}{2} \\ \frac{\sqrt{3}}{2} & \frac{1}{2} \end{pmatrix},$$

$$\begin{pmatrix} -\frac{1}{2} & -\frac{\sqrt{3}}{2} \\ \frac{\sqrt{3}}{2} & -\frac{1}{2} \end{pmatrix}, \quad \begin{pmatrix} -\frac{1}{2} & \frac{\sqrt{3}}{2} \\ -\frac{\sqrt{3}}{2} & -\frac{1}{2} \end{pmatrix}, \quad \begin{pmatrix} -\frac{1}{2} & -\frac{\sqrt{3}}{2} \\ -\frac{\sqrt{3}}{2} & \frac{1}{2} \end{pmatrix}.$$

$$(3.10)$$

Consequently, the linear symmetry group of an equilateral triangle is isomorphic to the symmetric group $\mathbb{S}^3$. On the other hand, the isosceles triangle with vertices $\widetilde{\Delta} = \{(1,0),(0,1),(0,-1)\}$ has only two linear symmetries: the identity and the reflection $y \mapsto -y$ through the x-axis. If we enlarge our class of transformations to include affine maps, then the isosceles triangle $\widetilde{\Delta}$ admits six affine symmetries, which we represent as 3×3 matrices as in Exercise 3.37:

$$
\begin{pmatrix} 1 & 0 & 0 \\ 0 & 1 & 0 \\ 0 & 0 & 1 \end{pmatrix}, \quad
\begin{pmatrix} 1 & 0 & 0 \\ 0 & -1 & 0 \\ 0 & 0 & 1 \end{pmatrix}, \quad
\begin{pmatrix} -\frac{1}{2} & \frac{1}{2} & \frac{1}{2} \\ \frac{3}{2} & \frac{1}{2} & -\frac{1}{2} \\ 0 & 0 & 1 \end{pmatrix},
$$

$$
\begin{pmatrix} -\frac{1}{2} & -\frac{1}{2} & \frac{1}{2} \\ \frac{3}{2} & -\frac{1}{2} & -\frac{1}{2} \\ 0 & 0 & 1 \end{pmatrix}, \quad
\begin{pmatrix} -\frac{1}{2} & \frac{1}{2} & \frac{1}{2} \\ -\frac{3}{2} & \frac{1}{2} & \frac{1}{2} \\ 0 & 0 & 1 \end{pmatrix}, \quad
\begin{pmatrix} -\frac{1}{2} & -\frac{1}{2} & \frac{1}{2} \\ -\frac{1}{2} & \frac{3}{2} & -\frac{1}{2} \\ 0 & 0 & 1 \end{pmatrix}.
$$

Indeed, the affine symmetry group of *any* triangle is isomorphic to $\mathbb{S}^3$.

For the square with vertices $S = \{(1,0),(0,1),(-1,0),(0,-1)\}$, the linear and affine symmetry groups coincide and consist of eight linear isometries: the identity; rotations through $90°$, $180°$, and $270°$; reflections through the two coordinate axes; and reflections through the two lines making $45°$ angles with the axes. The reader should verify that these transformations form a non-abelian group, known as the *dihedral group* D_4. Note that there are no linear (or affine) symmetries realizing every possible permutation of the four vertices; for instance, we cannot leave two adjacent vertices fixed and interchange the other two. There are, however, nonlinear transformations that realize such permutations. For example, the projective transformation[†]

$$
(x,y) \longmapsto \left(\frac{x-y+1}{2(x+y)}, \frac{-x+y+1}{2(x+y)} \right)
$$

fixes $(1,0)$ and $(0,1)$, while interchanging $(-1,0)$ and $(0,-1)$. (Since projective transformations map lines to lines, they also preserve the edges of the square.) In fact, there are 24 projective symmetries of a square, which form a group isomorphic to $\mathbb{S}^4$, realized as a group of 3×3 matrices.

Exercise 3.44. Prove that a generic quadrilateral has no nontrivial affine symmetries but always has 24 projective symmetries. Show

---

[†] See Chapter 10 for details.

that a triangle has an infinite number of projective symmetries. What is its isotropy group, that is, the subgroup leaving all three vertices fixed? Investigate the linear, affine, and projective symmetries of pentagons.

Given a transformation group G acting on a space X, the symmetry group of a single point $x \in X$ is known as its *isotropy subgroup*: $G_x = \{g \in G \mid g \cdot x = x\}$. The transformation group acts *freely* if all isotropy subgroups are all trivial: $G_x = \{e\}$ for all $x \in X$. This is equivalent to the statement that $g \cdot x = h \cdot x$ for any one point $x \in X$ if and only if $g = h$. Free actions should be contrasted with effective actions, where $g = h$ if and only if $g \cdot x = h \cdot x$ for *all* $x \in X$. For example, the rotation group SO(3) acts effectively on three-dimensional space $X = \mathbb{R}^3$, since the only rotation which leaves every point fixed is the identity; however, it does not act freely, since any nonzero point $0 \neq x \in \mathbb{R}^3$ is fixed by the rotations around the axis formed by the line passing through x and 0. And, of course, the origin is left fixed by all rotations. Similar remarks hold for SO(n) acting on $\mathbb{R}^n$ provided $n \geq 3$.

Given a transformation group G, we would like to characterize the possible G–invariant subsets. Trivially, X itself is invariant. Technically speaking, the empty set $\varnothing \subset X$ is also G–invariant, but this is uninteresting. Note that unions and intersections of G–invariant subsets are also G–invariant. The minimal G–invariant subsets are of particular importance.

Definition 3.45. An *orbit* of a transformation group is a minimal nonempty invariant subset. In particular, a *fixed point* is a G–invariant point $x_0 \in X$, so that $g \cdot x_0 = x_0$ for all $g \in G$.

Example 3.46. As a simple example, consider the standard action

$$(x, y) \longmapsto (x \cos \theta - y \sin \theta, x \sin \theta + y \cos \theta) \qquad (3.11)$$

of the rotation group SO(2) on $\mathbb{R}^2$. Any circle $\{x^2 + y^2 = r^2\}$ centered at the origin is a rotationally invariant subset of the plane. Since the circles are minimal — they contain no nonempty rotationally invariant subset — they are the orbits of SO(2). The only fixed point is the origin. Every other invariant subset, e.g., an annulus $\{a < x^2 + y^2 < b\}$, is necessarily a union of circles.

Example 3.47. For the usual linear action of GL(n) on $\mathbb{R}^n$, there are two orbits: the origin $\{0\}$ and the remainder $\mathbb{R}^n \setminus \{0\}$. The same holds for SL($n$) since we can still map any nonzero vector in $\mathbb{R}^n$ to any other nonzero vector by a matrix of determinant 1.

Proposition 3.48. *Given a transformation group acting on a space X, the orbit $\mathcal{O}_x$ through a point $x \in X$ is just the set of all images of x under arbitrary group transformations:* $\mathcal{O}_x = \{g \cdot x \mid g \in G\}$. *A subset $S \subset M$ is G–invariant if and only if it is the union of orbits.*

Exercise 3.49. Let $x, y \in X$ be points lying in the same orbit of G. Prove that their isotropy subgroups G_x and G_y are conjugate subgroups of G. More generally, given $Y \subset X$, show that the symmetry group of $g \cdot Y = \{g \cdot y \mid y \in Y\}$ is conjugate to that of Y.

A group action is called *transitive* if there is only one orbit, so for every $x, y \in X$ there exists at least one $g \in G$ such that $g \cdot x = y$. In this case, the only G–invariant subsets are the trivial ones $\varnothing$ and X. For example, the linear actions of the groups $\mathrm{GL}(n, \mathbb{R})$, $\mathrm{SL}(n, \mathbb{R})$, and $\mathrm{SO}(n)$ on $\mathbb{R}^n$ all induce transitive projective actions on the space $\mathbb{RP}^{n-1}$.

Exercise 3.50. Is the action of the real rotation group $\mathrm{SO}(2)$ on the complex projective plane $\mathbb{CP}^1$ given by "linear fractional rotations" $p \mapsto (p \cos\theta - \sin\theta)/(p \sin\theta + \cos\theta)$ transitive? effective? free? Describe the orbits.

Exercise 3.51. Prove that if G acts freely and transitively on X, then we can naturally identify $X \simeq G$, and the action is isomorphic to the left multiplication action of Example 3.31.

Equivalence and Canonical Forms

Given a transformation group G acting on X, we shall call two points $x, y \in X$ *equivalent* if there exists a group transformation mapping one to the other, so that $y = g \cdot x$ for some $g \in G$. In other words, two points are equivalent if and only if they lie in the same orbit. The *equivalence problem* for a transformation group is to find necessary and sufficient conditions for this to hold. A trivial case is when the group acts transitively, which implies that all points are equivalent. The solution to an equivalence problem typically requires the construction of suitable invariants which serve to distinguish the orbits of the group.

In this context, a *canonical form* of an element $x \in X$ just means a distinguished, "simple" representative $x_0 \in \mathcal{O}_x$ of the orbit containing x. Thus, a complete list of canonical forms can be identified with a list of orbits of the group, since each orbit must contain one (and, in an irredundant list, only one) canonical form, which thereby serves to

distinguish the orbit. Of course, there is no uniquely specified canonical form, and some choice, usually based on one's æsthetic judgment
of "simplicity", must be exercised. Note that the determination of a
complete system of canonical forms leads to an immediate solution to
the associated equivalence problem: Two objects are equivalent if and
only if they have the same canonical form.

For example, the well known Jordan canonical form of a matrix
corresponds to the conjugation action $X \mapsto AXA^{-1}$ of the complex
general linear group $\mathrm{GL}(n, \mathbb{C})$ on the space of $n \times n$ complex matrices
X. However, other less commonly known canonical forms, including the
older "rational canonical form", cf. [**219**], can also be advantageously
utilized.

The "coordinates" of the canonical form are particular invariants,
often referred to as the *moduli* for the given transformation group,
cf. [**156, 213**]. For the conjugation action of $\mathrm{GL}(n, \mathbb{C})$ on matrices, the
moduli are the eigenvalues of the matrix and the sizes of the different
Jordan blocks. Thus, moduli can be both continuous and discrete; in
regular cases, as discussed in Chapter 8, the moduli provide a complete
system of functionally independent invariants for the group action.

Example 3.52. For the rotation group $\mathrm{SO}(n)$ acting on $\mathbb{R}^n$, the
orbits are the spheres $\{\| x \| = \text{constant}\}$ centered at the origin. Hence,
two points in $\mathbb{R}^n$ are equivalent if and only if they have the same norm:
$\| x \| = \| y \|$. Thus, we can choose the canonical form of a vector $x \in \mathbb{R}^n$
to be a non-negative multiple of the first unit basis vector, $r\, e_1$, with
$r = \| x \| \geq 0$ being the single modulus for the group action. (Clearly,
there is no particular reason to choose the first basis vector for the
canonical form.) As for the actions of $\mathrm{GL}(n)$ and $\mathrm{SL}(n)$, each vector
$x \in \mathbb{R}^n$ has just two possible canonical forms: either 0 or, say, e_1.

Example 3.53. Consider the action of the group $\mathrm{GL}(2, \mathbb{R})$ of real
invertible 2×2 matrices on the space $\mathcal{Q}$ of real quadratic forms analyzed in Chapter 1. Since each quadratic form is uniquely determined
by its three coefficients, we can identify $\mathcal{Q} \simeq \mathbb{R}^3$. The induced action of
$\mathrm{GL}(2, \mathbb{R})$ on $\mathcal{Q}$ is explicitly given in (1.12). What are the orbits? According to the classification in Chapter 1, there are four different canonical
forms for real quadratic forms, and hence there are four distinct orbits.
These are almost completely distinguished by the discriminant; the first
orbit consists of all those quadratic forms with positive discriminant;
the second consists of those with negative discriminant. There is a fixed
point consisting of the zero quadratic form; the final orbit consists of

all not identically zero quadratic forms with zero discriminant. In terms of the coordinates (a, b, c) provided by the coefficients of the quadratic form (1.7), the orbits are

$$\mathcal{O}_+ = \{ac > b^2\}, \qquad \mathcal{O}_0 = \{ac = b^2, \; a^2 + b^2 + c^2 \neq 0\},$$
$$\mathcal{O}_- = \{ac < b^2\}, \qquad \mathcal{O}_* = \{a = b = c = 0\}.$$

The isotropy subgroup for a given quadratic form is its (linear) symmetry group and consists of all linear transformations that leave the form unchanged. The positive definite canonical form $x^2 + y^2$ was already considered in Example 3.35 — its symmetry group is the orthogonal subgroup $O(2) \subset GL(2)$; the same holds true for the negative definite canonical form $-x^2 - y^2$. The symmetry group of the indefinite canonical form $x^2 - y^2$ consists of hyperbolic rotations $\begin{pmatrix} \pm \cosh t & \pm \sinh t \\ \pm \sinh t & \pm \cosh t \end{pmatrix}$, where the product of signs in each row of the matrix must be the same. The symmetry group of the alternative indefinite quadratic form xy is the conjugate subgroup containing all matrices $\begin{pmatrix} \lambda & 0 \\ 0 & \lambda^{-1} \end{pmatrix}$ and $\begin{pmatrix} 0 & \mu \\ \mu^{-1} & 0 \end{pmatrix}$, for $\lambda, \mu \neq 0$. One can distinguish the symmetry groups of definite versus indefinite forms by their topology: the symmetry group of a definite quadratic form is compact, while that of an indefinite form is not. As for the degenerate forms $\pm x^2$, they have the same symmetry group, consisting of all matrices of the form $\begin{pmatrix} \pm 1 & 0 \\ \gamma & \delta \end{pmatrix}$. Unlike the definite forms, the symmetry group depends on two parameters. Finally the trivial zero form has the entire four-parameter group $GL(2)$ as a symmetry group. Thus, one can almost distinguish between the various orbits by the underlying structure of their isotropy subgroups.

Exercise 3.54. Use the canonical forms for cubic polynomials presented in Chapter 2 to determine the orbits of the action on $GL(2, \mathbb{R})$ on the space of cubic forms. Determine the symmetry groups of the canonical cubic forms.

Exercise 3.55. Consider the induced action of the rotation group $SO(2)$ on the space of quadratic forms $\mathcal{Q}$, obtained by restricting the general linear action to pure rotations. In other words, we use the transformation rules (1.12) with $\alpha = \delta = \cos\theta$, $-\beta = \gamma = \sin\theta$. Prove that a complete list of canonical forms is provided by the diagonal forms $Q_0(x, y) = \lambda x^2 + \mu y^2$. Is the list redundant, that is, can we map one diagonal quadratic form to another? Use this to describe the orbits.

The preceding examples are particular cases of the general theory of quadratic forms on finite-dimensional vector spaces. According to Sylvester's Law of Inertia, [**75**; §X.2, **219**; p. 89], the action $S \mapsto ASA^T$ of the general linear group $\mathrm{GL}(n, \mathbb{R})$ on the space of real symmetric $n \times n$ matrices S has a discrete set of orbits, which are classified by the matrix's *signature*, meaning the number of positive, zero, and negative eigenvalues. Canonical forms are provided by the diagonal matrices with $0 \leq p \leq n$ entries equal to $+1$, followed by $0 \leq q \leq n - p$ entries equal to -1, followed by $n - p - q$ zeros on the diagonal. In particular, the orbit containing the identity matrix consists of all symmetric, positive definite matrices.

Theorem 3.56. *Two symmetric matrices are equivalent under the action $S \mapsto ASA^T$, $A \in \mathrm{GL}(n, \mathbb{R})$, if and only if they have the same signature.*

If we restrict to the orthogonal subgroup $\mathrm{O}(n) \subset \mathrm{GL}(n, \mathbb{R})$, then we can still diagonalize any symmetric matrix S, leading to a canonical form $D = \mathrm{diag}(\lambda_1, \ldots \lambda_n)$. Note that this canonical form is not quite uniquely determined, since we can rearrange the order of the entries λ_i. Thus, in this case, the moduli are provided by symmetric functions of the eigenvalues.

Remark: We can uniquely identify each symmetric matrix S with a quadratic form $Q(\mathbf{x}) = x^T S x$ on $\mathbb{R}^n$, and so Sylvester's Theorem and its orthogonal counterpart describe the canonical forms for multivariate quadratic forms under linear and orthogonal transformations.

Chapter 4

Representations and Invariants

While general transformation groups play a ubiquitous role in geometry, the most important for invariant theoretic purposes are the linear versions (along with their projective counterparts). Linear group actions are referred to as "representations", so as to distinguish them from the more general nonlinear transformation groups. Representation theory is a vast field of mathematical research, and we shall only have room to survey its most elementary aspects, concentrating on those which are relevant to our subject. More complete treatments can be found in numerous reference texts, including [56, 144, 223, 231]. Representation theory has an amazing range of applications in physics and mathematics, including special function theory, [152, 212, 225], quantum mechanics and particle physics, [46, 136, 232, 238], solid state physics, [130], probability, [145], harmonic and Fourier analysis, [212, 228], number theory, [79, 118, 145], combinatorics, [143, 201], bifurcation theory, [82, 190], and many other fields. A quote of I. M. Gel'fand, "...all of mathematics is some kind of representation theory...", [132], is perhaps not as far-fetched as it might initially seem! The representations of a general (even nonlinear) transformation group on the scalar-valued functions defined on the space where the group acts are of particular importance. In classical invariant theory, our original transformation rules for inhomogeneous and homogeneous polynomials are very special cases of the general framework of group representations on function spaces. In this fashion, we are led back to our primary subject, renewed and reinvigorated by an understanding of the requisite mathematical theory.

Representations

For simplicity we shall state the basic concepts in representation theory in the context of real group actions on real vector spaces; the corresponding complex version is an immediate adaptation and it is left to the reader to fill in the details.

Definition 4.1. A *representation* of a group G is defined by a group homomorphism $\rho: G \to \mathrm{GL}(V)$ to the group of invertible linear transformations on a vector space V.

Thus, according to (3.3), any representation must satisfy the following basic rules:

$$\rho(g \cdot h) = \rho(g) \cdot \rho(h), \qquad \rho(e) = \mathbb{1}, \qquad \rho(g^{-1}) = \rho(g)^{-1}, \qquad (4.1)$$

for all $g, h \in G$. The vector space on which the representation acts will be called the *representation space*; it is not necessarily finite-dimensional, although, in view of our particular applications, we can safely avoid the many analytical complications inherent in the infinite-dimensional context.

Example 4.2. Let $G = \mathbb{R}$ be the additive group of real numbers. According to (3.4), for any fixed $n \times n$ matrix J, the matrix exponential $\rho_J(t) = e^{tJ}$ defines an n-dimensional representation of $\mathbb{R}$. Particular cases appeared in Example 3.18.

Example 4.3. A well-known representation of the symmetric group $\mathbb{S}^n$ is provided by the $n \times n$ permutation matrices. To be specific, a permutation $\pi \in \mathbb{S}^n$ is represented by the linear transformation A_π which permutes the entries of vectors $x = (x_1, \ldots, x_n) \in \mathbb{R}^n$ accordingly: $A_\pi(x) = (x_{\pi(1)}, \ldots, x_{\pi(n)})$. It is easy to see that the map $\rho: \pi \mapsto A_\pi$ defines a group monomorphism from $\mathbb{S}^n$ to $\mathrm{GL}(n)$ and hence determines a representation of the symmetric group. For example, the symmetric group $\mathbb{S}^3$, as given in (3.6), is represented by the following six matrices:

$$\begin{pmatrix} 1 & 0 & 0 \\ 0 & 1 & 0 \\ 0 & 0 & 1 \end{pmatrix}, \qquad \begin{pmatrix} 0 & 1 & 0 \\ 1 & 0 & 0 \\ 0 & 0 & 1 \end{pmatrix}, \qquad \begin{pmatrix} 0 & 0 & 1 \\ 0 & 1 & 0 \\ 1 & 0 & 0 \end{pmatrix},$$

$$\begin{pmatrix} 1 & 0 & 0 \\ 0 & 0 & 1 \\ 0 & 1 & 0 \end{pmatrix}, \qquad \begin{pmatrix} 0 & 0 & 1 \\ 1 & 0 & 0 \\ 0 & 1 & 0 \end{pmatrix}, \qquad \begin{pmatrix} 0 & 1 & 0 \\ 0 & 0 & 1 \\ 1 & 0 & 0 \end{pmatrix}.$$

See Example 3.43 for several other interesting representations of the symmetric group.

Example 4.4. Let $G = \mathrm{GL}(n, \mathbb{R})$ be the general linear group. The simplest possible representation is the trivial one-dimensional representation that assigns to each matrix $A \in \mathrm{GL}(n)$ the real number 1. Slightly more interesting is the one-dimensional determinantal representation

$\rho_d(A) = \det A$. More generally, any power $\rho_d^k(A) = (\det A)^k$ of the determinant also forms a one-dimensional representation of GL(n). An evident n-dimensional representation is provided by the identity representation $\rho_e(A) = A$ acting on $V = \mathbb{R}^n$. A second n-dimensional representation is the so-called *contragredient* or *dual* representation $\rho_c(A) = A^{-T}$ considered in Exercise 3.19. Higher dimensional representations can be constructed by a variety of techniques. For example, the conjugation action $X \mapsto AXA^{-1}$ defines a representation of GL(n) on the space $\mathcal{M}_{n \times n} \simeq \mathbb{R}^{n^2}$ of $n \times n$ matrices, as does the action $X \mapsto AXA^T$ arising in the theory of quadratic forms.

The preceding representations of the general linear group are particular instances of tensorial operations that can be applied to arbitrary representations. These include the operations of duality, direct sum, tensor product, and symmetric product. The simplest case is the direct sum $\rho \oplus \sigma$ of two representations ρ, σ of G on V, W, respectively, which acts in the obvious fashion on the sum $V \oplus W$ of the representation spaces.

If V is a real, finite-dimensional vector space, we let V^* denote the dual vector space, which is defined as the space of real-valued linear maps $\omega: V \to \mathbb{R}$. It is not difficult to see that V^* is a vector space of the same dimension as V. In terms of a basis $\{e_1, \ldots, e_n\}$ of V, the corresponding dual basis $\{\varepsilon_1, \ldots, \varepsilon_n\}$ of V^* is defined so that $\varepsilon_i(e_j) = \delta_j^i$, where δ_j^i is the usual Kronecker delta, which has the value 1 if $i = j$ and 0 otherwise. The dual of V^* can be identified with V again: $(V^*)^* \simeq V$; the identification takes $v \in V$ to the linear map $\tilde{v} \in (V^*)^*$ given by $\tilde{v}(\omega) = \omega(v)$ for $\omega \in V^*$. If $T: V \to V$ is any linear transformation, then there is an induced dual transformation $T^*: V^* \to V^*$, defined so that $T^*(\omega)(v) = \omega(T^{-1}v)$. In terms of the dual bases, if the linear transformation T has $n \times n$ matrix form A, then the dual map T^* has contragredient matrix form A^{-T}. Thus, given any representation ρ of a group G on V, we can construct the dual or contragredient representation ρ^* of G on V^*, where $\rho^*(g) = \rho(g)^*$ is obtained by applying the inverse transpose operation to the representation matrices.

Remark: In invariant theory, the vectors in $V = \mathbb{R}^n$ which transform according to the identity representation of GL(n) are known as *contravariant vectors*. In other words, a contravariant vector is one that is subject to our basic transformation rule $\bar{\mathbf{x}} = A\mathbf{x}$ for each $A \in$ GL(n), cf. (2.4). Vectors in the dual space $V^* \simeq \mathbb{R}^n$, which transform according to the contragredient representation, are known (perhaps perversely, although we are adhering to the classical terminology) as *covariant vec-*

tors. The transformation rule for covariant vectors is $\mathbf{a} = A^T \bar{\mathbf{a}}$ or $\bar{\mathbf{a}} = A^{-T} \mathbf{a}$. For example, under contravariant transformations of its argument, the coefficients $\mathbf{a}$ of a linear form $Q(\mathbf{x}) = \mathbf{a} \cdot \mathbf{x} = \sum_i a_i x_i$ form the components of a covariant vector; see (4.5).

Next, recall that the tensor product $V \otimes W$ of two vector spaces can be identified with the space of linear maps $T : V^* \to W$ from the dual space of V to W. If $v \in V$ and $w \in W$, then there is an induced element $v \otimes w \in V \otimes W$, defined as the rank one linear transformation that takes $\omega \in V^*$ to the scalar multiple $\omega(v)\, w \in W$. If $\{e_1, \ldots, e_n\}$ is a basis of V and $\{f_1, \ldots, f_m\}$ a basis of W, then the tensor products $e_i \otimes f_j$ form a basis of $V \otimes W$, which has dimension mn. (In terms of the given bases and their duals, $e_i \otimes f_j$ corresponds to the linear transformation whose matrix form has 1 in the $(j, i)^{\text{th}}$ position.) In particular, $V \otimes \mathbb{R} \simeq V$, where we identify $v \otimes c$ with the scalar product cv. Given linear maps $T : V \to V$ and $U : W \to W$, we can define their tensor product $T \otimes U : V \otimes W \to V \otimes W$ in the obvious manner: $(T \otimes U)(v \otimes w) = T(v) \otimes U(w)$. In this way, the tensor product $\rho \otimes \sigma$ of representations ρ of G on V and σ of G on W defines a representation on the tensor product space $V \otimes W$. For example, the tensor product $\rho_e \otimes \rho_e$ of the standard representation of $\mathrm{GL}(n)$ with itself acts on the space $\mathbb{R}^n \otimes \mathbb{R}^n \simeq \mathcal{M}_{n \times n}$, which we identify with the space of $n \times n$ matrices. The reader can verify that this representation can be identified with the representation $X \mapsto A X A^T$ mentioned above. Similarly, the conjugation representation $X \mapsto A X A^{-1}$ is the same as the tensor product $\rho_e \otimes \rho_c$ of the standard representation with its dual.

Example 4.5. Consider the particular case when $G = \mathrm{GL}(2)$. An important example is provided by the tensor product of the contragredient representation

$$\rho_c \begin{pmatrix} \alpha & \beta \\ \gamma & \delta \end{pmatrix} = \frac{1}{\alpha\delta - \beta\gamma} \begin{pmatrix} \delta & -\gamma \\ -\beta & \alpha \end{pmatrix}$$

with the determinantal representation $\rho_d \begin{pmatrix} \alpha & \beta \\ \gamma & \delta \end{pmatrix} = \alpha\delta - \beta\gamma$. The representation space can be identified as $\mathbb{R}^2 \otimes \mathbb{R} \simeq \mathbb{R}^2$, and hence this tensor product defines the two-dimensional representation

$$(\rho_c \otimes \rho_d) \begin{pmatrix} \alpha & \beta \\ \gamma & \delta \end{pmatrix} = \begin{pmatrix} \delta & -\gamma \\ -\beta & \alpha \end{pmatrix} = \tilde{A}.$$

However, the latter representation is, in fact, isomorphic to the identity representation $\rho_e(A) = A$. Indeed, introduction of the alternative basis

$\widetilde{e}_1 = e_2$, $\widetilde{e}_2 = -e_1$ produces

$$\widetilde{A}\,\widetilde{e}_1 = \widetilde{A}\,e_2 = -\beta e_1 + \alpha e_2 = \alpha\,\widetilde{e}_1 + \beta\,\widetilde{e}_2,$$
$$\widetilde{A}\,\widetilde{e}_2 = -\widetilde{A}\,e_1 = -\delta e_1 + \gamma e_2 = \gamma\,\widetilde{e}_1 + \delta\,\widetilde{e}_2,$$

and hence, in terms of the new basis, $\widetilde{A}$ has the same matrix form as A. In other words, in a two-dimensional vector space, contravariant vectors can be obtained from their covariant counterparts by multiplication by the determinantal representation.

Remark: This result is peculiar to GL(2) — there is no combination of contragredient and determinantal representations of GL(n) that reproduces the identity representation when $n > 2$. As a consequence, the invariant theory of GL(n) for $n > 2$ is rather more involved than the two-dimensional case. See Chapter 10 for further details.

Remark: The tensor product $V \otimes \mathbb{C}$ of a real vector space with the complex numbers defines a complex vector space known as the *complexification* of V. The complexification process, applied to real representations, proves to be a particularly powerful tool in the general theory.

Remark: The elements of the tensor powers of the vector space V are known as *contravariant tensors*, while the tensor powers of the dual V^* are known as *covariant tensors*. Binary forms can be regarded as symmetric covariant tensors. Elements of tensor products of one or more copies of the identity representation with one or more copies of the contragredient representation are known as *mixed tensors*. Example 4.5 implies that, in the two-dimensional situation, every type of tensorial representation can be obtained by tensoring a covariant tensor representation with by a suitable power of the determinantal representation, which amounts to an adjustment of the overall weight of the tensor.

Irreducibility

As we have just seen, many complicated group representations can be built up from simpler representations using the basic tensor operations. The simplest nontrivial cases are the irreducible representations, meaning those which admit no nontrivial subrepresentations.

Definition 4.6. Let ρ define a representation of a group G on a vector space V. A subspace $W \subset V$ is an *invariant subspace* if it has the property that $\rho(g)W \subset W$ for all $g \in G$.

Note that the restriction of ρ to an invariant subspace W induces a subrepresentation of G on W. Trivial invariant subspaces, valid for any representation, are $W = \{0\}$ and $W = V$.

Definition 4.7. A *reducible* representation is one that contains a nontrivial invariant subspace $\{0\} \neq W \subsetneq V$. An *irreducible* representation is one that has no nontrivial invariant subspaces.

Example 4.8. The direct sum $\rho \oplus \sigma$ of two representations is reducible since each appears as a subrepresentation therein.

Example 4.9. The identity representation ρ_e of $\mathrm{GL}(V)$ on V is clearly irreducible. However, none of its tensor powers, acting on the tensor product spaces $\otimes^k V = V \otimes \cdots \otimes V$, are irreducible. For example, if we identify $V \simeq \mathbb{R}^n$, then, as we remarked, the second tensor power $V \otimes V \simeq \mathcal{M}_{n \times n} \simeq \mathbb{R}^{n^2}$ can be identified with the space of $n \times n$ matrices, with representation $X \mapsto AXA^T$. Clearly the subspaces consisting of all symmetric matrices, $\odot^2 V = \{S^T = S\}$, and all skew-symmetric matrices, $\wedge^2 V = \{K^T = -K\}$, are invariant subspaces, and so $\otimes^2 V = \odot^2 V \oplus \wedge^2 V$ decomposes into a direct sum of symmetric and skew components, both of which form irreducible subrepresentations. Let us prove irreducibility in the symmetric case. Suppose $\{0\} \neq W \subset \odot^2 V$ is an invariant subspace, and let $0 \neq S \in W$ be a nonzero symmetric matrix therein. Using Sylvester's Theorem 3.56, we can diagonalize $ASA^T = S_0 = \mathrm{diag}(\pm 1, \ldots, \pm 1, 0, \ldots, 0) \in W$. Moreover, choosing $D = \mathrm{diag}(\lambda_1, \ldots, \lambda_n)$ shows that $DS_0 D^T = \mathrm{diag}(\pm \lambda_1^2, \ldots, \pm \lambda_k^2, 0, \ldots, 0)$ must lie in W. The permutation matrices constructed in Example 4.3 will act on such diagonal matrices by permuting the entries, so the nonzero entries can be placed anywhere on the diagonal. Since W is a subspace, we can take linear combinations of such diagonal matrices and arrive at the conclusion that W contains all diagonal matrices. But then Theorem 3.56 shows that every symmetric matrix is in W, proving the result. The skew-symmetric version is similar and left to the interested reader. (For instance, one can utilize the canonical forms for skew-symmetric matrices, cf. [**75**; §XI.4].)

The higher tensor powers of a vector space include the symmetric and skew-symmetric subspaces as irreducible subrepresentations, but these are not an exhaustive list. One indication of this fact is to note that $\otimes^3 V$ has dimension n^3, while the symmetric and completely skew-symmetric subspaces have respective dimensions $\frac{1}{6} n(n+1)(n+2)$ and $\frac{1}{6} n(n-1)(n-2)$, whose sum is less than n^3. The full classification of the irreducible subrepresentations of the tensor representations of $\mathrm{GL}(V)$ can be found in [231], for instance, and lead to the theory of symmetry classes of tensors, classified by Young diagrams, which are also intimately related to the representation theory of the symmetric group $\mathbb{S}^n$. The resulting theory has important applications not only in mathematics, including combinatorics and geometry, cf. [143, 231], but also in quantum mechanics and chemistry, cf. [232, 238].

Exercise 4.10. Prove that if $\rho \colon G \to \mathrm{GL}(n)$ is any n-dimensional representation, then its determinant $\det \rho \colon G \to \mathbb{R} \setminus \{0\}$ defines a one-dimensional representation.

In many of the most important cases — for instance, when the group is finite, or compact, or the general tensor representations of $\mathrm{GL}(n)$ — finite-dimensional reducible representations can always be decomposed into a direct sum of irreducible subrepresentations, cf. [231]. For such groups, then, the irreducible representations form the fundamental building blocks, and it suffices to study their properties in order to understand completely general representations. The classification of irreducible representations has been the focus of major research efforts for over a half century; see [144, 223, 231] for example. Unfortunately, space precludes us from looking at anything beyond the particular examples that form the focus of this book.

Example 4.11. A simple counterexample to decomposability is provided by the reducible two-dimensional representation

$$\rho(A) = \begin{pmatrix} 1 & \log|\det A| \\ 0 & 1 \end{pmatrix}$$

of the general linear group $\mathrm{GL}(n, \mathbb{R})$. This representation leaves the subspace spanned by the vector $(1,0)^T$ invariant, but there is no complementary invariant subspace, and hence ρ cannot be written as the direct sum of two subrepresentations.

Exercise 4.12. Prove that if a representation of G acts transitively on $V \setminus \{0\}$, then it is irreducible. Is the converse true?

Exercise 4.13. A representation is called *unitary*[†] if its image is contained in the group of norm-preserving linear transformations of an

---

[†] The term comes from the complex version, in which the representation maps G to the unitary group $\mathrm{U}(V)$ consisting of all norm-preserving linear transformations of the complex Hermitian inner product space V.

inner product space V, so that, in the real case, $\rho\colon G \to \mathrm{O}(V)$. Prove that any finite-dimensional unitary representation decomposes into a direct sum of irreducible subrepresentations. *Hint*: Prove that the orthogonal complement to an invariant subspace is also invariant.

The decomposability of the representations of finite groups is based on Exercise 4.13 combined with the following observation.

Proposition 4.14. *Every finite-dimensional representation of a finite group is equivalent to a unitary representation.*

Proof: Let us choose any inner product $\langle x \, ; y \rangle$ on the representation space V. We then average over the group[†] to define a new inner product

$$\langle x \, ; y \rangle_G = \frac{1}{\#G} \sum_{g \in G} \langle \rho(g)x \, ; \rho(g)y \rangle \tag{4.2}$$

on V. It is easy to check that ρ is unitary with respect to the new group-invariant inner product. *Q.E.D.*

Remark: The same result holds for compact Lie groups (see Chapter 8), where one replaces the sum in (4.2) by an integral based on the invariant Haar measure, cf. [**212, 231**]. Thus, Proposition 4.14 combined with Exercise 4.13 implies that any representation of a finite group or compact Lie group decomposes into a direct sum of irreducible subrepresentations.

Remark: A matrix group $G \subset \mathrm{GL}(n)$ is called *linearly reductive*, [**156, 213**], if every representation that depends rationally on the matrix entries of $A \in G$ decomposes into a sum of irreducible representations. As we remarked, finite groups, compact Lie groups, and the groups $\mathrm{GL}(n)$ and $\mathrm{SL}(n)$ are all linearly reductive. The Hilbert Basis Theorem 2.42 holds for all linearly reductive groups, cf. [**156**], and hence one can argue that they form the optimal class of groups for generalizing the full range of methods and results in classical invariant theory.

Function Spaces

One easy way to turn a nonlinear group action into a linear representation is through its induced action on the functions defined on the space. Let G be a transformation group acting on a space X. Then there is a

---

[†] See (4.8) for a more detailed discussion of this process.

naturally induced representation of G on the linear space $\mathcal{F} = \mathcal{F}(X)$ of real-valued functions $F: X \to \mathbb{R}$. A group element $g \in G$ will map the function F to the transformed function $\overline{F} = g \cdot F$, which is defined by

$$\overline{F}(\bar{x}) = F(g^{-1} \cdot \bar{x}) = F(x). \tag{4.3}$$

The introduction of the inverse g^{-1} in (4.3) ensures that the action defines a group homomorphism: $g \cdot (h \cdot F) = (g \cdot h) \cdot F$ for all $g, h \in G$, and $F \in \mathcal{F}$. The representation of G on the (infinite dimensional) function space $\mathcal{F}$ will typically include a wide variety of important subrepresentations, including representations on subspaces of polynomials, continuous functions, smooth (differentiable) functions, analytic functions, normalizable (L^2) functions, and so on. This basic construction clearly extends to the space $\mathcal{F}^k(X)$ consisting of all vector-valued functions $F: X \to \mathbb{R}^k$.

For our invariant theoretic purposes, the most important examples are the standard representation of the general linear group $\mathrm{GL}(2, \mathbb{R})$ acting on $\mathbb{R}^2$, and its complex counterpart $\mathrm{GL}(2, \mathbb{C})$ acting on $\mathbb{C}^2$. Concentrating on the real version, the induced representation (4.3) on the space of real-valued functions $Q: \mathbb{R}^2 \to \mathbb{R}$ is explicitly given by

$$\overline{Q}(\bar{x}, \bar{y}) = \overline{Q}(\alpha\, x + \beta\, y, \gamma\, x + \delta\, y) = Q(x, y), \qquad \begin{pmatrix} \alpha & \beta \\ \gamma & \delta \end{pmatrix} \in \mathrm{GL}(2). \tag{4.4}$$

In particular, if Q is a homogeneous polynomial, so is $\overline{Q}$, and hence the space $\mathcal{P}^{(n)} \subset \mathcal{F}(\mathbb{R}^2)$ of homogeneous polynomials of degree n forms an invariant subspace, and the representation (2.5) reduces to our original transformation rules (2.5) for binary forms. We let $\rho_n = \rho_{n,0}$ denote the induced finite-dimensional representation of $\mathrm{GL}(2)$ on $\mathcal{P}^{(n)}$. We can uniquely identify each homogeneous polynomial $Q \in \mathcal{P}^{(n)}$ with the $(n+1)$-tuple $\mathbf{a} = (a_0, a_1, \dots, a_n) \in \mathbb{R}^{n+1}$, where the a_i are its coefficients relative to the basis of $\mathcal{P}^{(n)} \simeq \mathbb{R}^{n+1}$ provided by the scaled monomials $\binom{n}{i} x^i y^{n-i}$, for $i = 0, \dots, n$. The explicit action of $\mathrm{GL}(2)$ on the coefficients a_i was given in (2.6). For example, the coefficients of a general linear polynomial $Q(x, y) = ax + by$ will transform according to

$$\begin{pmatrix} a \\ b \end{pmatrix} = \begin{pmatrix} \alpha & \gamma \\ \beta & \delta \end{pmatrix} \begin{pmatrix} \bar{a} \\ \bar{b} \end{pmatrix}, \qquad A = \begin{pmatrix} \alpha & \beta \\ \gamma & \delta \end{pmatrix} \in \mathrm{GL}(2), \tag{4.5}$$

and so form a covariant vector. Therefore, the representation ρ_1 on the space $\mathcal{P}^{(1)}$ can be identified with the contragredient representation of $\mathrm{GL}(2)$. On the space of quadratic forms $Q(x, y) = ax^2 + 2bxy + cy^2$, the induced representation ρ_2 acts on the coefficients as in (1.12), which we identify with the second symmetric power of the contragredient representation, cf. Example 4.9.

A particularly important class of representation is obtained by tensoring the polynomial representations with powers of the determinantal representation. Under a group element, the function $\overline{Q}(\bar{x}, \bar{y})$ maps to $Q(x, y)$, where

$$\overline{Q}(\bar{x}, \bar{y}) = (\alpha\delta - \beta\gamma)^k \, Q(x, y) = (\alpha\delta - \beta\gamma)^k \, \overline{Q}(\alpha\, x + \beta\, y, \gamma\, x + \delta\, y). \tag{4.6}$$

The restriction of the representation (4.6) to the space $\mathcal{P}^{(n)}$ of homogeneous polynomials serves to define the fundamental representation $\rho_{n,k}$ of *weight* k and *degree* n, cf. (2.32). In particular, according to Example 4.5, the representation $\rho_{1,1}$ of weight 1 and degree 1 is isomorphic to the identity representation of GL(2).

Theorem 4.15. *The representation* $\rho_{n,k}$ *of* GL(2) *is irreducible.*

Proof: Let $\{0\} \neq W \subset \mathcal{P}^{(n)}$ be a nonzero invariant subspace. We first show that if $Q(x, y)$ is any polynomial in W, as in (2.1), then every monomial appearing in Q with nonzero coefficient also lies in W; that is, $x^i y^{n-i} \in W$ provided $a_i \neq 0$. This follows immediately from the invariance of W under the unimodular scaling $(x, y) \mapsto (\lambda x, \lambda^{-1} y)$, which changes Q into

$$\widehat{Q}_\lambda(x, y) = \sum_{i=0}^{n} \binom{n}{i} a_i \lambda^{2i-n} x^i y^{n-i} = \lambda^{-n} \sum_{i=0}^{n} \binom{n}{i} a_i \lambda^{2i} x^i y^{n-i}. \tag{4.7}$$

Now $\widehat{Q}_\lambda \in W$ for each $\lambda \in \mathbb{R}$. But $\lambda^n \widehat{Q}_\lambda$ is a polynomial in the parameter λ, hence this will hold if and only if each coefficient of each power of λ itself lies in W, proving the assertion.

Thus, we have proved that any subspace invariant under the unimodular scaling subgroup must be spanned by monomials. Next, suppose $x^k y^{n-k} \in W$ is any monomial. We apply the linear transformation $(x, y) \mapsto (x + \beta y, y)$, which maps it to $\sum_{i=0}^{k} \binom{k}{i} \beta^i x^i y^{n-i}$, which, as before, lies in W, and hence each of its constituent monomials, $x^k y^{n-k}$, $x^{k-1} y^{n-k+1}, \ldots, y^n$, also lies in W. Application of the linear transformation $(x, y) \mapsto (x, \gamma x + y)$ shows that the monomials $x^n, \ldots, x^k y^{n-k}$ also lie in W. Therefore W contains every monomial of degree n, and hence $W = \mathcal{P}^{(n)}$, which proves irreducibility.   *Q.E.D.*

Remark: In fact, the representations $\rho_{n,k}$ provide a complete list of all irreducible finite-dimensional representations of GL(2). Moreover, any tensor representation can be decomposed into a direct sum of copies of the $\rho_{n,k}$; see [231] for a proof.

Note that our proof utilized only unimodular linear transformations, proving that $\rho_{n,k}$ remains irreducible when restricted to the special linear group SL(2). On the other hand, if $n \geq 2$, the restriction of the representation $\rho_{n,k}$ to the rotation subgroup SO(2) $\subset$ GL(2) is no longer irreducible. For example, the representation of SO(2) on the space of real quadratic polynomials $\mathcal{P}^{(2)}$ decomposes into the sum of two irreducible subrepresentations, a trivial one on the one-dimensional subspace spanned by $x^2 + y^2$, and a two-dimensional representation on the subspace spanned by $x^2 - y^2$ and xy. Finally, we remark that the preceding proof shows that the subgroup of upper triangular unimodular matrices is *not* linearly reductive since each of the subspaces $W_k = \text{Span}\{x^k y^{n-k}, x^{k-1}y^{n-k+1}, \ldots, y^n\}$ is invariant, but clearly not irreducible (unless $k = 0$).

Exercise 4.16. Decompose the restriction of $\rho_{n,k}$ to the subgroup SO(2) $\subset$ GL(2) into irreducible subrepresentations. *Hint*: Note that the subspaces $\mathcal{H}_m = \{(x^2+y^2)^m Q(x,y) \mid Q \in \mathcal{P}^{(n-2m)}\}$ are rotationally invariant.

Exercise 4.17. The (complex) one-dimensional representations of the circle group SO(2) $\simeq S^1$ are given by $\rho_n(\theta) = e^{in\theta}$, where θ is the angular coordinate on S^1 and $n \in \mathbb{Z}$. It can be proved that any other (complex) representation can be decomposed into a direct sum of these irreducible representations. For example, prove that the standard representation $\rho(\theta) = \begin{pmatrix} \cos\theta & -\sin\theta \\ \sin\theta & \cos\theta \end{pmatrix}$ decomposes into the direct sum $\rho_1 \oplus \rho_{-1}$. Similarly, show that the representation $\rho^{(2)}$ of SO(2) on the space of complex quadratic polynomials breaks up into the direct sum $\rho_2 \; \rho_0 \oplus \rho_{-2}$. *Note*: The subspace $W = \text{Span}\{x^2 - y^2, xy\} \subset \mathcal{P}^{(2)}$ mentioned above is irreducible over the reals but decomposes into two invariant one-dimensional subspaces over the complexes.

Remark: The appearance of the Fourier modes $e^{in\theta}$ in this case is not an accident, but forms a special case of a general theory, due to Peter and Weyl, [179], of Fourier analysis on compact Lie groups. Details, including applications to the theory of special functions, can be found in [212].

Invariant Functions

In general, an invariant is defined as a real-valued function that is unaffected by group transformations. The determination of a complete set of invariants of a given group action is a problem of supreme importance for the study of equivalence and canonical forms. For example, in sufficiently regular cases, the orbits, and hence the canonical forms, for a group action are completely characterized by its invariants; see Theorem 8.17.

Definition 4.18. Let G be a transformation group acting on a space X. An *invariant* is a real-valued function $I: X \to \mathbb{R}$ which satisfies $I(g \cdot x) = I(x)$ for all transformations $g \in G$.

In other words, an invariant function is a fixed point for the induced representation of G on the function space $\mathcal{F}(X)$.

Proposition 4.19. *Let $I: X \to \mathbb{R}$. The following conditions are equivalent:*
(a)   *I is a G-invariant function.*
(b)   *I is constant on the orbits of G.*
(c)   *All level sets $\{I(x) = c\}$ are G-invariant subsets of X.*

In particular, constant functions are trivially G-invariant. If G acts transitively on X, then these are the only invariants.

Example 4.20. For the standard action (3.11) of the rotation group SO(2) on $\mathbb{R}^2$, the radius $r = \sqrt{x^2 + y^2}$ is an invariant function, as is any function $f(r)$ thereof. Indeed, Proposition 4.19 immediately implies that these are the only invariants for the rotation group. Note that in this case the invariant function r serves to distinguish the orbits: two points lie in the same circular orbit if and only if they have the same value for the radial invariant.

Example 4.21. Even though there are several orbits, there are no continuous, nonconstant invariant functions for the action of GL(2) on the space $\mathcal{Q}$ of real quadratic forms described in Example 3.53, and hence the orbits are *not* distinguished by continuous invariants. However, we can certainly find discontinuous invariant functions, e.g. the sign of the discriminant,[†] which will serve the purpose.

[†] The discriminant itself is an SL(2) invariant, but is not GL(2)–invariant.

A fundamental problem is to determine *all* possible (continuous, polynomial, analytic, etc.) invariants of a group of transformations. Note that if $I_1(x), \ldots, I_k(x)$ are invariants, and $H(y_1, \ldots, y_k)$ is any function, then $I(x) = H(I_1(x), \ldots, I_k(x))$ is also invariant. Therefore, we only need to find a complete set of functionally independent invariants, having the property that any other invariant can be written as a function thereof. In the case of polynomials, one can ask for more detailed information: algebraically independent invariants, as in the definition of a Hilbert basis, rationally independent invariants, and so on.

For finite groups, one can introduce a method of "group averaging" to generate invariants. Let G be a finite group with $\#G$ elements that acts on a space X. The *symmetrization* or *Reynolds operator*

$$\varsigma = \frac{1}{\#G} \sum_{g \in G} g \tag{4.8}$$

averages functions over the group and so defines a projection from the space $\mathcal{F}(X)$ of all functions on X to the subspace of G-invariant functions. In other words, if $F: X \to \mathbb{R}$ is any function, then $I(x) = \varsigma \cdot F(x) = (\#G)^{-1} \sum F(g \cdot x)$ is an invariant; moreover, if I is already invariant, then $\varsigma \cdot I = I$.

Example 4.22. Consider the representation of the symmetric group $\mathbb{S}^n$ on $\mathbb{R}^n$ given by the permutation matrices, as in Example 4.3. A function $I: \mathbb{R}^n \to \mathbb{R}$ is invariant under $\mathbb{S}^n$ if and only if it satisfies

$$I(x_{\pi(1)}, \ldots, x_{\pi(n)}) = I(x_1, \ldots, x_n), \qquad \text{for every} \qquad \pi \in \mathbb{S}^n. \tag{4.9}$$

The invariants of the permutation group $\mathbb{S}^n$ are known as *symmetric functions*. According to (4.8), the associated symmetrization operator

$$\varsigma = \frac{1}{n!} \sum_{\pi \in \mathbb{S}^n} \pi \tag{4.10}$$

defines the projection onto the subspace of all symmetric functions. Of particular importance are the elementary symmetric polynomials

$$\sigma_1(x_1, \ldots, x_n) = x_1 + \cdots + x_n,$$
$$\sigma_2(x_1, \ldots, x_n) = x_1 x_2 + x_1 x_3 + \cdots + x_{n-1} x_n = \sum_{i<j} x_i x_j,$$
$$\vdots$$
$$\sigma_n(x_1, \ldots, x_n) = x_1 x_2 \cdots x_n.$$

In general,

$$\sigma_k(x_1, \ldots, x_n) = \binom{n}{k} \varsigma(x_1 x_2 \cdots x_k) = \sum_{i_1 < i_2 < \cdots < i_k} x_{i_1} x_{i_2} \cdots x_{i_k}. \quad (4.11)$$

The Fundamental Theorem for the symmetric group states that the elementary symmetric polynomials form a complete system of invariants.

Theorem 4.23. *If $I(x)$ is a symmetric polynomial, then I can be written uniquely in terms of the elementary symmetric polynomials: $I(x) = P(\sigma_1(x), \cdots, \sigma_n(x))$.*

Proof: Let us order the monomials $x^K = x_1^{k_1} \cdots x_n^{k_n}$ *lexicographically*, so that $x^K < x^J$ whenever the first nonvanishing difference $j_1 - k_1, j_2 - k_2, \ldots$ is positive. The last monomial in the elementary symmetric polynomial σ_k in the lexicographic ordering is $x_1 x_2 \cdots x_k$. Given a symmetric polynomial $I(x)$, let $a_K x^K$, $a_K \neq 0$, be its last monomial. Since $I(x)$ is symmetric, the permuted monomials $x^{\pi(K)}$ also occur with the same nonvanishing coefficient a_K; therefore, $x^{\pi(K)} < x^K$, which implies that $k_1 \geq k_2 \geq \cdots \geq k_n$. On the other hand, the last monomial in the power product of elementary symmetric polynomials $S^{\overline{K}} = \sigma_1^{k_1-k_2} \sigma_2^{k_2-k_3} \cdots \sigma_{n-1}^{k_{n-1}-k_n} \sigma_n^{k_n}$ is also x^K. The difference $\tilde{I} = I - a_K S^{\overline{K}}$ is also symmetric and has a last monomial of lower order. A straightforward induction completes the demonstration that I can be written as a polynomial in the elementary symmetric polynomials. In order to prove uniqueness, it suffices to note that each power product $S^{\overline{K}}$ of elementary symmetric polynomials has a different last monomial, namely x^K.   *Q.E.D.*

Exercise 4.24. The *power sum symmetric polynomials* are obtained by symmetrizing $(x_1)^k$, so $P_k(x) = \varsigma((x_1)^k) = \sum (x_i)^k$. Use the method of proof of Theorem 4.23 to express the P_k in terms of the elementary symmetric polynomials for k small. Can you extend your result to general k? See [**143**] for details.

Exercise 4.25. Prove that the coefficients c_k in a polynomial (2.16) of degree n can be written in terms of the elementary symmetric polynomials of its (complex) roots p_k, as in (2.17). Specifically,

$$c_k = (-1)^{n-k} c_n \sigma_k(p_1, \ldots, p_n), \qquad k = 1, \ldots, n.$$

Remark: Extensions of the Fundamental Theorem 4.23 to more general symmetric functions are discussed in [**14, 16, 81**].

Joint Invariants

Joint invariants appear when a transformation group acts simultaneously on several different spaces or, more typically, on multiple copies of the same space. More specifically, suppose G is a fixed group which acts on the spaces $X_1, \ldots, X_m$. Then there is a naturally induced action of G on the Cartesian product space $X_1 \times \cdots \times X_m$ given by $g \cdot (x_1, \ldots, x_m) = (g \cdot x_1, \ldots, g \cdot x_m)$ for $x_i \in X_i$, $g \in G$.

Definition 4.26. A *joint invariant* is merely an invariant function $J: X_1 \times \cdots \times X_m \to \mathbb{R}$ for a Cartesian product action of a group G. In other words, $J(g \cdot x_1, \ldots, g \cdot x_m) = J(x_1, \ldots, x_m)$ for all $g \in G$, $x_i \in X_i$.

When the spaces $X_i = X$ are identical, with identical actions of G, we shall call a joint invariant $J(x_1, \ldots, x_m)$ depending on m points $x_i \in X$ an m–*fold joint invariant*. Any ordinary invariant $I(x)$ induces trivial m–fold joint invariants, namely, $J_i(x_1, \ldots, x_m) = I(x_i)$ for any $1 \leq i \leq m$. Typically, one is really only interested in "genuine" joint invariants, which are not trivially obtained from joint invariants depending on fewer than m points. Vice versa, if $J(x_1, \ldots, x_m)$ is an m–fold joint invariant, then its *trace*

$$I(x) = J(x, x, \ldots, x), \qquad x \in X, \qquad (4.12)$$

produces an ordinary invariant on X.

Example 4.27. Consider the Euclidean group E(2) acting on the plane $X = \mathbb{R}^2$. Since the action is transitive, there are no ordinary invariants. On the Cartesian product $X^{(2)} = X \times X$, there is a single joint invariant, namely, the distance $d(\mathbf{x}_1, \mathbf{x}_2) = \| \mathbf{x}_1 - \mathbf{x}_2 \|$ between two points $\mathbf{x}_1, \mathbf{x}_2 \in \mathbb{R}^2$. On higher Cartesian powers, $X^{(m)} = X \times \cdots \times X$, the two-fold joint invariant induces a variety of joint invariants:

$$d_{ij} = d(\mathbf{x}_i, \mathbf{x}_j) = \| \mathbf{x}_i - \mathbf{x}_j \| \qquad (4.13)$$

for any $i < j$. A fundamental result states that these are the only joint invariants — any other joint invariant $I(\mathbf{x}_1, \ldots, \mathbf{x}_m)$ can be written as a function of the interpoint distances. See Example 8.29, and also Weyl, [**231**; Theorem 2.9.A], for more details.

Example 4.28. A similar result holds for the equi-affine group, SA(2), consisting of area-preserving affine transformations, as introduced in Example 3.36. In this case there are no (continuous) ordinary or two-fold joint invariants, so the first example of a joint invariant arises on $X^{(3)}$, and is given by a_{123}, where[†]

$$a_{ijk} = A(\mathbf{x}_i, \mathbf{x}_j, \mathbf{x}_k) = \tfrac{1}{2}(\mathbf{x}_j - \mathbf{x}_i) \wedge (\mathbf{x}_j - \mathbf{x}_k) \qquad (4.14)$$

denotes the (signed) area of the triangle with corners $\mathbf{x}_i, \mathbf{x}_j, \mathbf{x}_k$. Again, in Example 8.31 we prove that every other joint invariant can be written as a function of the three-fold area invariants (4.14). Note that the area a_{ijk} is only a proper Euclidean joint invariant, since a reflection will reverse its sign; however, its square is Euclidean–invariant. According to the preceding example, then, we should be able to write the latter in terms of the distances between the vertices of the triangle. Of course, the answer is provided by the well-known semi-perimeter formula $(a_{ijk})^2 = s(s - d_{ij})(s - d_{ik})(s - d_{jk})$, where $s = \tfrac{1}{2}(d_{ij} + d_{ik} + d_{jk})$ is the semi-perimeter of the triangle with vertices $\mathbf{x}_i, \mathbf{x}_j, \mathbf{x}_k$.

For the full affine group $A(2)$, the area of a triangle is multiplied by the determinant of the linear part of the transformation and hence forms a "relative joint invariant" — see below. The ratio $r_{ijkl} = a_{ijk}/a_{ijl}$ between two such areas provides a genuine joint affine invariant. Again, all joint affine invariants are expressible as functions of these ratios.

Exercise 4.29. Determine the orbits of the action of the groups $SA(2)$ and $A(2)$ acting on $\mathbb{R}^2 \times \mathbb{R}^2$ and $\mathbb{R}^2 \times \mathbb{R}^2 \times \mathbb{R}^2$.

Example 4.30. The next example is of crucial importance for understanding how the group of linear transformations affects the geometry of the roots of polynomials. Consider the joint projective action of $SL(2, \mathbb{C})$ on the m-fold Cartesian product $\mathbb{CP}^1 \times \cdots \times \mathbb{CP}^1$ given by

$$(p_1, \ldots, p_m) \longmapsto \left(\frac{\alpha p_1 + \beta}{\gamma p_1 + \delta}, \ldots, \frac{\alpha p_m + \beta}{\gamma p_m + \delta} \right), \quad \begin{pmatrix} \alpha & \beta \\ \gamma & \delta \end{pmatrix} \in SL(2, \mathbb{C}).$$

Let us concentrate on the open subset $\widetilde{X}^{(m)} = \{p_i \neq p_j, i \neq j\}$ consisting of distinct m-tuples of points. For $m \leq 3$, the action is transitive on $\widetilde{X}^{(m)}$, and there are no nonconstant two- or three-fold invariants. Indeed, we can map any three distinct points (p_1, p_2, p_3) on the Riemann sphere to any desired canonical form, e.g., $(0, 1, \infty)$, by a suitable linear fractional transformation. The *cross-ratio*

$$[\, p_1, p_2, p_3, p_4\,] = \frac{(p_1 - p_2)(p_3 - p_4)}{(p_1 - p_3)(p_2 - p_4)} \qquad (4.15)$$

[†] The symbol $\wedge$ denotes the standard scalar cross product between planar vectors: $(x, y) \wedge (x', y') = xy' - x'y$.

is four-fold joint invariant, as can be verified directly. (If one of the points is infinite, (4.15) is computed in a consistent manner.) Now, only the first three points can be fixed, so a canonical form for four points could be $(0, 1, \infty, z)$, where the value of the projective modulus z is fixed by the cross-ratio $[0, 1, \infty, z] = 1/(1-z)$. In Example 8.34 we shall prove that every joint projective invariant can be written as a function of the cross-ratios of the points taken four at a time.

Exercise 4.31. Discuss the action of $SL(2, \mathbb{R})$ on $\mathbb{RP}^1 \times \cdots \times \mathbb{RP}^1$, and on $\mathbb{CP}^1 \times \cdots \times \mathbb{CP}^1$.

Example 4.32. A joint invariant for the standard action of the symmetric group $\mathbb{S}^n$ on vectors in $\mathbb{R}^n$, as in Example 4.3, is known as a *multi-symmetric function*. The theory of multi-symmetric polynomials is less extensively investigated than the more elementary theory of symmetric polynomials, although it has been the subject of intermittent research activity, beginning with Junker, [**121, 122**], and Netto, [**158**; §377–386]. See [**147, 80**], for connections with combinatorics, [**5**] for recent applications to algebraic topology, and [**172**] for applications to dissipative decompositions of partial differential equations.

We consider the Cartesian product representation of $\mathbb{S}^n$ on the space $\mathbb{R}^n \times \cdots \times \mathbb{R}^n \simeq \mathbb{R}^{mn}$ consisting of m copies of $\mathbb{R}^n$. A function $Q(\mathbf{x}^1, \ldots, \mathbf{x}^m)$, where $\mathbf{x}^i = (x_1^i, \ldots, x_n^i) \in \mathbb{R}^n$, is called a multi-symmetric function if $\pi \cdot Q = Q$ for every $\pi \in \mathbb{S}^n$, which acts by simultaneous permutation of the components of the vectors $\mathbf{x}_i$. Alternatively, one can view a multi-symmetric function $\widetilde{Q}(\widetilde{\mathbf{x}}_1, \ldots, \widetilde{\mathbf{x}}_n)$ as depending on n vectors $\widetilde{\mathbf{x}}_k = (x_k^1, \ldots, x_k^m) \in \mathbb{R}^m$ and require symmetry under permutations of the vectors: $\widetilde{Q}(\widetilde{\mathbf{x}}_{\pi(1)}, \ldots, \widetilde{\mathbf{x}}_{\pi(n)}) = \widetilde{Q}(\widetilde{\mathbf{x}}_1, \ldots, \widetilde{\mathbf{x}}_n)$.

We can evidently construct multi-symmetric functions by applying the symmetrization operator (4.10) to ordinary functions. In particular, the *elementary multi-symmetric polynomials* are found by applying ς to the multi-linear monomials:

$$\sigma_I(\mathbf{x}^1, \ldots, \mathbf{x}^m) = \binom{n}{k} \varsigma(x_1^{i_1} x_2^{i_2} \cdots x_k^{i_k}), \qquad \begin{array}{l} I = (i_1, \ldots, i_k), \\ 1 \leq i_\kappa \leq n. \end{array} \tag{4.16}$$

For example, in the case $m = n = 2$, the multi-symmetric functions depend on two sets of two variables and satisfy the symmetry condition $Q(x_2, x_1; y_2, y_1) = Q(x_1, x_2; y_1, y_2)$. The linear multi-symmetric functions are

$$\sigma_1 = x_1 + x_2, \qquad \sigma_2 = y_1 + y_2.$$

There are three quadratic elementary multi-symmetric functions:

$$\sigma_{11} = x_1 x_2, \quad \sigma_{12} = \tfrac{1}{2}(x_1 y_2 + x_2 y_1), \quad \sigma_{22} = y_1 y_2.$$

As before, the Fundamental Theorem states that the elementary multi-symmetric polynomials form a generating set.

Theorem 4.33. *Any multi-symmetric polynomial can be written as a polynomial in the elementary multi-symmetric polynomials.*

The proof of this result is similar to that of Theorem 4.23; see [80, 158]. There is, however, one crucial difference between the two versions — the formula expressing a multi-symmetric polynomial in terms of the elementary ones is *not* unique, owing to the existence of nontrivial syzygies. The simplest such syzygy is in the case $n = 2$, where

$$4(\sigma_{ij}\sigma_{kl} - \sigma_{ik}\sigma_{jl}) = \sigma_i\sigma_j\sigma_{kl} + \sigma_k\sigma_l\sigma_{ij} - \sigma_i\sigma_k\sigma_{jl} - \sigma_j\sigma_l\sigma_{ik}. \quad (4.17)$$

Thus, in the case $m = n = 2$, there is one such syzygy,

$$4\left[\sigma_{11}\sigma_{22} - (\sigma_{12})^2\right] = (\sigma_2)^2\sigma_{11} - 2\sigma_1\sigma_2\sigma_{12} + (\sigma_1)^2\sigma_{22}. \quad (4.18)$$

It can be shown, [63], that (4.17) forms a complete list of syzygies when $n = 2$. Higher order syzygies are more complicated, and still not completely classified,[†] although particular cases appear in [121, 122, 172].

Multiplier Representations

Although induced actions of transformation groups on functions provide us with a wide variety of important representations, these are not quite general enough for our purposes. Consider the linear fractional action (2.7) of the group $GL(2, \mathbb{C})$ on the projective line $\mathbb{CP}^1$. According to the preceding construction, (4.3), this induces the representation

$$\bar{F}(\bar{p}) = \bar{F}\left(\frac{\alpha p + \beta}{\gamma p + \delta}\right) = F(p)$$

[†] A personal story: In the first version of [172], unaware of the work of Junker, Cheri Shakiban and I ran several computer algebra computations of multi-symmetric syzygies. Using the specialized MACAULAY computer algebra package on a workstation, we were just able to treat the cases when $m, n \le 3$, but this came close to the limits of the machine at that time. However, the referee of our paper pointed out Junker's thesis (written under Hilbert), which successfully treated cases like $m = 2, n = 7$, and $m = 3, n = 4$. So much for the power of symbolic computing!

on the space of scalar-valued functions on $\mathbb{CP}^1$, written in terms of the projective coordinate $p = x/y$. However, with the exception of the constants, which are homogeneous of degree 0, this representation does not naturally contain any of the fundamental polynomial representations $\rho_{n,k}$ given by (4.6). Indeed, each homogeneous polynomial (2.1) has an inhomogeneous representative given by (2.2). The induced action of GL(2) corresponding to the representation $\rho_{n,k}$ is readily seen to be

$$
\begin{aligned}
Q(p) &= (\alpha\delta - \beta\gamma)^k \, (\gamma p + \delta)^n \, \overline{Q}(\bar{p}) \\
&= (\alpha\delta - \beta\gamma)^k \, (\gamma p + \delta)^n \, \overline{Q}\left(\frac{\alpha p + \beta}{\gamma p + \delta}\right).
\end{aligned}
\tag{4.19}
$$

The linear action (4.19) defines a more general kind of representation of the projective group on the space of functions over $\mathbb{CP}^1$, which is known as a "multiplier representation", and the prefactor $(\alpha\delta - \beta\gamma)^k(\gamma p + \delta)^n$, or, rather, its reciprocal, is called the *multiplier*. The general definition[†] follows, cf. [**17, 153, 231**].

Definition 4.34.   Let G be a group acting on a space X. A *multiplier representation* of G is a representation $\overline{F} = g \cdot F$ on the space of real-valued functions $\mathcal{F}(X)$ of the particular form

$$
\overline{F}(\bar{x}) = \overline{F}(g \cdot x) = \mu(g, x) \, F(x), \qquad g \in G, \quad F \in \mathcal{F}.
\tag{4.20}
$$

The condition that (4.20) actually defines a representation of the group G requires that the multiplier μ satisfy a certain algebraic identity, which follows directly from the group law $g \cdot (h \cdot F) = (g \cdot h) \cdot F$.

Lemma 4.35.   *A function* $\mu \colon G \times X \to \mathbb{C} \setminus \{0\}$ *is a multiplier for a transformation group* G *acting on a space* X *if and only if it satisfies the multiplier equation*

$$
\begin{aligned}
\mu(g{\cdot}h, x) &= \mu(g, h{\cdot}x)\,\mu(h, x), \\
\mu(e, x) &= 1,
\end{aligned}
\qquad \text{for all} \qquad
\left\{
\begin{aligned}
&g, h \in G, \\
&x \in X.
\end{aligned}
\right.
\tag{4.21}
$$

Note that if $\mu(g, x)$ and $\widehat{\mu}(g, x)$ are multipliers for a group G, so is their product $\mu(g, x) \cdot \widehat{\mu}(g, x)$, as is any power $\mu(g, x)^k$.

Remark: If a multiplier does not depend on the point x, so $\mu = \mu(g)$, then the multiplier equation (4.21) reduces to

$$
\mu(g \cdot h) = \mu(g)\,\mu(h), \quad \text{for all} \quad g, h \in G, \qquad \text{and} \qquad \mu(e) = 1.
$$

[†] This definition of multiplier representation is *not* the same as that appearing in the work of Mackey, [**144**]; the latter are also known as projective representations, [**100**], and will not play any role in our discussion.

In other words, μ defines a one-dimensional representation of the group G. Weyl, [231], only permits such multipliers, but (4.19) underscores the need to allow x dependence in our treatment.

Example 4.36. Consider the usual projective action (2.7) of the general linear group $GL(2)$. By the preceding remark, the determinantal representation $\mu_{0,1}(A) = \alpha\delta - \beta\gamma$, for $A = \begin{pmatrix} \alpha & \beta \\ \gamma & \delta \end{pmatrix}$, automatically defines a multiplier. More interestingly, the function

$$\mu_1\left(\begin{pmatrix} \alpha & \beta \\ \gamma & \delta \end{pmatrix}, p\right) = \gamma p + \delta$$

also defines a multiplier.[†] Indeed, $\mu_1(\mathbb{1}, p) = 1$, and

$$\mu_1\left(\begin{pmatrix} \alpha\widetilde{\alpha} + \beta\widetilde{\gamma} & \alpha\widetilde{\beta} + \beta\widetilde{\delta} \\ \gamma\widetilde{\alpha} + \delta\widetilde{\gamma} & \gamma\widetilde{\beta} + \delta\widetilde{\delta} \end{pmatrix}, p\right) = (\gamma\widetilde{\alpha} + \delta\widetilde{\gamma})p + (\gamma\widetilde{\beta} + \delta\widetilde{\delta})$$

$$= \gamma(\widetilde{\alpha}p + \widetilde{\beta}) + \delta(\widetilde{\gamma}p + \widetilde{\delta}) = \left[\gamma\left(\frac{\widetilde{\alpha}p + \widetilde{\beta}}{\widetilde{\gamma}p + \widetilde{\delta}}\right) + \delta\right](\widetilde{\gamma}p + \widetilde{\delta})$$

$$= \mu_1\left(\begin{pmatrix} \alpha & \beta \\ \gamma & \delta \end{pmatrix}, \frac{\widetilde{\alpha}p + \widetilde{\beta}}{\widetilde{\gamma}p + \widetilde{\delta}}\right) \cdot \mu_1\left(\begin{pmatrix} \widetilde{\alpha} & \widetilde{\beta} \\ \widetilde{\gamma} & \widetilde{\delta} \end{pmatrix}, p\right),$$

as required by (4.21). Since we can multiply multipliers, the functions

$$\mu_{n,k}\left(\begin{pmatrix} \alpha & \beta \\ \gamma & \delta \end{pmatrix}, p\right) = (\alpha\delta - \beta\gamma)^{-k}(\gamma p + \delta)^{-n} \tag{4.22}$$

also define multipliers for the projective group. The restriction of this multiplier representation to the space $\mathcal{P}^n$ of polynomials of degree $\leq n$ coincides with the fundamental representation $\rho_{n,k}$ defined by (4.19).

There is a trivial way to obtain multipliers from the ordinary representation (4.3) of a transformation group G on the function space $\mathcal{F}(X)$. Suppose we multiply every function by a fixed nonvanishing function $\eta: X \to \mathbb{C} \setminus \{0\}$, and set $F^*(x) = \eta(x) F(x)$ for each $F \in \mathcal{F}$. The function η is sometimes known as a *gauge factor*.[‡] In terms of the

---

[†] Note that μ_1 vanishes at the point $p_0 = -\gamma/\delta$, but the projective group transformation maps p_0 to ∞, so consistency is maintained.

[‡] This terminology comes from physics and has its origins in Weyl's speculative attempt, [233], to unify electromagnetism and gravity. Our use of the term "gauge" is closer in spirit to Weyl's original coinage since, unlike the modern definition appearing in quantum electrodynamics, cf. [232, 22], our gauge factors $\eta(x)$ are not restricted to be of modulus 1. See [85, 169] for further developments and applications.

new choice of gauge, then, the standard representation (4.3) takes the modified form

$$\overline{F}^*(\bar{x}) = \frac{\eta(\bar{x})}{\eta(x)} F^*(x), \qquad \text{where} \qquad \bar{x} = g \cdot x, \quad \overline{F}^* = g \cdot F^*.$$

The function $\mu(g, x) = \eta(g \cdot x)/\eta(x)$ is readily seen to satisfy the multiplier equation (4.21) and so defines a *trivial multiplier*.

Definition 4.37. Two multipliers $\mu, \tilde{\mu} \colon G \times X \to \mathbb{C}$ are *gauge equivalent* if they are related by the formula

$$\tilde{\mu}(g, x) = \frac{\eta(g \cdot x)}{\eta(x)} \mu(g, x), \qquad (4.23)$$

for some nonzero function $\eta \colon X \to \mathbb{C} \setminus \{0\}$.

Equivalent multipliers prescribe the same underlying multiplier representation, up to multiplication by a function. For example, it can be shown, cf. [169], that the multipliers (4.22) provide a complete list of inequivalent multipliers for the projective action of the group GL(2).

Exercise 4.38. Consider the standard linear action of the group SL(2) on $\mathbb{R}^2$ given by matrix multiplication. Prove that, for $k \in \mathbb{R}$,

$$\mu_k \left(\begin{pmatrix} \alpha & \beta \\ \gamma & \delta \end{pmatrix} ; x, y \right) = \exp \left[\frac{k\gamma}{y(\gamma x + \delta y)} \right], \qquad \alpha\delta - \beta\gamma = 1, \quad (4.24)$$

defines a multiplier. In fact, these give a complete list of inequivalent multipliers for this group action, cf. [84], [169; Exercise 3.25]. Does the multiplier (4.24) extend to the full general linear group GL(2)?

Although we shall primarily deal with scalar multipliers, the definition readily extends to vector-valued functions. On the function space $\mathcal{F}^k(X)$, a multiplier will be a matrix-valued function $\mu \colon G \times X \to \text{GL}(k)$ that satisfies the multiplier equation (4.21); the order of the factors is now important in the first condition, and we replace 1 by the $k \times k$ identity matrix $\mathbb{1}$ in the second. In particular, any representation $\rho \colon G \to \text{GL}(k)$ defines a matrix multiplier $\mu = \rho(g)$ that does not depend on the spatial coordinates x.

Exercise 4.39. Suppose G acts smoothly on $X \subset \mathbb{R}^n$, with explicit transformation formulae $g \cdot x = w(g, x)$. Prove that the Jacobian matrix $J(g, x) = \partial w(g, x)/\partial x$ defines a matrix-valued multiplier. Use this to conclude that the Jacobian determinant $\mu(g, x) = \det J(g, x)$ forms a scalar multiplier for any transformation group. Determine the Jacobian multiplier for the linear fractional action of GL(2).

Relative Invariants

Earlier we characterized an invariant of a group of transformations as a fixed point of the induced representation on the space of functions: $g \cdot I = I$ for all $g \in G$. The analog of an invariant for a multiplier representation is known as a relative invariant.

Definition 4.40. Let $\mu: G \times X \to \mathbb{R}$ be a multiplier for a transformation group action. A *relative invariant* of *weight* μ is a function $R: X \to \mathbb{R}$ which satisfies

$$R(g \cdot x) = \mu(g, x) R(x). \tag{4.25}$$

It is not hard to see that, as long as $R \not\equiv 0$ satisfies (4.25), the weight function μ must necessarily be a multiplier for the group action. For clarity, ordinary invariants are, occasionally, referred to as *absolute invariants*. For example, the invariants appearing in classical invariant theory are absolute invariants for the unimodular group SL(2), but only relative invariants for the full group GL(2). The relevant multiplier is the determinantal power $\mu_k(A) = (\det A)^{-k}$, where k determines the weight of the invariant. Thus, (2.19) is a particular case of the general condition (4.25) for a relative invariant.

If R and S are relative invariants corresponding to the *same* multiplier μ, then any linear combination $c_1 R + c_2 S$ is also a relative invariant of weight μ. (However, this certainly does not hold if R and S are relative invariants corresponding to different multipliers!) If R has weight μ and S has weight ν, then their product $R \cdot S$ is a relative invariant for the product multiplier $\mu \cdot \nu$. In particular, the ratio R/S of two relative invariants having the *same* weight is an absolute invariant of the group. Therefore, once we know one relative invariant of a given weight μ, we can easily provide a complete list of all such relative invariants.

Proposition 4.41. *Let μ be a scalar multiplier for a transformation group G. If $R_0(x)$ is a nonvanishing relative invariant of weight μ, then every other relative invariant of weight μ has the form $R(x) = I(x)R_0(x)$, where I is any absolute invariant.*

Remark: Proposition 4.41 does *not* guarantee the existence of a nontrivial relative invariant. In the case of matrix multipliers, one typically requires several different relative invariants to form a basis (with respect to the absolute invariants) for the space of vector-valued relative invariants. See [**68**] for a general theorem that characterizes the precise number of relative invariants for "regular" matrix multipliers.

The classical theory can also be fitted into this general framework. A classical invariant $I(\mathbf{a})$ is a relatively invariant function $I: \mathcal{P}^{(n)} \to \mathbb{C}$ defined on the space of homogeneous polynomials of degree n under the induced representation of GL(2). A covariant $J(\mathbf{a}, \mathbf{x})$ can be viewed as a homogeneous function $J: \mathcal{P}^{(n)} \times \mathbb{C}^2 \to \mathbb{C}$ which is a *joint relative invariant* for the Cartesian product action with respect to the determinantal multiplier of weight k. In the case of projective coordinates, where $J: \mathcal{P}^{(n)} \times \mathbb{CP}^2 \to \mathbb{C}$, one replaces the determinantal multiplier by a suitable fundamental multiplier of the form (4.22) for appropriate values of k, the weight, and n, the degree, of the covariant.

The simplest example is the "tautologous" function

$$Q(\mathbf{a}, \mathbf{x}) = \sum_{i=0}^{n} \binom{n}{i} a_i \, x^i y^{n-i}, \qquad (4.26)$$

which defines a function $Q: \mathcal{P}^{(n)} \times \mathbb{C}^2 \to \mathbb{C}$. In other words, $Q(\mathbf{a}, \mathbf{x}) = Q(\mathbf{x})$ coincides with the original binary form, now considered as a function of both its coefficients and its arguments. The function Q forms a relative invariant whose weight equals the original weight of the binary form: $Q(\overline{\mathbf{a}}, \overline{\mathbf{x}}) = (\alpha\delta - \beta\gamma)^{-m} \, Q(\mathbf{a}, \mathbf{x})$. In practice, we shall revert to our old notation and, at the slight risk of confusion, identify the function Q with the (general) binary form Q.

A fixed point for a matrix multiplier representation will form a vector-valued relative invariant. These are much less common in classical invariant theory, but the following particular example plays an important role in establishing the symbolic method presented in the next chapter.

Exercise 4.42. Let G be any transformation group acting on $\mathbb{R}^n$. Prove that if $I(x)$ is any sufficiently smooth absolute invariant, then its gradient $\nabla I: \mathbb{R}^n \to \mathbb{R}^n$ defines a vector-valued relative invariant for the Jacobian multiplier representation of Exercise 4.39. (In differential geometric terms, this means that the Jacobian multiplier governs the induced action of a transformation group on the space of differential one-forms, cf. [**169**].)

An important case occurs when $G = $ GL(2) acts linearly on $\mathbb{R}^2$, and so the Jacobian multiplier coincides with the identity representation. Of course, there are no nonconstant absolute invariants, and so this case is not particularly interesting. However, consider the Cartesian product action on $\mathcal{P}^{(n)} \times \mathbb{R}^2$, and suppose $J(\mathbf{a}, \mathbf{x})$ is any absolute covariant. Then

$\nabla J = (J_x, J_y)^T$ will be a vector-valued relative invariant for the identity multiplier; in other words,

$$\alpha \bar{J}_{\bar{x}} + \gamma \bar{J}_{\bar{y}} = J_x, \qquad \beta \bar{J}_{\bar{x}} + \delta \bar{J}_{\bar{y}} = J_y. \qquad (4.27)$$

Therefore, the gradient operator satisfies

$$\nabla \longmapsto A^{-T} \nabla = \overline{\nabla}, \qquad \text{under the map} \qquad \mathbf{x} \longmapsto \overline{\mathbf{x}} = A\mathbf{x}, \qquad (4.28)$$

which indicates that it transforms like a covariant vector! Of course, one can establish this formula by a simple direct calculation; nevertheless, it is of interest to understand how it fits into the general theory of relative invariants and multiplier representations. More generally, if J is a covariant of weight k, then ∇J will be a relative invariant for the matrix-valued multiplier $\mu(A) = (\det A)^k \cdot A$ corresponding to the tensor product of the kth power of the determinantal representation with the identity representation.

Exercise 4.43. Determine what the gradient operator looks like in projective coordinates. What is the appropriate matrix multiplier?

Chapter 5

Transvectants

Of all the available methods for computing invariants and covariants, the most incisive are those based on certain invariant differential operators. The Hessian of a binary form, and the Jacobian of a pair of forms, are just special cases of a general method of constructing covariants known as "transvection". Not only does this process provide another method for constructing explicit covariants and invariants, but, in fact, *every* covariant and invariant of any system of binary forms arises using these fundamental operations. These ideas have their origins in Cayley's old theory of "hyperdeterminants", [38, 39], and culminated in their general formulation by the German school of invariant theorists led by Aronhold, [12], Clebsch, [49], and Gordan, [89]. The transvection process is fundamental in the implementation of Gordan's algorithm, [86, 92], for constructing a complete system of independent covariants.

In our presentation, we shall concentrate on the formulation of transvectants based on invariant differential operators or "invariant processes" rather than the more traditional symbolic approach. This point of view has several advantages. First, the formulae for the transvection processes are explicitly given in terms of derivatives of the polynomials, rather than the more shadowy "umbral" or symbolic form. As such, one can immediately apply such invariant processes to more general functions and thereby simultaneously implement an "invariant theory of smooth or analytic functions" without additional work. In fact, the homogeneous polynomials will be seen to play a somewhat anomalous, and computationally more difficult, role in the general theory. Second, the use of differential operators makes clear the fascinating and mostly unexplored connections between classical invariant theory and a wide variety of recent developments in mathematics and mathematical physics, including solitons and integrable systems, quantum groups, quantization of Poisson brackets, and the calculus of variations. Finally, the differential operator formalism actually subsumes the more traditional symbolic method, in a sense to be made more precise in the following chapter. The one disadvantage is that such methods are, strictly speaking, outside the traditional algebraic sphere and are only directly effective in fields of characteristic zero.

The Omega Process

The transvection method is based on an important invariant differential operator originally introduced by Cayley, [**37**]. As usual, we work exclusively with binary forms here, leaving the higher dimensional generalizations until Chapter 10.

The key idea is that invariants and covariants of binary forms can be often recovered by reduction or trace, (4.12), of certain naturally defined joint invariants. In the simplest case, we consider the joint action of $GL(2, \mathbb{C})$ on the Cartesian product space $\mathbb{C}^2 \times \mathbb{C}^2$, given by simultaneous linear transformations $(\mathbf{x}_\alpha, \mathbf{x}_\beta) \mapsto (A\mathbf{x}_\alpha, A\mathbf{x}_\beta)$ for $\mathbf{x}_\alpha, \mathbf{x}_\beta \in \mathbb{C}^2$. We shall find it convenient to label the variables in each copy of $\mathbb{C}^2$ by a different "symbolic letter" $\alpha, \beta, \gamma, \dots$; see below for details. For any integer k, there is an induced determinantal multiplier representation ρ_k of weight k, whose action on functions is given by

$$\overline{F}(\overline{\mathbf{x}}_\alpha, \overline{\mathbf{x}}_\beta) = (\det A)^k F(\mathbf{x}_\alpha, \mathbf{x}_\beta), \qquad \text{when} \qquad \begin{aligned} \overline{\mathbf{x}}_\alpha &= A\mathbf{x}_\alpha, \\ \overline{\mathbf{x}}_\beta &= A\mathbf{x}_\beta. \end{aligned} \qquad (5.1)$$

Cayley's invariant process is given by a simple second order differential operator that acts on the two-fold relative invariants.

Definition 5.1. The second order differential operator

$$\Omega_{\alpha\beta} = \det \mathbf{\Omega}_{\alpha\beta} = \det \begin{vmatrix} \dfrac{\partial}{\partial x_\alpha} & \dfrac{\partial}{\partial y_\alpha} \\ \dfrac{\partial}{\partial x_\beta} & \dfrac{\partial}{\partial y_\beta} \end{vmatrix} = \frac{\partial^2}{\partial x_\alpha \partial y_\beta} - \frac{\partial^2}{\partial x_\beta \partial y_\alpha} \qquad (5.2)$$

is known as the *omega process* with respect to the variables $\mathbf{x}_\alpha = (x_\alpha, y_\alpha)$ and $\mathbf{x}_\beta = (x_\beta, y_\beta)$.

If we subject the variables to a simultaneous linear transformation, $\mathbf{x}_\alpha \mapsto A\mathbf{x}_\alpha$, $\mathbf{x}_\beta \mapsto A\mathbf{x}_\beta$, then the matrix differential operator $\mathbf{\Omega}_{\alpha\beta}$ whose determinant defines the omega process is covariantly transformed: $\mathbf{\Omega}_{\alpha\beta} \mapsto A^{-T}\mathbf{\Omega}_{\alpha\beta}$. This observation follows immediately from the covariant transformation rules (4.28) for the gradient operator. Taking the determinant, we find that, under a simultaneous linear transformation, the omega process is merely multiplied by the determinant:

$$\Omega_{\alpha\beta} \longmapsto (\det A)^{-1} \Omega_{\alpha\beta}. \qquad (5.3)$$

This is the fundamental property that facilitates the construction of general covariants for (systems of) binary forms.

Definition 5.2. The rth order *transvectant* of a pair of smooth functions $Q(x,y)$, $R(x,y)$ is the function

$$(Q,\,R)^{(r)} = (\Omega_{\alpha\beta})^r \left[Q(x_\alpha, y_\alpha)\, R(x_\beta, y_\beta) \right] \Big|_{\substack{x=x_\alpha=x_\beta \\ y=y_\alpha=y_\beta}}. \qquad (5.4)$$

The low order cases are computed as follows. First,

$$(Q,\,R)^{(0)} = Q(x_\alpha, y_\alpha) \cdot R(x_\beta, y_\beta) \Big|_{\substack{x=x_\alpha=x_\beta \\ y=y_\alpha=y_\beta}} = Q(x,y)\, R(x,y),$$

so the zeroth transvectant is just the product of the two functions, whose covariance was already noted. Next,

$$(Q,\,R)^{(1)} = \Omega_{\alpha\beta} \left[Q(x_\alpha, y_\alpha)\, R(x_\beta, y_\beta) \right] \Big|_{\substack{x=x_\alpha=x_\beta \\ y=y_\alpha=y_\beta}}$$

$$= \left[\frac{\partial Q}{\partial x}(x_\alpha, y_\alpha) \frac{\partial R}{\partial y}(x_\beta, y_\beta) - \frac{\partial Q}{\partial y}(x_\alpha, y_\alpha) \frac{\partial R}{\partial x}(x_\beta, y_\beta) \right] \Big|_{\substack{x=x_\alpha=x_\beta \\ y=y_\alpha=y_\beta}}$$

$$= Q_x(x,y) R_y(x,y) - Q_y(x,y) R_x(x,y),$$

so that the first transvectant of two functions is the same as their Jacobian determinant. The second transvectant is computed similarly; the result is the polarized Hessian covariant (2.35),

$$(Q,\,R)^{(2)} = Q_{xx} R_{yy} - 2 Q_{xy} R_{xy} + Q_{yy} R_{xx}. \qquad (5.5)$$

This operation is less familiar but does appear as the basic "Monge–Ampère bracket" in the von Kármán equations that model plate mechanics in elasticity, cf. [11; p. 596]. In particular, specializing to $Q = R$ reduces (5.5) to the Hessian covariant

$$H[Q] = \tfrac{1}{2}(Q,\,Q)^{(2)} = Q_{xx} Q_{yy} - Q_{xy}^2, \qquad (5.6)$$

realized as one-half the second transvectant of the form with itself. In fact, there is a simple explicit formula for the rth transvectant, which is proved by a direct expansion of $(\Omega_{\alpha\beta})^r$ via the Binomial Theorem:

$$(Q,\,R)^{(r)} = \sum_{i=0}^{r} (-1)^i \binom{r}{i} \frac{\partial^r Q}{\partial x^{r-i} \partial y^i} \frac{\partial^r R}{\partial x^i \partial y^{r-i}}. \qquad (5.7)$$

Exercise 5.3. Prove that the invariant i of the binary quartic, (2.29), is a multiple of the fourth transvectant of the quartic with itself: $i = \frac{1}{1152}(Q,\,Q)^{(4)}$. Further, prove that the second invariant (2.30) of a binary quartic is a multiple of the fourth transvectant of Q with its Hessian: $j = c\,(Q,\,H)^{(4)} = \frac{c}{2}\,(Q,\,(Q,\,Q)^{(2)})^{(4)}$. By direct computation, determine the numerical factor c.

The rth transvectant $(Q,\,R)^{(r)}$ is symmetric or skew-symmetric under interchange of Q and R depending on whether r is even or odd:

$$(Q, R)^{(r)} = (-1)^r\,(R, Q)^{(r)}. \tag{5.8}$$

In particular, any odd transvectant $(Q,\,Q)^{(2j+1)}$ of a form with itself automatically vanishes. Clearly, if Q and R are homogeneous polynomials of respective degrees n and m, then their r^{th} transvectant $(Q,\,R)^{(r)}$ is a homogeneous polynomial of degree $n + m - 2r$. In particular, if r is greater than the degree of Q or the degree of R, then $(Q,\,R)^{(r)} \equiv 0$. The fact that the rth transvectant (5.4) of Q and R is a joint covariant of Q, R follows directly from the covariance of the omega process, (5.3).

Theorem 5.4. *If Q and R have respective weights k and j, then their r^{th} transvectant $(Q,\,R)^{(r)}$ is a joint covariant of weight $k + j + r$. If, in addition, Q and R are homogeneous functions of respective degrees n and m, then $(Q,\,R)^{(r)}$ is homogeneous of degree $n + m - 2r$.*

Remark: The omega process is, interestingly, related to the Hirota bilinear operator $\mathbb{D}_x$, which arises in the analysis of soliton equations, [**109, 159**]. Specifically, if we apply Ω to functions of the form $Q(x, y) = f(x)e^y$, $R(x, y) = g(x)e^y$, we recover the Hirota operator applied to f, g. Therefore, the rth transvectant of such functions is given by

$$e^{-2y}\,(f(x)e^y, g(x)e^y)^{(r)} = \mathbb{D}_x^r\,[\,f \cdot g\,]. \tag{5.9}$$

With the basic operations at hand, one can now build up a very large collection of covariants (and invariants) by taking successive transvectants. Starting with the binary form Q of degree n, one first forms the quadratic covariants $S_i = (Q, Q)^{(2i)}$, which are nonzero provided $2i \leq n$. If the degree $n = 2m$ of Q is even, then the last of these, S_m, is an invariant. Next, one can take transvectants of the S_i's with the form itself, yielding further covariants $T_{i,j} = (Q, S_i)^{(j)}$, or, more generally, among the S_i's themselves: $V_{i,j,k} = (S_i, S_j)^{(k)}$. And so on. Moreover, yet more invariants and covariants can be produced by composing these transvectants. In a similar fashion, both iterated transvectants and composition can be used to produce large quantities of joint covariants and invariants for arbitrary systems of binary forms. Thus, the main problem is not producing covariants but, rather, systematizing their construction and determining which ones are fundamental.

The First Fundamental Theorem of Invariant Theory, as proved in Chapter 6, states that *all* polynomial covariants and invariants of any system of binary forms can be expressed as linear combinations of iterated transvectants. Consequently, every joint polynomial covariant of the binary forms $Q_1, \ldots, Q_m$ can be written as a homogeneous constant coefficient differential polynomial in the forms. Of course, not all of these transvectants are independent, and so the key problem is to construct a fundamental system of algebraically independent covariants.

Exercise 5.5. Express the discriminant of a cubic in terms of iterated transvectants.

Projective Coordinates

Since transvectants preserve the class of homogeneous functions, one can re-express them in terms of the projective coordinate $p = x/y$. The resulting ordinary differential polynomials turn out to be quite interesting in their own right and facilitate proofs of some of the basic transvectant properties. We begin with the basic projective transvectant formula, cf. [96, 174].

Theorem 5.6. *Let $Q(p)$ and $R(p)$ be polynomials of degrees n and m, respectively. For $r \leq \min\{n, m\}$, the rth transvectant of Q and R is the polynomial*

$$(Q, R)^{(r)} = r! \sum_{k=0}^{r} (-1)^k \binom{n-r+k}{k} \binom{m-k}{r-k} Q^{(r-k)}(p) R^{(k)}(p).$$

$$(5.10)$$

Proof: Formula (5.10) is proved inductively by differentiating the formula (2.3) that relates a homogeneous polynomial and its projective counterpart. The details are left to the reader.   *Q.E.D.*

Example 5.7. Let Q have degree n and R have degree m. The simplest cases of (5.10) are

$$(Q, R)^{(0)} = QR, \tag{5.11}$$
$$(Q, R)^{(1)} = mQ'R - nQR',$$
$$(Q, R)^{(2)} = m(m-1)Q''R - 2(m-1)(n-1)Q'R' + n(n-1)QR'',$$
$$(Q, R)^{(3)} = m(m-1)(m-2)Q'''R - 3(m-1)(m-2)(n-2)Q''R' +$$
$$+ 3(m-2)(n-1)(n-2) - n(n-1)(n-2)QR''.$$

In particular, the Hessian of a polynomial Q of degree n is

$$H[Q] = \tfrac{1}{2}(Q, Q)^{(2)} = n(n-1) \left[QQ'' - \frac{n-1}{n}(Q')^2 \right]. \qquad (5.12)$$

Formula (5.12) provides a simple direct proof of Proposition 2.23, which states that a polynomial has zero Hessian if and only if it is the power of a linear form. Indeed, for $n \geq 2$, we have $H \equiv 0$ if and only if Q satisfies the second order ordinary differential equation

$$Q\,Q'' = \frac{n-1}{n}(Q')^2. \qquad (5.13)$$

This can be straightforwardly integrated. If $Q\,Q' \neq 0$, then

$$\log Q' = \int \frac{Q''}{Q'}\,dp = \frac{n-1}{n}\int \frac{Q'}{Q}\,dp = \frac{n-1}{n}\log Q + c,$$

and hence

$$Q' = k\,Q^{(n-1)/n},$$

for some constant k. One further integration produces the general solution to (5.13),

$$Q(p) = (ap+b)^n,$$

demonstrating that Q is the nth power of a linear polynomial. (The singular solutions $Q \equiv$ constant are included in the general formula.)

Theorem 5.4 holds unchanged in the projective context. In particular, if Q is a polynomial of weight k and degree n, and R has weight j and degree m, then the rth transvectant $(Q, R)^{(r)}$ will be a polynomial covariant of weight $k+j+r$ and degree $n+m-2r$. The latter fact does *not* (at least to me) seem obvious from the explicit formula (5.10).

Exercise 5.8. Let Q be a polynomial of weight k and degree n. Prove that the iterated transvectant $T = (Q, H)^{(1)} = \tfrac{1}{2}(Q, (Q, Q)^{(2)})^{(1)}$, which is a covariant of weight $3k+3$ and degree $3n-6$, has the following formula in projective coordinates:

$$T = -n^2(n-1) \left[Q^2Q''' - 3\frac{(n-2)}{n}QQ'Q'' + 2\frac{(n-1)(n-2)}{n^2}(Q')^3 \right]. \qquad (5.14)$$

Use this result to prove that

$$U = (Q, T)^{(1)} = n^3(n-1)\,V - 3\frac{(n-2)}{(n-1)}H^2, \qquad (5.15)$$

where

$$V = Q^3 Q'''' - 4 \frac{(n-3)}{n} Q^2 Q' Q'' + 6 \frac{(n-2)(n-3)}{n^2} Q Q'^2 Q'' +$$
$$+ \frac{(n-1)(n-2)(n-3)}{n^3} (Q')^4. \tag{5.16}$$

Explain why V is a covariant and determine its weight and order.

Remark: Formula (5.10) plays an important role in the theory of automorphic and modular forms. By definition, an automorphic form is a polynomial $Q(p)$ which forms a relative invariant for the restriction of a fundamental multiplier representation $\rho_{n,k}$ to a discrete subgroup $\Gamma \subset GL(2)$, that is, the polynomial satisfies (4.19) for all matrices $A \in \Gamma$. The particularly important case of modular forms occurs when $\Gamma = SL(2, \mathbb{Z})$. Theorem 5.6 implies that any transvectant of two automorphic forms is also automorphic. For the Hessian covariant, this result was apparently first noted by Rankin, [**182**]; it was generalized to higher transvectants by H. Cohen, [**52**]. However, the connection with classical transvectants demonstrates that the result holds for *any* — not only discrete — subgroup of GL(2). Modular forms play a crucial role in Wiles' proof of Fermat's Last Theorem, cf. [**236, 185**].

Remark: In [**175**], the rth order transvectant (5.10) is called the "Lie derivative of order r". The authors study their invariance under more general change of variables. Applications to the Adler–Gel'fand–Dikii bracket, which plays a crucial role in the theory of soliton systems, [**6, 78**], and, more recently, quantum groups and generalizations of the Virasoro algebra are under investigation, [**83**]. See also [**119**] for applications in Fourier analysis and physics, where transvectants are interpreted as generalized Hankel operators.

Partial Transvectants

As we remarked earlier, one can compute transvectants, not just of homogeneous polynomials, but of any sufficiently differentiable functions. They will continue to retain their basic covariance properties. However, in contrast to the case of homogeneous polynomials, or even homogeneous functions, transvectants do *not* form the most general covariant differential polynomials of arbitrary functions, and we must introduce a slight generalization, which we call a "partial transvectant", which essentially coincides with Cayley's original hyperdeterminant operator,

[**38**, **39**]. See [**37**; p. 585, note 14] for Cayley's own remarks on how they generalize transvectants. Grace and Young, [**92**; §81] call them generalized transvectants.

For this purpose, and in order to effectively compute identities among successive transvectants, it is helpful to introduce a more transparent notation. Given functions $P(x,y)$, $Q(x,y)$, we let $P \otimes Q$ denote the joint function $P(x_\alpha, y_\alpha) Q(x_\beta, y_\beta)$ defined on $\mathbb{C}^2 \times \mathbb{C}^2$. Equating the arguments (x_α, y_α) and (x_β, y_β) of the product $P \otimes Q$ will be identified as the *trace* operation

$$\operatorname{tr}(P \otimes Q) \equiv P(x_\alpha, y_\alpha)Q(x_\beta, y_\beta)\bigg|_{\substack{x=x_\alpha=x_\beta \\ y=y_\alpha=y_\beta}} = P(x,y)\,Q(x,y). \quad (5.17)$$

Thus, the kth transvectant (5.4) of P, Q can be simply written as

$$(P,\,Q)^{(k)} = \operatorname{tr}(\Omega_{\alpha\beta})^k (P \otimes Q). \quad (5.18)$$

More generally, the three-fold function $P(x_\alpha, y_\alpha)Q(x_\beta, y_\beta)R(x_\gamma, y_\gamma)$ will be denoted by $P \otimes Q \otimes R$, so that $\operatorname{tr}(P \otimes Q \otimes R) = P(x,y)Q(x,y)R(x,y)$, and so on. Different symbolic letters $\alpha, \beta, \gamma, \ldots$ will be used to label the different variables appearing in the "tensor product".

Given a product $\mathbf{Q} = Q_\alpha \otimes \cdots \otimes Q_\varepsilon$ of several functions $Q_\alpha, \ldots, Q_\varepsilon$, some or all of which may be the same, there are various "omega processes" which can be applied to it. Indeed, for each pair of variables, e.g., $\mathbf{x}_\alpha = (x_\alpha, y_\alpha)$, $\mathbf{x}_\beta = (x_\beta, y_\beta)$, there is a corresponding omega process, $\Omega_{\alpha\beta}$, each of which produces a (simultaneous) covariant of the functions $Q_\alpha, Q_\beta, \ldots$. Note that $\Omega_{\beta\alpha} = -\Omega_{\alpha\beta}$, while $\Omega_{\alpha\alpha} = 0$.

Definition 5.9. An rth order *partial transvectant* of a system of homogeneous functions $Q_\alpha, \ldots, Q_\varepsilon$ is a differential polynomial expression of the form

$$\operatorname{tr} \mathcal{D}\, \mathbf{Q} = \operatorname{tr} \left\{ \prod_{\nu=1}^{r} \Omega_\nu \right\} Q_\alpha \otimes \cdots \otimes Q_\varepsilon, \quad (5.19)$$

where the differential operator $\mathcal{D} = \prod \Omega_\nu$ is an r-fold product of various omega processes, each of whose (distinct) labels are chosen from those symbolic letters $\alpha, \beta, \ldots, \varepsilon$ appearing in $\mathbf{Q}$.

Example 5.10. If P, Q, R are functions, then their possible first order partial transvectants are

$$\operatorname{tr} \Omega_{\alpha\beta}(P \otimes Q \otimes R) = (P_x Q_y - P_y Q_x)R,$$
$$\operatorname{tr} \Omega_{\alpha\gamma}(P \otimes Q \otimes R) = (P_x R_y - P_y R_x)Q,$$
$$\operatorname{tr} \Omega_{\beta\gamma}(P \otimes Q \otimes R) = (Q_x R_y - Q_y R_x)P.$$

The labels α, β, γ correspond to P, Q, R, respectively. Another interesting example is the second order partial transvectant

$$\operatorname{tr} \Omega_{\alpha\gamma} \Omega_{\beta\gamma} (P \otimes Q \otimes R) = P_x Q_x R_{yy} - (P_x Q_y + P_y Q_x) R_{xy} + P_y Q_y R_{xx},$$

which, in the special case $P = Q = R$, has been proposed as an edge detector in digital images, cf. [142].

Exercise 5.11. Prove that when Q is homogeneous, the partial transvectant $\operatorname{tr} \Omega_{\alpha\gamma} \Omega_{\beta\gamma} (Q \otimes Q \otimes Q) = Q_x^2 Q_{yy} - 2Q_x Q_y Q_{xy} + Q_y^2 Q_{xx}$ coincides with a multiple of the product QH, where $H = \frac{1}{2}(Q, Q)^{(2)}$.

Since, under simultaneous transformation of the $\mathbf{x}_\alpha$, $\mathbf{x}_\beta$, $\ldots$, the omega processes are all invariant under GL(2) (up to a determinantal factor), every r^{th} order partial transvectant of a system of functions provides a joint covariant of that system.

Theorem 5.12. *Suppose that each function Q_ν has weight k_ν. Then the rth order partial transvectant* (5.19) *is a covariant of weight $k = k_\alpha + \cdots + k_\varepsilon + r$. If each Q_ν is homogeneous of degree n_ν, then the rth order partial transvectant* (5.19) *is also homogeneous of degree $n = n_\alpha + \cdots + n_\varepsilon - 2r$.*

Remark: In general, partial transvectants are *not* expressible in terms of ordinary transvectants; however, in most cases, if the functions Q_κ are homogeneous, then every partial transvectant is equal to a linear combination of ordinary (iterated) transvectants; see Theorem 7.10.

Identities among iterated partial (and ordinary) transvectants can be effectively computed. Consider an rth order transvectant

$$(Q, R)^{(r)} = \operatorname{tr} (\Omega_{\alpha\lambda})^r (Q_\alpha \otimes R), \tag{5.20}$$

between a function Q_α and a partial transvectant

$$R = \operatorname{tr} \mathcal{D} (Q_\beta \otimes \cdots \otimes Q_\varepsilon). \tag{5.21}$$

In (5.20) we are adopting a new label λ to index R in the tensor product $Q_\alpha \otimes R$, so as not to be confused with the indices $\beta, \ldots, \varepsilon$ in the partial transvectant formula (5.21) for R itself. In essence, this is the primary reason for our alphabetical labeling convention, as opposed to labeling by, say, positive integers, which soon becomes unwieldy in such compound formulae. The basic rule for iterated transvectants is as follows:

$$(Q_\alpha, R)^{(r)} = \operatorname{tr} (\Omega_{\alpha\lambda})^r \big[Q_\alpha \otimes \operatorname{tr} \mathcal{D} (Q_\beta \otimes \cdots \otimes Q_\varepsilon) \big] \tag{5.22}$$

$$= \operatorname{tr} \big(\Omega_{\alpha\beta} + \Omega_{\alpha\gamma} + \cdots + \Omega_{\alpha\varepsilon} \big)^r \mathcal{D} (Q_\alpha \otimes Q_\beta \otimes \cdots \otimes Q_\varepsilon).$$

Note that we can algebraically expand the right-hand side of (5.22) into a linear combination of partial transvectants.

Example 5.13. The Jacobian of P with the second transvectant of Q and R is given as a linear combination of two partial transvectants:

$$(P, (Q, R)^{(2)})^{(1)} = \operatorname{tr} \Omega_{\alpha\lambda} \left[P \otimes \operatorname{tr} (\Omega_{\beta\gamma})^2 (Q \otimes R) \right]$$
$$= \operatorname{tr} (\Omega_{\alpha\beta} + \Omega_{\alpha\gamma})(\Omega_{\beta\gamma})^2 (P \otimes Q \otimes R).$$

Therefore $(P, (Q, R)^{(2)})^{(1)}$ is the sum of the two partial transvectants

$$\operatorname{tr} \Omega_{\alpha\beta}(\Omega_{\beta\gamma})^2 (P \otimes Q \otimes R) = P_x \left(Q_{xxy} R_{yy} - 2 Q_{xyy} R_{xy} + Q_{yyy} R_{xx} \right) -$$
$$- P_y \left(Q_{xxx} R_{yy} - 2 Q_{xxy} R_{xy} + Q_{xyy} R_{xx} \right),$$
$$\operatorname{tr} \Omega_{\alpha\gamma}(\Omega_{\beta\gamma})^2 (P \otimes Q \otimes R) = P_x \left(Q_{xx} R_{yyy} - 2 Q_{xy} R_{xyy} + Q_{yy} R_{xxy} \right) -$$
$$- P_y \left(Q_{xx} R_{xyy} - 2 Q_{xy} R_{xxy} + Q_{yy} R_{xxx} \right).$$
$$(5.23)$$

In particular, if $P = Q = R$, then the two partial transvectants are equal, and we recover our earlier formula (2.27) for the transvectant $T = (Q, H)^{(1)} = \frac{1}{2} (Q, (Q, Q)^{(2)})^{(1)}$.

Similar computations will provide formulae for the composition of covariants defined by ordinary or partial transvectants. A simple example will suffice to illustrate the basic idea. Consider the composed covariant $i \circ H(Q)$ discussed earlier for a binary quartic, which, up to multiple, can be identified with the iterated transvectant $((Q, Q)^{(2)}, (Q, Q)^{(2)})^{(4)}$. We compute using a variant of the basic rule (5.22),

$$((Q, Q)^{(2)}, (Q, Q)^{(2)})^{(4)} =$$
$$= \operatorname{tr} (\Omega_{\lambda\mu})^4 \left\{ \left[\operatorname{tr} (\Omega_{\alpha\beta})^2 (Q \otimes Q) \right] \otimes \left[\operatorname{tr} (\Omega_{\gamma\delta})^2 (Q \otimes Q) \right] \right\}$$
$$= \operatorname{tr} \left[\Omega_{\alpha\gamma} + \Omega_{\alpha\delta} + \Omega_{\beta\gamma} + \Omega_{\beta\delta} \right]^4 (\Omega_{\alpha\beta})^2 (\Omega_{\gamma\delta})^2 (Q \otimes Q \otimes Q \otimes Q).$$

The last expression can clearly be expanded into a linear combination of partial transvectants. In general, then, if $R = \operatorname{tr} \mathcal{D}(Q_{\alpha_1} \otimes \cdots \otimes Q_{\alpha_k})$ and $S = \operatorname{tr} \mathcal{E}(Q_{\beta_1} \otimes \cdots \otimes Q_{\beta_l})$ are any two partial transvectants, then their r^{th} transvectant is given as

$$(R, S)^{(r)} = \operatorname{tr} \left(\sum_{\kappa=1}^{k} \sum_{\nu=1}^{l} \Omega_{\alpha_\kappa \beta_\nu} \right)^r \mathcal{D} \cdot \mathcal{E} \left(Q_{\alpha_1} \otimes \cdots \otimes Q_{\beta_l} \right). \quad (5.24)$$

The Fundamental Theorem on iterated transvectants is a straightforward consequence of the basic rules (5.22) and (5.24).

Theorem 5.14. *Every iterated partial transvectant can be expressed as a linear combination of partial transvectants.*

The Scaling and Polarization Processes

Besides the omega process, there is a second important invariant differential operator, namely, the *scaling process*

$$\sigma = \mathbf{x} \cdot \nabla = x \, \partial_x + y \, \partial_y. \qquad (5.25)$$

Combining the covariant transformation rules (4.28) for the gradient with the contravariant transformation rules (2.4) for the coordinates, we see that σ is an *invariant operator* for the general linear group. Consequently, given any function $Q(x, y)$, the function $\sigma(Q) = xQ_x + yQ_y$ is covariant having the same weight as Q. Of course, if Q is a homogeneous function of degree n, then Euler's formula (2.13) implies $\sigma(Q) = nQ$, so the scaling process just rescales homogeneous functions, and will only produce something new when applied to inhomogeneous functions. The scaling process can be slightly generalized.

Definition 5.15. The *polarization process* corresponding to variables $\mathbf{x}_\alpha = (x_\alpha, y_\alpha)$ and $\mathbf{x}_\beta = (x_\beta, y_\beta)$ is the differential operator

$$\sigma_{\alpha\beta} = x_\alpha \frac{\partial}{\partial x_\beta} + y_\alpha \frac{\partial}{\partial y_\beta}. \qquad (5.26)$$

In particular $\sigma_{\alpha\alpha} = \sigma_\alpha = x_\alpha \partial_{x_\alpha} + y_\alpha \partial_{y_\alpha}$ is just the scaling process applied to the αth function in a tensor product $\mathbf{Q}$. Note that

$$\sigma\big[\operatorname{tr}(Q_\alpha \otimes \cdots \otimes Q_\varepsilon)\big] = \operatorname{tr}(\sigma_\alpha + \cdots + \sigma_\varepsilon)(Q_\alpha \otimes \cdots \otimes Q_\varepsilon). \qquad (5.27)$$

Proposition 5.16. *Let* $Q_\alpha, \ldots, Q_\varepsilon$ *be covariants having respective weights* $i_\alpha, \ldots, i_\varepsilon$. *Then any polarization* $\operatorname{tr} \sigma_{\alpha\beta}(Q_\alpha \otimes \cdots \otimes Q_\varepsilon)$ *is a covariant of weight* $i = i_\alpha + \cdots + i_\varepsilon$.

Actually, this is completely trivial, since upon taking the trace the variables $\mathbf{x}_\alpha$ are all set equal, and so the initial index in $\sigma_{\alpha\beta}$ is, in a sense, irrelevant. The First Fundamental Theorem will imply that every differential polynomial covariant — meaning every covariant which can be expressed as a differential polynomial — of any system of functions can be expressed in terms of iterated partial transvectants and scalings.

Exercise 5.17. Prove the following commutation formulae for the polarization and omega processes:

$$\sigma_{\alpha\beta}\, \sigma_{\gamma\varepsilon} - \sigma_{\gamma\varepsilon}\, \sigma_{\alpha\beta} = \delta^\beta_\gamma\, \sigma_{\alpha\varepsilon} - \delta^\alpha_\varepsilon\, \sigma_{\gamma\beta},$$

$$\Omega_{\alpha\beta}\, \sigma_{\gamma\varepsilon} - \sigma_{\gamma\varepsilon}\, \Omega_{\alpha\beta} = \delta^\beta_\gamma\, \Omega_{\alpha\varepsilon} - \delta^\alpha_\gamma\, \Omega_{\beta\varepsilon}, \qquad (5.28)$$

where δ_α^β denotes the Kronecker delta. Note also that the omega operators all mutually commute.

The Poisson and Moyal Brackets

The explicit forms (5.7) of the transvectants appear, strikingly, in the definition of the "Moyal bracket" which arises in quantum mechanics as the essentially unique deformation of the classical Poisson bracket, [**154, 224, 19**]. On a two-dimensional phase space $X = \mathbb{R}^2$, the *Poisson bracket* between two functions (or "observables") $P(x, y)$, $Q(x, y)$ is the same as their Jacobian,

$$\{P, Q\} = (P, Q)^{(1)} = P_x Q_y - P_y Q_x. \tag{5.29}$$

The Poisson bracket plays a fundamental role in the Hamiltonian theory of classical mechanical systems, cf. [**1, 168**].

Exercise 5.18. Prove that the Poisson bracket (5.29) satisfies the following basic properties:

Linearity: $\qquad \{P + R, Q\} = \{P, Q\} + \{R, Q\},$

Skew-Symmetry: $\quad \{P, Q\} = -\{Q, P\},$ $\qquad\qquad$ (5.30)

Leibniz Rule: $\qquad \{P, Q \cdot R\} = \{P, Q\} \cdot R + Q \cdot \{P, R\},$

Jacobi Identity: $\quad \{P, \{Q, R\}\} + \{R, \{P, Q\}\} + \{Q, \{R, P\}\} = 0.$

Our iterated transvectant formula (5.22) can be regarded as a generalization of the Leibniz and Jacobi properties, which correspond to the zero[th] and first order transvectants between Q and R.

Regarding classical mechanics as a limiting case of quantum mechanics, as Planck's constant $\hbar \to 0$, motivates one to investigate the deformations of the Poisson bracket operation. We shall use t (which is typically a multiple of $\hbar$) to denote the deformation parameter.

Definition 5.19. Let $P(x, y)$, $Q(x, y)$ be smooth functions, and t a scalar parameter. The *star product* of P and Q is the formal series

$$P \star_t Q = \mathrm{tr}\,(\exp t\,\Omega_{\alpha\beta})(P \otimes Q) = \sum_{n=0}^{\infty} \frac{t^n}{n!}\,(P, Q)^{(n)}. \tag{5.31}$$

Note that the series (5.31) terminates if P, Q are polynomials. The covariance properties of the transvectants imply that the star product is invariant under the unimodular group SL(2). More generally, transforming by $A \in \mathrm{GL}(2)$ has the same effect as scaling the parameter t

by the determinantal factor $\det A$. The star product is the essentially unique deformation of the multiplicative product $(P, Q) \mapsto P \cdot Q$.

Proposition 5.20. *The star product is associative:*

$$P \star_t (Q \star_t R) = (P \star_t Q) \star_t R. \qquad (5.32)$$

Proof: We compute using (5.22) and the fact that the omega process operators mutually commute:

$$
\begin{aligned}
P \star_t (Q \star_t R) &= \mathrm{tr} \left(\exp t\, \Omega_{\alpha\lambda} \right) \left[P \otimes \mathrm{tr} \left(\exp t\, \Omega_{\beta\gamma} \right) (Q \otimes R) \right] \\
&= \mathrm{tr} \left[\exp t\, (\Omega_{\alpha\beta} + \Omega_{\alpha\gamma}) \cdot (\exp t\, \Omega_{\beta\gamma})(P \otimes Q \otimes R) \right] \\
&= \mathrm{tr} \left[\exp t\, (\Omega_{\alpha\beta} + \Omega_{\alpha\gamma} + \Omega_{\beta\gamma})(P \otimes Q \otimes R) \right].
\end{aligned}
$$

The final expression clearly equals $(P \star_t Q) \star_t R$ (and also $R \star_t (P \star_t Q)$, $Q \star_t (R \star_t P)$, etc.). *Q.E.D.*

Exercise 5.21. The associativity (5.32) of the star product implies a number of interesting iterated transvectant identities, found by equating the various powers of the deformation parameter t. The first of these is a Jacobian identity

$$(P,\, Q \cdot R)^{(1)} + P \cdot (Q,\, R)^{(1)} = (P \cdot Q,\, R)^{(1)} + (P,\, Q)^{(1)} \cdot R, \qquad (5.33)$$

which follows from the Leibniz rule (5.30). Write out the transvectant identities provided by the second and third order terms in (5.32).

Since even transvectants are symmetric while odd ones are skew-symmetric, $P \star_t Q = Q \star_{(-t)} P$. The *Moyal bracket* is the "odd" part of the star product[†]

$$[\![P, Q]\!]_t = \frac{P \star_t Q - Q \star_t P}{2t} = \mathrm{tr} \left(\frac{\sinh t\, \Omega_{\alpha\beta}}{t} \right) P \otimes Q, \qquad (5.34)$$

which, by the associativity of the star product, automatically satisfies the Jacobi identity, (5.30), and provides a quantum deformation of the Poisson bracket: $[\![P, Q]\!]_t = t\, \{P, Q\} + \mathrm{O}(t^3)$. Unfortunately, space precludes a more detailed development of these fascinating connections.

[†] Some authors replace t by $\sqrt{-1}\, t$, converting the hyperbolic sinh to a trigonometric sine.

Chapter 6

Symbolic Methods

In the classical approach to invariant theory, the symbolic method originally introduced by Aronhold, [**12**], and Clebsch, [**48**], plays a pivotal role. By appealing directly to the omega process and related differential operators, we have effectively bypassed this classical algebraic technique. In fact, when restricted to homogeneous polynomials, our approach can be directly reduced to the classical method. The connection between the two is made via a transform which, like the Fourier transform of classical analysis, changes questions about derivatives and differential polynomials into questions about ordinary algebraic polynomials, thus making them amenable to the powerful techniques of commutative algebra and invariant theory. Unfortunately, unlike the Fourier transform, which plays a vital role in the solution of linear partial differential equations, the nonlinear transform does not appear to be of much use for actually solving nonlinear differential equations. Nevertheless, its power for studying algebraic and geometric properties has been amply documented.

A special case of this transform was first introduced by Gel'fand and Dikii, [**77**], in their study of integrable soliton equations, inverse scattering problems, and the calculus of variations. It was generalized by Shakiban, [**192, 193**], who used it to apply the invariant theory of finite groups to the study of conservation laws of differential equations. The connections between the transform method and the classical symbolic method were first recognized, in slightly different form, in [**163, 164**]. The present version is essentially the same as that used by Ball, Currie, and Olver, [**15, 163**], to classify null Lagrangians, with applications to non-convex variational problems arising in nonlinear elasticity. Recent applications include the work of Anderson and Pohjanpelto, [**9, 10**], on symmetries and conservation laws in the calculus of variations, and Sanders and Wang, [**188**], in which methods of Diophantine approximation theory have been applied to classify integrable evolution equations. In this chapter, however, we shall concentrate just on its role in classical invariant theory.

The Fourier Transform

We begin our discussion with the simple case of a linear differential polynomial, for which the transform (and the symbolic method as discussed below) reduces to the classical Fourier transform. Consider an analytic function $Q(\mathbf{x}) = Q(x, y)$ depending on two scalar variables. Each linear differential operator acting on Q is defined by a polynomial function of the partial derivative operators $\nabla = (\partial_x, \partial_y)$, with $\mathbf{x}$-dependent coefficients:

$$\mathcal{D}[Q] = T(\mathbf{x}, \nabla)\, Q = \sum_{j,k} a_{jk}(\mathbf{x})\, \partial_x^j \partial_y^k Q. \qquad (6.1)$$

The coefficients $a_{jk}(\mathbf{x})$ can, in principle, be any analytic functions, but usually we restrict our attention to differential operators with polynomial coefficients.

The *Fourier transform* of (6.1) is simply found by replacing the derivatives ∂_x and ∂_y by *symbolic variables*[†] ξ, η. (A similar device appears in the foundations of Schrödinger's theory of quantum mechanics, cf. [**191, 232, 59, 136**], where the partial derivative operators ∂_x, ∂_y are interpreted as the momenta ξ, η "dual" to the position variables x, y.) Therefore, the *transform* or *symbolic form* (initially, the terms will be interchangeable) of the linear differential polynomial (6.1) is the algebraic polynomial

$$\widehat{\mathcal{D}} = T(x, y; \xi, \eta) = \sum_{j,k} a_{jk}(x, y)\, \xi^j \eta^k. \qquad (6.2)$$

We will consistently use hats to denote transforms. For example, the symbolic form of the scaling process $\sigma = x\partial_x + y\partial_y$ is just $\widehat{\sigma} = x\xi + y\eta$. In the transform space, the variables x, y, ξ, η all mutually commute, so that we could also write the transform of σ as $\xi x + \eta y$. On the other hand, differential operators *do not* commute with multiplication operators. Therefore, it is critically important that we place all the coefficient functions to the left of the differentiations *before* performing the transform.[‡] So $\xi x = x\xi$ is the transform of $xQ_x = x\partial_x Q$, *not* of $\partial_x(xQ) = xQ_x + Q$, which has transform $x\xi + 1$. Our convention

---

[†] In the analytical theory of Fourier transforms, we would multiply ξ and η by $i = \sqrt{-1}$. This additional complex factor will not play any role in our more formal approach, and hence, for simplicity, we shall omit it.

[‡] This is similar to the process of "normal ordering" in the theory of second quantization, [**21**].

has the important implication that the product of two transforms is *not* (except for constant coefficient operators) the same as the product of the corresponding differential operators. For example, the symbolic square $(x\xi + y\eta)^2$ does not represent the square of the scaling process, since

$$\sigma^2 = (x\partial_x + y\partial_y)^2 = x^2\partial_x^2 + 2xy\partial_x\partial_y + y^2\partial_y^2 + x\partial_x + y\partial_y,$$

which has symbolic representation $(x\xi + y\eta)^2 + x\xi + y\eta$. This particular computation readily generalizes, and the result will be used later.

For this purpose, we adopt a useful notation for "falling factorial powers" advocated by Knuth, [**131**; p. 414]. Given a real or complex number z and a non-negative integer r, define

$$z^{\underline{r}} = z(z-1)(z-2)\cdots(z-r+1). \tag{6.3}$$

By convention $z^{\underline{0}} = 1$. The falling factorial powers (6.3) are also referred to as "Pochhammer symbols" and have the more common notation $(z)_r$, although as Knuth remarks, Pochhammer, [**180**], actually reserved the latter notation for binomial coefficients! Knuth's notation is more suggestive, especially in view of the falling factorial version of the binomial identity

$$(z+w)^{\underline{r}} = \sum_{i=0}^{r} \binom{r}{i} z^{\underline{i}} \, w^{\underline{r-i}}. \tag{6.4}$$

Lemma 6.1. *The symbolic polynomial* $(x\xi + y\eta)^r$ *represents the "falling factorial power"* $\sigma^{\underline{r}} = \sigma(\sigma - 1)\cdots(\sigma - r + 1)$ *of the scaling process. If Q is a homogeneous function of degree n, so $\sigma Q = nQ$ according to Euler's formula, then*

$$\sigma^{\underline{r}} Q = n^{\underline{r}} Q = n(n-1)\cdots(n-r+1)Q. \tag{6.5}$$

In particular, if n is a positive integer, then $\sigma^{\underline{r}} Q = 0$ when $r > n$.

Equation (6.4) is the bivariate case of the general falling factorial multinomial identity

$$(z_1 + z_2 + \cdots + z_m)^{\underline{r}} = \sum_{I} \binom{r}{I} \mathbf{z}^{\underline{I}}. \tag{6.6}$$

In (6.6), the sum is over all multi-indices $I = (i_1, \ldots, i_m)$ with $|I| = i_1 + \cdots + i_m = r$, and $\mathbf{z}^{\underline{I}}$ is the multi-index version of the falling factorial power:

$$\mathbf{z}^{\underline{I}} = z_1^{\underline{i_1}} z_2^{\underline{i_2}} \cdots z_m^{\underline{i_m}}, \qquad \text{where} \qquad \mathbf{z} = (z_1, \ldots, z_m). \tag{6.7}$$

The General Transform

We shall now extend the Fourier transform to general differential polynomials. We begin with the most straightforward case. The *transform* of a constant coefficient multilinear differential polynomial involving *distinct* functions $Q_\alpha(x,y), \ldots, Q_\varepsilon(x,y)$ is defined by Fourier transforming each individual term, but using a different pair $\boldsymbol{\xi}_\alpha = (\xi_\alpha, \eta_\alpha), \ldots, \boldsymbol{\xi}_\varepsilon = (\xi_\varepsilon, \eta_\varepsilon)$ of symbolic variables to represent the derivatives of the different functions. For example, the Jacobian $J = Q_x R_y - Q_y R_x$ has transform

$$\widehat{J} = \xi_\alpha \eta_\beta - \eta_\alpha \xi_\beta, \tag{6.8}$$

in which $\boldsymbol{\xi}_\alpha = (\xi_\alpha, \eta_\alpha)$ represents derivatives of $Q = Q_\alpha$, and $\boldsymbol{\xi}_\beta = (\xi_\beta, \eta_\beta)$ those of $R = Q_\beta$. In general, if $P = \mathcal{D}[Q_\alpha, \ldots, Q_\varepsilon]$ is any multilinear differential polynomial with transform $\widehat{P} = T(\boldsymbol{\xi}_\alpha, \ldots, \boldsymbol{\xi}_\varepsilon)$, then we can recover P by substituting derivatives for the symbolic variables in T. We can conveniently represent this using our earlier tensor product and trace notation (5.17), where $\mathbf{Q} = Q_\alpha \otimes \cdots \otimes Q_\varepsilon$ represents the product $Q_\alpha(\mathbf{x}_\alpha) \cdots Q_\varepsilon(\mathbf{x}_\varepsilon)$. Any multilinear differential polynomial can then be recovered by applying a differential operator to $\mathbf{Q}$ and setting all the variables equal:

$$\mathcal{D}[Q_\alpha, \ldots, Q_\varepsilon] = \operatorname{tr} T(\nabla_\alpha, \ldots, \nabla_\varepsilon)(Q_\alpha \otimes \cdots \otimes Q_\varepsilon), \tag{6.9}$$

where $\nabla_\alpha = \left(\partial_{x_\alpha}, \partial_{y_\alpha}\right)$ denotes the gradient operator with respect to the αth pair of variables. For example, replacing the symbolic variables by the corresponding derivatives changes the Jacobian transform (6.8) into the omega process $\Omega_{\alpha\beta}$, and so (6.9) in this case reduces to the basic transvectant formula for the Jacobian covariant

$$J = Q_x R_y - Q_y R_x = \operatorname{tr} \Omega_{\alpha\beta}(Q \otimes R) = \operatorname{tr} \left\{ \frac{\partial^2}{\partial x_\alpha \partial y_\beta} - \frac{\partial^2}{\partial x_\beta \partial y_\alpha} \right\} Q \otimes R.$$

This computation immediately generalizes to higher order transvectants:

Proposition 6.2. *The rth transvectant $(Q, R)^{(r)}$ has as its transform the rth power $(\xi_\alpha \eta_\beta - \eta_\alpha \xi_\beta)^r$.*

Exercise 6.3. Determine the transforms of the partial transvectants in Examples 5.10 and 5.13 directly from the definition.

If some of the functions $Q_\alpha, \ldots, Q_\varepsilon$ coincide, then a similar transforming process can be established, but there are some ambiguities that need to be resolved. For instance, consider the quadratic differential polynomial $\mathcal{D}[Q] = Q_x Q_y$. If we regard $\mathcal{D}[Q] = \mathcal{D}[Q,Q]$ as a bilinear differential polynomial with two equal arguments, then we can represent it as

$$Q_x Q_y = \text{tr} \; \frac{\partial^2}{\partial x_\alpha \partial y_\beta} \, Q \otimes Q,$$

leading to the transform expression $T = \xi_\alpha \eta_\beta$. On the other hand, we can just as easily write the two factors in the opposite order, so $\mathcal{D}[Q] = Q_y Q_x$, leading to the alternative transform $\tilde{T} = \xi_\beta \eta_\alpha$. Either polynomial is a valid candidate for the transform of $\mathcal{D}[Q]$, since both reduce to $\mathcal{D}[Q]$ upon equating the variables $\mathbf{x}_\alpha = \mathbf{x}_\beta = \mathbf{x}$ as in (6.9). One way to resolve the ambiguity is to symmetrize by averaging the different transforms obtained by writing the factors in various orders; in our simple case, this would give the symmetric symbolic expression $T^\varsigma = \frac{1}{2}(\xi_\alpha \eta_\beta + \eta_\alpha \xi_\beta)$. It is not hard to see that the symmetric transform is unique. As a second example, consider the Hessian $H = Q_{xx} Q_{yy} - Q_{xy}^2$. We can write H in symbolic form either as $\widehat{H}_1 = \xi_\alpha^2 \eta_\beta^2 - \xi_\alpha \eta_\alpha \xi_\beta \eta_\beta$ or as $\widehat{H}_2 = \eta_\alpha^2 \xi_\beta^2 - \xi_\alpha \eta_\alpha \xi_\beta \eta_\beta$, depending on which order we write the factors in H. (Of course, the second summand is unchanged.) The symmetric transform is again found by averaging,

$$\widehat{H}^\varsigma = \tfrac{1}{2}(\xi_\alpha^2 \eta_\beta^2 + \eta_\alpha^2 \xi_\beta^2) - \xi_\alpha \eta_\alpha \xi_\beta \eta_\beta = \tfrac{1}{2}(\xi_\alpha \eta_\beta - \eta_\alpha \xi_\beta)^2, \qquad (6.10)$$

thereby recovering, in symbolic form, its omega process representation.

This serves to justify the general definition of the transform of an arbitrary homogeneous differential polynomial. We state it here for constant coefficient differential polynomials depending on a single function Q, although, as we will indicate later, extensions to variable coefficient differential polynomials involving several different functions are straightforward.

Definition 6.4. Let $Q(\mathbf{x})$ be an analytic function and let $\mathcal{D}[Q]$ be a constant coefficient homogeneous differential polynomial of degree m. A *transform* of $\mathcal{D}[Q]$ is any polynomial $T(\boldsymbol{\xi}_\alpha, \ldots, \boldsymbol{\xi}_\varepsilon)$, depending on m symbolic variables, which satisfies

$$\mathcal{D}[Q] = \text{tr} \, T(\nabla_\alpha, \ldots, \nabla_\varepsilon) \, (Q \otimes \cdots \otimes Q).$$

Each permutation $\pi \in \mathbb{S}^m$ acts on a symbolic polynomial by permuting its arguments. If $T(\boldsymbol{\xi}_\alpha, \ldots, \boldsymbol{\xi}_\varepsilon)$ is a transform for a differential polynomial $\mathcal{D}$, then, clearly, so is any permuted version $\pi \cdot T = T(\boldsymbol{\xi}_{\pi(\alpha)}, \ldots, \boldsymbol{\xi}_{\pi(\varepsilon)})$. The polynomial T is called *symmetric* if it is unchanged by the action of the permutations: $\pi \cdot T = T$ for all permutations

$\pi \in \mathbb{S}^m$. According to Chapter 4, symmetric transforms are, in fact, multi-symmetric polynomials. Any other transform T can always be made symmetric by applying the symmetrization operator (4.10), which defines a projection onto the subspace of all multi-symmetric polynomials. A basic, elementary result is that the symmetric transform is unique, [192].

Theorem 6.5. *Any homogeneous constant coefficient differential polynomial $\mathcal{D}[Q]$ has a unique symmetric transform.*

Proof: Given $I = (k, l)$, $k, l \in \mathbb{N}$, let $Q_I = \partial^{k+l}Q/\partial x^k \partial y^l$ denote the corresponding partial derivative of the function $Q(x, y)$. Each constant coefficient differential polynomial is a linear combination of monomials

$$Q_{\mathcal{I}} = Q_{I_\alpha} Q_{I_\beta} \cdots Q_{I_\varepsilon}. \tag{6.11}$$

One transform of the monomial (6.11) is provided by the power product

$$\boldsymbol{\xi}^{\mathcal{I}} = \boldsymbol{\xi}_\alpha^{I_\alpha} \, \boldsymbol{\xi}_\beta^{I_\beta} \cdots \boldsymbol{\xi}_\varepsilon^{I_\varepsilon},$$

where $\boldsymbol{\xi}_\alpha^I = (\xi_\alpha)^k (\eta_\alpha)^l$ for $I = (k, l)$. Consequently, the symmetric transform of $Q_{\mathcal{I}}$ is

$$\varsigma(\boldsymbol{\xi}^{\mathcal{I}}) = \frac{1}{m!} \sum_{\pi \in \mathbb{S}^m} \boldsymbol{\xi}_{\pi(\alpha)}^{I_\alpha} \, \boldsymbol{\xi}_{\pi(\beta)}^{I_\beta} \cdots \boldsymbol{\xi}_{\pi(\varepsilon)}^{I_\varepsilon}, \tag{6.12}$$

where m is the degree of $Q_{\mathcal{I}}$. Each monomial in the summation (6.12) can be obtained as a transform of $Q_{\mathcal{I}}$ by permuting the factors in (6.11); moreover, this is the only way such a summand can arise. This immediately proves the uniqueness of the symmetric transform (6.12). *Q.E.D.*

As a consequence of Theorem 6.5, the transform sets up a linear isomorphism between the space of homogeneous differential polynomials and the corresponding space of multi-symmetric polynomials. Although this implies that we could work exclusively with multi-symmetric symbolic forms without loss of generality (and perhaps with some gain in precision), practical computational considerations inspire us to retain the possibility of using non-symmetric symbolic forms.

Example 6.6. Let $H = \frac{1}{2}(Q, Q)^{(2)}$ denote the Hessian, and consider the covariant $T = (Q, H)^{(1)}$, whose explicit formula appears in (2.27). Multiplying out and transforming each term using the indicated order of factors leads to one possible transform, namely,

$$\widehat{T}_0 = \left[\xi_\alpha(\eta_\beta + \eta_\gamma) - \eta_\alpha(\xi_\beta + \xi_\gamma) \right] (\xi_\beta^2 \eta_\gamma^2 - \xi_\beta \eta_\beta \xi_\gamma \eta_\gamma).$$

Symmetrizing over all permutations of $\{\alpha, \beta, \gamma\}$ leads to the complete symmetric transform

$$\widehat{T}^\varsigma = \tfrac{1}{6}\big[\,(\xi_\alpha \eta_\beta - \eta_\alpha \xi_\beta + \xi_\alpha \eta_\gamma - \eta_\alpha \xi_\gamma)(\xi_\beta \eta_\gamma - \eta_\beta \xi_\gamma)^2 +$$

$$+ (\eta_\alpha \xi_\gamma - \xi_\alpha \eta_\gamma + \eta_\beta \xi_\gamma - \xi_\beta \eta_\gamma)(\xi_\alpha \eta_\beta - \eta_\alpha \xi_\beta)^2 + \qquad (6.13)$$

$$+ (\eta_\alpha \xi_\beta - \xi_\alpha \eta_\beta + \xi_\beta \eta_\gamma - \eta_\beta \xi_\gamma)(\xi_\alpha \eta_\gamma - \eta_\alpha \xi_\gamma)^2\,\big].$$

However, a more compact and useful, but non-symmetric transform is found by taking just one of the preceding summands, for instance,

$$\widehat{T} = (\xi_\alpha \eta_\beta - \eta_\alpha \xi_\beta)(\xi_\beta \eta_\gamma - \eta_\beta \xi_\gamma)^2. \qquad (6.14)$$

The latter formula has a direct correspondence with our earlier omega process-based formula (2.27). The general result is stated in Proposition 6.13.

If our differential polynomial $\mathcal{D}[Q]$ depends explicitly on $\mathbf{x} = (x, y)$, then its transform will be a polynomial $T(\mathbf{x}, \boldsymbol{\xi}_\alpha, \boldsymbol{\xi}_\beta, \boldsymbol{\xi}_\gamma, \boldsymbol{\xi}_\delta, \boldsymbol{\xi}_\varepsilon)$ in the relevant symbolic variables with $\mathbf{x}$–dependent coefficients. For example, the differential polynomial $x^3 y\, Q_x Q_{yy}$ has the simple transform $x^3 y\, \xi_\alpha (\eta_\beta)^2$ and symmetric transform $\tfrac{1}{2} x^3 y \big[\xi_\alpha (\eta_\beta)^2 + \xi_\beta (\eta_\alpha)^2\big]$. As with the original Fourier transform, it is important to place any $\mathbf{x}$–dependent coefficients in $\mathcal{D}[Q]$ to the left of any differentiations before replacing derivatives by their symbolic counterparts. The uniqueness Theorem 6.5 is then readily seen to hold in this more general context.

Remark: The reader may be wondering why we are not replacing the coordinates $\mathbf{x}$ in the differential polynomial by one or more symbolic variables $\mathbf{x}_\alpha, \ldots, \mathbf{x}_\varepsilon$ occurring in the product $Q \otimes \cdots \otimes Q$. This is indeed possible, and perhaps more natural; however, there are two disadvantages. First, since the trace equates all the symbolic coordinates, $\mathbf{x}_\alpha = \cdots \mathbf{x}_\varepsilon = \mathbf{x}$, we do not gain anything by using different symbols for the transforms of the coefficients, and the computations are unnecessarily complicated. A second disadvantage is that the uniqueness result of Theorem 6.5 is not valid in this more general context, since many (multi-symmetric) polynomials in the symbolic variables, e.g., $(x_\alpha y_\beta - x_\beta y_\alpha)^2$, have zero trace.

An important observation is that *every* function $F(a_0, \ldots, a_n, x, y)$ depending on the coefficients a_i of a degree n binary form (2.1) can be expressed in symbolic form, since F can be written (in many ways) as a differential polynomial. Indeed, the evident formula

$$n!\, a_i = \partial_x^i \partial_y^{n-i} Q, \qquad i = 0, \ldots, n, \qquad (6.15)$$

serves to express the coefficients of Q in terms of its nth order partial derivatives. One can then directly transform the resulting differential polynomial; of course, such a representation is by no means uniquely determined.

Example 6.7. Consider the quadratic invariant i for a binary quartic, as given in (2.29). The corresponding differential polynomial is

$$i = a_0 a_4 - 4a_1 a_3 + 3a_2^2 = \tfrac{1}{576} \left(Q_{xxxx} Q_{yyyy} - 4Q_{xxxy} Q_{xyyy} + 3(Q_{xxyy})^2 \right).$$

An immediate transform is

$$\widehat{i} = \tfrac{1}{576} \left(\xi_\alpha^4 \eta_\beta^4 - 4\xi_\alpha^3 \eta_\alpha \xi_\beta \eta_\beta^3 + 3\xi_\alpha^2 \eta_\alpha^2 \xi_\beta^2 \eta_\beta^2 \right),$$

but the symmetric transform factors

$$\widehat{i}^\varsigma = \tfrac{1}{1152} (\xi_\alpha \eta_\beta - \xi_\beta \eta_\alpha)^4. \tag{6.16}$$

We thus recover its transvectant version, $i = \tfrac{1}{1152}(Q, Q)^{(4)}$, as indicated in Exercise 5.3.

Example 6.8. Consider the Hessian (2.24) of a binary cubic (2.21). Replacing each coefficient by the associated derivative (6.15), we find a variable coefficient differential polynomial representation

$$H = x^2 (Q_{xxx} Q_{xyy} - Q_{xxy}^2) + xy(Q_{xxx} Q_{yyy} - Q_{xxy} Q_{xyy}) + \\ + y^2 (Q_{xxy} Q_{yyy} - Q_{xyy}^2). \tag{6.17}$$

The symmetric transform of (6.17) factors into

$$\widehat{H} = \tfrac{1}{2} (x\xi_\alpha + y\eta_\alpha)(x\xi_\beta + y\eta_\beta)(\xi_\alpha \eta_\beta - \xi_\beta \eta_\alpha)^2, \tag{6.18}$$

providing an alternative symbolic form for the cubic Hessian covariant.

Exercise 6.9. Use Euler's formula (2.13) to prove the equality of the two differential polynomials (2.24), (6.17) for binary cubics. Generalize the Hessian formulae (6.17), (6.18) to binary forms of degree n.

These results can be readily extended to homogeneous differential polynomials depending on several functions. For example, a constant coefficient differential polynomial $\mathcal{D}[Q, R]$ which has degree 3 in Q and degree 2 in R has transform $T(\boldsymbol{\xi}_\alpha, \boldsymbol{\xi}_\beta, \boldsymbol{\xi}_\gamma, \boldsymbol{\xi}_\delta, \boldsymbol{\xi}_\varepsilon)$ satisfying

$$\mathcal{D}[Q, R] = \operatorname{tr} T(\nabla_\alpha, \nabla_\beta, \nabla_\gamma, \nabla_\delta, \nabla_\varepsilon) \, Q \otimes Q \otimes Q \otimes R \otimes R.$$

Permutations of the symbolic letters $\{\alpha, \beta, \gamma, \delta, \varepsilon\}$ which do not affect the tensor product $Q \otimes Q \otimes Q \otimes R \otimes R$, namely, the "isotropy subgroup" $\mathbb{S}^3 \times \mathbb{S}^2 \subset \mathbb{S}^5$ generated by the permutations of the subsets $\{\alpha, \beta, \gamma\}$

and $\{\delta, \varepsilon\}$, will give potentially different symbolic forms. Again the symmetric symbolic form obtained by symmetrizing over the isotropy subgroup is unique.

Brackets

As the reader may have noticed, certain particular symbolic expressions keep reappearing in the transform formulae for covariants. Since they play a ubiquitous role in the theory, it is worth introducing a special, classical notation for these so-called bracket factors.

Definition 6.10. (a) A *bracket factor of the first kind* is the pairing

$$(\alpha \mathbf{x}) = \xi_\alpha x + \eta_\alpha y \tag{6.19}$$

between symbolic and coordinate variables.

(b) A *bracket factor of the second kind* is the 2×2 determinant

$$[\alpha \beta] = \xi_\alpha \eta_\beta - \eta_\alpha \xi_\beta = \det \begin{vmatrix} \xi_\alpha & \eta_\alpha \\ \xi_\beta & \eta_\beta \end{vmatrix}, \tag{6.20}$$

depending on two symbolic variables.

A bracket factor of the first kind (6.19) is nothing but the transform of the scaling process σ given by (5.25). Similarly, a bracket factor of the second kind (6.20) is just the transform of the omega process $\Omega_{\alpha\beta}$, as presented in Proposition 6.2.

In order to understand why the bracket factors appear so often, we investigate their behavior under linear transformations. The symbolic variables $\boldsymbol{\xi}_\alpha = (\xi_\alpha, \eta_\alpha)$ represent the gradient operator $\nabla = (\partial_x, \partial_y)$. According to (4.28), if we change variables linearly, $\overline{\mathbf{x}} = A \mathbf{x}$, then the gradient transforms contragrediently: $\overline{\nabla} = A^{-T} \nabla$. Therefore, under a linear transformation A we have the general transformation rules

$$\overline{\mathbf{x}} = A \mathbf{x}, \qquad \overline{\boldsymbol{\xi}}_\alpha = A^{-T} \boldsymbol{\xi}_\alpha, \tag{6.21}$$

for the variables appearing in any transform expression. In other words, while the usual coordinates $\mathbf{x}$ define contravariant vectors, the symbolic variables define dual, covariant vectors.

Proposition 6.11. *Under a linear transformation* (6.21), *the two bracket factors* (6.19) *and* (6.20) *transform as follows:*

$$(\alpha \mathbf{x}) \longmapsto (\alpha \mathbf{x}), \qquad [\alpha \beta] \longmapsto (\det A)^{-1} [\alpha \beta]. \tag{6.22}$$

In other words, bracket factors of the first kind have weight 0, while those of the second kind are of weight +1 under linear transformations.

Definition 6.12. A *bracket polynomial* is a symbolic expression which can be written as a polynomial in the bracket factors of the first and second kinds.

Remark: One can also introduce *bracket factors of the third kind*:

$$\llbracket \alpha\,\beta \rrbracket = x_\alpha y_\beta - y_\alpha x_\beta = \det \begin{vmatrix} x_\alpha & x_\beta \\ y_\alpha & y_\beta \end{vmatrix}, \qquad (6.23)$$

depending on two different contravariant vectors. They transform with weight -1, so $\llbracket \alpha\,\beta \rrbracket \mapsto (\det A)\,\llbracket \alpha\,\beta \rrbracket$. Moreover, once we allow several different contravariant vectors into the picture, then we can construct various bracket factors of the first kind, e.g., $(\alpha\,\lambda) = \xi_\alpha x_\lambda + \eta_\alpha y_\lambda$. The bracket factors of the third kind all vanish upon equating the contravariant variables via our trace operation. Thus, in light of our earlier comments on the use of only one set of contravariant variables $\mathbf{x}$ for transform computations, we shall not have much occasion to use the third types of bracket factors here.

Remark: We could, in fact, adopt a common notation for both brackets, since those of the first kind can be written as a 2×2 determinant $(\alpha\,\mathbf{x}) = \det \begin{vmatrix} \xi_\alpha & \eta_\alpha \\ -y & x \end{vmatrix}$. This artifice relies on the identification, in the two-dimensional case, of covariant with contravariant vectors, as described in Example 4.5, and can be used to make general bracket computations more uniform. However, there are good reasons not to rely on this identification in general. First, the transformation rules, (6.22), for the brackets under GL(2) — as opposed to SL(2) — are different. Second, this artifice is particular to the case of binary forms and does not extend to forms in three or more variables, because one can no longer identify co- and contravariant vectors; see Chapter 10.

Each of the covariants discussed above has a transform which is a bracket polynomial. For example, the symmetric transform (6.10) of the Hessian is the simple bracket polynomial $\widehat{H}^\varsigma = \frac{1}{2}[\alpha\,\beta]^2$; its (cubic) alternative (6.17) still has its transform (6.18) being expressed as a bracket polynomial, namely, $\frac{1}{2}[\alpha\,\beta]^2\,(\alpha\,\mathbf{x})(\beta\,\mathbf{x})$. Similarly, the covariant T in (2.27) has symbolic form (6.14) given by the bracket polynomial $\widehat{T} = [\alpha\,\beta]\,[\beta\,\gamma]^2$. (Its symmetric symbolic form (6.13) is also a bracket polynomial, obtained by symmetrizing $\widehat{T}$ over permutations of α, β, γ.) A key result is that any covariant obtained through (partial) transvection has symbolic form given by a bracket polynomial.

Proposition 6.13. *The transform of the partial transvectant*

$$\mathcal{D}[\mathbf{Q}] = \text{tr} \left\{ \prod_{\nu=1}^{r} \Omega_{\alpha_\nu \beta_\nu} \right\} (Q_\alpha \otimes \cdots \otimes Q_\varepsilon)$$

is given by the same product, $\widehat{\mathcal{D}[Q]} = \prod_\nu [\, \alpha_\nu \, \beta_\nu \,]$, of bracket factors of the second kind.

The First Fundamental Theorem of classical invariant theory states that *every* differential polynomial covariant can be written in symbolic form as a bracket polynomial.

Theorem 6.14. *Let $J = \mathcal{D}[\mathbf{Q}]$ be a homogeneous, differential polynomial covariant depending on the functions $\mathbf{Q} = (Q_\alpha, \ldots, Q_\varepsilon)$. Then the symmetric transform of J can be written as a bracket polynomial. Conversely, if S is any homogeneous bracket polynomial, then S is a symbolic form of a covariant differential polynomial.*

Corollary 6.15. *Every differential polynomial covariant of a system of functions of two variables can be expressed in terms of partial transvectants and scaling processes.*

A proof of the First Fundamental Theorem will appear shortly. As we remarked earlier, cf. (6.15), the classical invariants and covariants of binary forms can always be written in differential polynomial form and so are automatically included in the result. Indeed, if the functions are homogeneous, then the bracket factors of the first kind correspond to scaling processes, and hence their only effect is to multiply the expression by some overall factor. Thus, a more refined version of Theorem 6.14 is the following:

Theorem 6.16. *If $Q_\alpha, \ldots, Q_\varepsilon$ are homogeneous functions, then every invariant or covariant has a transform which is a homogeneous polynomial in the bracket factors of the second kind; conversely, every such bracket polynomial is the transform of a homogeneous covariant.*

Exercise 6.17. Determine bracket polynomials representing the discriminants of the binary cubic and quartic. See Gordan, [**88**], for the symbolic forms of more general resultants and discriminants.

Exercise 6.18. The kth *Gundelfinger covariant* of a binary form Q of degree n is the $(k+1) \times (k+1)$ determinant

$$G_k[Q] = \det \left[\frac{\partial^{2k} Q}{\partial^{2k-i-j} x \, \partial^{i+j} y} \right]. \tag{6.24}$$

In particular, $G_0[Q] = Q$, while $G_1[Q] = H[Q]$ is the Hessian. Prove that $G_k[Q]$ is a covariant and determine its weight and a symbolic bracket expression. See [96, 134] for applications to the problem of representing binary forms as sums of powers.

Syzygies

According to the First Fundamental Theorem 6.14, we can explicitly write down all the polynomial invariants and covariants of any system of binary forms merely by writing down all the bracket polynomials. However, there is considerable redundancy in this procedure. First, since any homogeneous polynomial function of a collection of covariants of a binary form is also a covariant, we should eliminate these reducible covariants from our "minimal" list or Hilbert basis of covariants. Even more fundamental is the fact that the symbolic form of a given covariant does *not* have a unique bracket polynomial representative, owing to the presence of certain relations among the bracket factors themselves, which we discuss next.

There are three fundamental identities involving the bracket factors. These are referred to as "syzygies" and take the following form:

$$[\alpha\,\beta] = -[\beta\,\alpha], \tag{6.25}$$

$$[\alpha\,\beta]\,(\gamma\,\mathbf{x}) + [\gamma\,\alpha]\,(\beta\,\mathbf{x}) + [\beta\,\gamma]\,(\alpha\,\mathbf{x}) = 0, \tag{6.26}$$

$$[\alpha\,\beta]\,[\gamma\,\delta] + [\gamma\,\alpha]\,[\beta\,\delta] + [\beta\,\gamma]\,[\alpha\,\delta] = 0. \tag{6.27}$$

The reader can easily verify each of these identities directly from the formulae for the bracket factors. Indeed, under our artificial identification of the two types of bracket factors, the second and third syzygies are two versions of the same simple determinantal identity. In Chapter 7 we shall see how each of these syzygies can be applied to simplify bracket polynomials and, ultimately, derive the Hilbert basis for the covariants of a binary form of a given degree.

Remark: The three syzygies can be identified as the transforms of the following basic differential operator identities for the omega and scaling processes:

$$\Omega_{\alpha\beta} = -\Omega_{\beta\alpha}, \tag{6.28}$$

$$\sigma_\gamma\Omega_{\alpha\beta} + \sigma_\alpha\Omega_{\beta\gamma} + \sigma_\beta\Omega_{\gamma\alpha} = 0, \tag{6.29}$$

$$\Omega_{\alpha\beta}\Omega_{\gamma\delta} + \Omega_{\beta\gamma}\Omega_{\alpha\delta} + \Omega_{\gamma\alpha}\Omega_{\beta\delta} = 0. \tag{6.30}$$

The Second Fundamental Theorem of classical invariant theory says that any other syzygy between bracket polynomials can be deduced from the three basic ones.

Theorem 6.19. *Every polynomial identity among the different bracket factors is obtained as a linear combination of the three basic syzygies* (6.25), (6.26), (6.27).

A proof of the Second Fundamental Theorem will be provided later in this chapter.

Example 6.20. In view of the evident matrix identity

$$\begin{pmatrix} [\alpha\lambda] & [\alpha\mu] & [\alpha\nu] \\ [\beta\lambda] & [\beta\mu] & [\beta\nu] \\ [\gamma\lambda] & [\gamma\mu] & [\gamma\nu] \end{pmatrix} = \begin{pmatrix} \xi_\alpha & \eta_\alpha \\ \xi_\beta & \eta_\beta \\ \xi_\gamma & \eta_\gamma \end{pmatrix} \begin{pmatrix} -\eta_\lambda & -\eta_\mu & -\eta_\nu \\ \xi_\lambda & \xi_\mu & \xi_\nu \end{pmatrix},$$

the determinant of the left-hand 3×3 matrix of bracket factors of the second kind must vanish. This provides an interesting, nontrivial syzygy,

$$\left[\!\!\left[\begin{matrix} \alpha & \beta & \gamma \\ \lambda & \mu & \nu \end{matrix} \right]\!\!\right] \equiv \det \begin{vmatrix} [\alpha\lambda] & [\alpha\mu] & [\alpha\nu] \\ [\beta\lambda] & [\beta\mu] & [\beta\nu] \\ [\gamma\lambda] & [\gamma\mu] & [\gamma\nu] \end{vmatrix} = 0, \qquad (6.31)$$

which depends on six symbolic letters, some of which may coincide. The Second Fundamental Theorem 6.19 implies that (6.31) must be a linear combination of the basic syzygies

$$[\!\![\, \alpha\beta\gamma\delta \,]\!\!] \equiv [\,\alpha\beta\,]\,[\,\gamma\delta\,] + [\,\gamma\alpha\,]\,[\,\beta\delta\,] + [\,\beta\gamma\,]\,[\,\alpha\delta\,]. \qquad (6.32)$$

Indeed, expanding the determinant along its first row and then adding and subtracting the extra terms in $[\,\beta\gamma\,]\,[\!\![\,\alpha\lambda\mu\nu\,]\!\!]$ show that

$$\left[\!\!\left[\begin{matrix} \alpha & \beta & \gamma \\ \lambda & \mu & \nu \end{matrix} \right]\!\!\right] = [\,\alpha\lambda\,]\,[\!\![\,\beta\mu\gamma\nu\,]\!\!] + [\,\alpha\mu\,]\,[\!\![\,\beta\nu\gamma\lambda\,]\!\!] + \\ + [\,\alpha\nu\,]\,[\!\![\,\beta\lambda\gamma\mu\,]\!\!] + [\,\beta\gamma\,]\,[\!\![\,\alpha\lambda\mu\nu\,]\!\!]. \qquad (6.33)$$

Example 6.21. Consider the bracket monomial $M_k = [\,\alpha\beta\,][\,\beta\gamma\,]^k$. If $k = 2j$ is even, M_k represents the covariant $T_j = \frac{1}{2}(Q, (Q, Q)^{(2j)})^{(1)}$. If $k = 2j + 1$ is odd, then we use the second syzygy (6.26) to write

$$[\,\alpha\beta\,]\,[\,\beta\gamma\,]^k\,(\gamma\mathbf{x}) + [\,\gamma\alpha\,]\,[\,\beta\gamma\,]^k\,(\beta\mathbf{x}) + [\,\beta\gamma\,]^{k+1}\,(\alpha\mathbf{x}) = 0.$$

When k is odd, skew-symmetry of the brackets, (6.25), implies that the first two terms can be obtained from each other by permuting the symbolic letters. (When k is even, the first two terms cancel each other and show that the third term represents a trivial covariant.) Moreover, if Q is homogeneous of degree n, then the bracket factors of the first kind represent scaling processes and so can be replaced by appropriate numerical multiples. Therefore, $M_k \simeq -\frac{n}{2(n-k)}[\beta\gamma]^{k+1}$ represents the reducible cubic covariant $-\frac{n}{2(n-k)}Q \cdot (Q, Q)^{(k+1)}$.

The Classical Symbolic Method

The transform method introduced above is a decidedly unorthodox approach to invariant theory. However, it directly parallels the classical symbolic method or umbral calculus for constructing invariants and covariants of binary forms, [**89, 92, 135**]. It is thus important, particularly for comparing our results with the classical literature, to point out the precise connection between the two approaches.

The motivating idea behind the symbolic method is that the theory of binary forms would be extremely simple if our degree n binary form $Q(x, y) = \sum \binom{n}{i} a_i x^i y^{n-i}$ were just the nth power of a linear form:

$$(\xi x + \eta y)^n = \sum_{i=0}^{n} \binom{n}{i} \xi^i \eta^{n-i} x^i y^{n-i}.$$

This would indeed be the case if its coefficients happened to be of the special form

$$a_i = \xi^i \eta^{n-i}. \tag{6.34}$$

In the symbolic method, one effectively "pretends" that the general polynomial Q is a power of a linear form. One then uses (6.34) to replace the coefficients $\mathbf{a} = (a_0, \ldots, a_n)$ of Q that occur in a polynomial $J(\mathbf{a}, \mathbf{x})$, leading to the "symbolic form" of J. The variables $\boldsymbol{\xi} = (\xi, \eta)$ are referred to as "symbolic variables". Symbolic variables by themselves have no real meaning; it is only when they appear in the particular power products (6.34) of degree n that they acquire an actual value in terms of the coefficients of our binary form.

Simple examples reveal that this naïve approach must be immediately modified so as to avoid ambiguities. For instance, in the case of a binary quadratic $a_2 x^2 + 2a_1 xy + a_0 y^2$, it would not distinguish between the monomials $a_0 a_2$ and a_1^2; according to the previous rule, both would degenerate to the same symbolic form $\xi^2 \eta^2$. (This is just another way of stating that a quadratic form is a perfect square if and only if its discriminant vanishes: $a_0 a_2 = a_1^2$.) The way to resolve this ambiguity is to use a *different* pair of symbolic variables $\boldsymbol{\xi}_\alpha = (\xi_\alpha, \eta_\alpha)$, indexed by distinct symbolic letters $\alpha, \beta, \ldots$, for each factor a_i. Thus, for the binary quadratic, in the monomial $a_0 a_2$ we replace a_0 by ξ_α^2 and a_2 by η_β^2, leading to the symbolic form $\xi_\alpha^2 \eta_\beta^2$; on the other hand, for the monomial a_1^2, we replace one factor a_1 by the product $\xi_\alpha \eta_\alpha$ and the second factor by $\xi_\beta \eta_\beta$, leading to the different symbolic form $\xi_\alpha \eta_\alpha \xi_\beta \eta_\beta$. Note that we can readily pass back and forth between the explicit formula for a polynomial depending on the coefficients of the binary form and its symbolic forms; for instance, in the case of a quadratic form, the symbolic polynomial $\xi_\alpha^2 \xi_\beta \eta_\beta \eta_\gamma^2 x y^2$ depending on three symbolic letters would represent the monomial $a_0 a_1 a_2 x y^2$. (The x's and y's are not affected.)

As with the earlier transform, there is *not* a uniquely determined symbolic form for a given polynomial $J(\mathbf{a}, \mathbf{x})$. For instance, if we write $a_0 a_2$, which has symbolic form $\xi_\alpha^2 \eta_\beta^2$, in reverse order as $a_2 a_0$, we obtain the symbolic form $\eta_\alpha^2 \xi_\beta^2$. However, all the different symbolic forms can be obtained one from the other merely by interchanging the symbolic letters. There is a unique symmetric symbolic form for any given polynomial, obtained by symmetrizing any symbolic representative by applying the operator (4.10) to it. For instance, the discriminant $\Delta = a_0 a_2 - a_1^2$ of a quadratic has symmetric symbolic form

$$\tfrac{1}{2}\left\{ (\xi_\alpha^2 \eta_\beta^2 - \xi_\alpha \eta_\alpha \xi_\beta \eta_\beta) + (\eta_\alpha^2 \xi_\beta^2 - \xi_\alpha \eta_\alpha \xi_\beta \eta_\beta) \right\} = \tfrac{1}{2}(\xi_\alpha \eta_\beta - \eta_\alpha \xi_\beta)^2$$

$$= \tfrac{1}{2}[\alpha\,\beta]^2,$$

where we adopt the same bracket notation (6.20) as before.

Remark: Unlike the transform, a symbolic polynomial will represent an actual polynomial in the coefficients of a binary form of degree n if and only if each symbolic letter $\alpha, \beta, \ldots$ occurs precisely n times in each symbolic monomial; that is, the symbolic polynomial is homogeneous of degree n in each of its symbolic variables $\boldsymbol{\xi}_\alpha, \boldsymbol{\xi}_\beta, \ldots$. The number of different symbolic letters represents the order of the symbolic polynomial, meaning its degree in the coefficients a_i.

Example 6.22. For the invariant $i = a_0 a_4 - 4 a_1 a_3 + 3 a_2^2$ of the binary quartic, cf. Example 2.24, we obtain one symbolic form immediately: $\xi_\alpha^4 \eta_\beta^4 - 4\,\xi_\alpha^3 \eta_\alpha \xi_\beta \eta_\beta^3 + 3\,\xi_\alpha^2 \eta_\alpha^2 \xi_\beta^2 \eta_\beta^2$. The symmetric form is

$$\tfrac{1}{2}\left(\xi_\alpha^4 \eta_\beta^4 - 4\,\xi_\alpha^3 \eta_\alpha \xi_\beta \eta_\beta^3 + 6\,\xi_\alpha^2 \eta_\alpha^2 \xi_\beta^2 \eta_\beta^2 - 4\,\xi_\alpha \eta_\alpha^3 \xi_\beta^3 \eta_\beta + \eta_\alpha^4 \xi_\beta^4 \right) =$$

$$= \tfrac{1}{2}(\xi_\alpha \eta_\beta - \eta_\alpha \xi_\beta)^4 = \tfrac{1}{2}[\alpha\,\beta]^4,$$

which, apart from a numerical multiple, agrees with its transform (6.16). Similarly, if we replace each coefficient in the explicit expression for the

Hessian (2.31) of a quartic by its symbolic form, we find that $\frac{1}{144}H$ has a symbolic form

$$(\xi_\alpha^4\xi_\beta^2\eta_\beta^2 - \xi_\alpha^3\eta_\alpha\xi_\beta^3\eta_\beta)\,x^4 + 2(\xi_\alpha^4\xi_\beta\eta_\beta^3 - \xi_\alpha\eta_\alpha^3\xi_\beta^2\eta_\beta^2)\,x^3y +$$
$$+ (\xi_\alpha^4\eta_\beta^4 + 2\xi_\alpha^3\eta_\alpha\xi_\beta\eta_\beta^3 - 3\xi_\alpha^2\eta_\alpha^2\xi_\beta^2\eta_\beta^2)\,x^2y^2 +$$
$$+ 2(\xi_\alpha^3\eta_\alpha\eta_\beta^4 - \xi_\alpha^2\eta_\alpha^2\xi_\beta\eta_\beta^3)\,xy^3 + (\xi_\alpha^2\eta_\alpha^2\eta_\beta^4 - \xi_\alpha\eta_\alpha^3\xi_\beta\eta_\beta^3)\,y^4.$$

Interchanging α and β, and averaging, we obtain a symmetric symbolic form for the Hessian:

$$\widehat{H} = 72\,(\xi_\alpha\eta_\beta - \eta_\alpha\xi_\beta)^2\,(x\,\xi_\alpha + y\,\eta_\alpha)^2\,(x\,\xi_\beta + y\,\eta_\beta)^2$$
$$= 72\,[\,\alpha\,\beta\,]^2\,(\alpha\,\mathbf{x})^2\,(\beta\,\mathbf{x})^2.$$

In this case, the $\mathbf{x}$–independent parts of the symbolic formula are the same (up to a factor) as the transform, which is $\frac{1}{2}[\,\alpha\,\beta\,]^2$.

This particular observation holds true in general. Apart from a multiplicative factor, the $\mathbf{x}$–independent transform version of a polynomial covariant coincides with the $\mathbf{x}$–independent factors of its classical symbolic form. The precise connection is embodied in the following theorem, which provides an explicit mechanism for changing a symbolic form into a transform and vice versa, and shows that the correspondence is "morally" one-to-one. The complication is due to the fact that the scaling process σ, when applied to a homogeneous polynomial, reduces to multiplication by the degree of homogeneity. Therefore, one can expect that many different differential polynomials will represent the same function of the coefficients of a homogeneous polynomial. In fact, a direct corollary of the theorem is that this is the only ambiguity: two differential polynomials give the same function if and only if they can be transformed into each other by replacing scaling processes by appropriate factors. See Example 6.8 for an example of this phenomenon.

Theorem 6.23. *Let $Q(\mathbf{x})$ be a binary form of degree n, as in (2.1). Let $J(\mathbf{a}, \mathbf{x})$ be any homogeneous polynomial of order m in the coefficients $\mathbf{a}$ of Q. Suppose that $S(\mathbf{x}, \boldsymbol{\xi}_\alpha, \ldots, \boldsymbol{\xi}_\varepsilon)$ is a symbolic form of J depending on m symbolic letters. Then the rescaled polynomial*

$$T(\mathbf{x}, \boldsymbol{\xi}_\alpha, \ldots, \boldsymbol{\xi}_\varepsilon) = \frac{1}{(n!)^m}\,S(\mathbf{x}, \boldsymbol{\xi}_\alpha, \ldots, \boldsymbol{\xi}_\varepsilon) \qquad (6.35)$$

is a transform of a differential polynomial $\mathcal{D}[Q]$, which, for the particular binary form, coincides with $J = \mathcal{D}[Q]$.

Conversely, given a homogeneous differential polynomial $\mathcal{D}[Q]$, suppose $T(\mathbf{x}, \boldsymbol{\xi}_\alpha, \ldots, \boldsymbol{\xi}_\varepsilon)$ is a transform, which we assume to be homoge-

neous[†] *of degree* l_κ *in each* $\boldsymbol{\xi}_\kappa$. *Then*

$$S(\mathbf{x}, \boldsymbol{\xi}_\alpha, \dots, \boldsymbol{\xi}_\varepsilon) = T(\mathbf{x}, \boldsymbol{\xi}_\alpha, \dots, \boldsymbol{\xi}_\varepsilon)\, n^{\underline{l_\alpha}}\, (\alpha\,\mathbf{x})^{n-l_\alpha} \cdots n^{\underline{l_\varepsilon}}\, (\varepsilon\,\mathbf{x})^{n-l_\varepsilon}$$
(6.36)

is a symbolic form of the function $J(\mathbf{a}, \mathbf{x}) = \mathcal{D}[Q]$.

Remark: While a symbolic form *must* be homogeneous of degree n in each symbolic letter, this is not necessary for transforms. The extra bracket factors $(\alpha\,\mathbf{x}), \dots, (\varepsilon\,\mathbf{x})$ appearing in the second formula (6.36) are needed so that the symbolic letters $\boldsymbol{\xi}_\alpha, \dots, \boldsymbol{\xi}_\varepsilon$ will all appear in the symbolic form S with the proper degree.

Remark: The theorem does not assume that either the transform or the symbolic form is symmetric. Since they have exactly the same symmetry properties under permutations of symbolic letters, we need only prove the formulae for one particular transform/symbolic form in order to conclude that it holds in general.

Proof: First, suppose $J(\mathbf{a}, \mathbf{x})$ is homogeneous of degree m in $\mathbf{a}$. By linearity, it suffices to consider each monomial in J separately. The monomial

$$M = a_{i_1} a_{i_2} \cdots a_{i_m} x^j y^k$$

has symbolic form

$$S = \xi_{\alpha_1}^{i_1} \eta_{\alpha_1}^{n-i_1} \xi_{\alpha_2}^{i_2} \eta_{\alpha_2}^{n-i_2} \cdots \xi_{\alpha_m}^{i_m} \eta_{\alpha_m}^{n-i_m} x^j y^k.$$

Substituting the formula (6.15) for the coefficients of Q in terms of derivatives demonstrates that M can be recovered by evaluating the differential polynomial

$$\mathcal{D}[Q] = \frac{x^j y^k}{(n!)^m} \prod_{\kappa=1}^{m} \frac{\partial^n Q}{\partial x^{i_\kappa} \partial y^{n-i_\kappa}}.$$

The transform of this differential monomial is clearly just $(n!)^{-m}$ times the symbolic form S, which proves (6.35) for these particular forms.

Conversely, a general differential monomial

$$\mathcal{D}[Q] = x^j y^k \prod_{\kappa=1}^{m} \frac{\partial^{l_\kappa} Q}{\partial x^{i_\kappa} \partial y^{j_\kappa}},$$

[†] If T is not homogeneous, we just break it up into homogeneous summands and apply the result separately to each.

where $l_\kappa = i_\kappa + j_\kappa$, has

$$T = x^j y^k \prod_{\kappa=1}^{m} \xi_{\alpha_\kappa}^{i_\kappa} \eta_{\alpha_\kappa}^{j_\kappa}$$

as a possible transform. Now,

$$\frac{\partial^{i+j} Q}{\partial x^i \partial y^j} = n^{\underline{l}} \sum_{\nu=0}^{n-l} \binom{n-l}{\nu} a_{i+\nu} x^\nu y^{n-l-\nu}, \qquad \text{where} \qquad l = i+j.$$

Therefore, we find

$$\mathcal{D}[Q] = x^j y^k \prod_{\kappa=1}^{m} \left\{ n^{\underline{l_\kappa}} \sum_{\nu=0}^{n-l_\kappa} \binom{n-l_\kappa}{\nu} a_{i_\kappa+\nu} x^\nu y^{n-l_\kappa-\nu} \right\}$$

has symbolic form

$$S = x^j y^k \prod_{\kappa=1}^{m} \left\{ n^{\underline{l_\kappa}} \sum_{\nu=0}^{n-l_\kappa} \binom{n-l_\kappa}{\nu} \xi_{\alpha_\kappa}^{i_\kappa+\nu} \eta_{\alpha_\kappa}^{n-i_\kappa-\nu} x^\nu y^{n-l_\kappa-\nu} \right\}$$

$$= x^j y^k \prod_{\kappa=1}^{m} n^{\underline{l_\kappa}} (x\xi_{\alpha_\kappa} + y\eta_{\alpha_\kappa})^{n-l_\kappa} \xi_{\alpha_\kappa}^{i_\kappa} \eta_{\alpha_\kappa}^{j_\kappa}.$$

Comparison with the earlier transform expression proves (6.36). *Q.E.D.*

Corollary 6.24. *Given two homogeneous polynomials P, Q having respective degrees m, n, then the symbolic form of their r^{th} transvectant $(P, Q)^{(r)}$ is*

$$r^{\underline{m}} r^{\underline{n}} [\alpha \beta]^r (\alpha \mathbf{x})^{m-r} (\beta \mathbf{x})^{n-r}. \qquad (6.37)$$

In the classical literature, the falling factorial prefactors $r^{\underline{m}} r^{\underline{n}}$ do not appear in the symbolic form of transvectants, which means that the classical versions are represented by suitably scaled differential polynomials, the scaling depending on the degree(s) of the form(s). However, I maintain that the present version, because it also applies to more general functions, is more consistent, and hence preferable — even though it does tend to introduce larger numerical coefficients into the final formulae.

Combining Theorems 6.14 and 6.23 leads us immediately to the classical version of the First Fundamental Theorem. Note in particular that, unlike the transform version, invariants are distinguished by their symbolic forms involving only bracket factors of the second kind.

Theorem 6.25. *Every covariant of a system of binary forms has a symbolic form given by a homogeneous bracket polynomial; conversely* *every homogeneous bracket polynomial is the symbolic form of a covariant. A bracket polynomial is the symbolic form of an invariant if and only if it contains only bracket factors of the second kind.*

Example 6.26. By direct computation, the discriminant of the binary cubic, (2.22), can be shown to have its symbolic form given by the bracket monomial $[\alpha\beta]^2 [\alpha\gamma] [\beta\delta] [\gamma\delta]^2$. This implies that the discriminant of a cubic can be expressed by the following partial transvectant formula, cf. Exercise 5.5:

$$\mathcal{D} = \tfrac{1}{1296} \operatorname{tr} (\Omega_{\alpha\beta})^2 \, \Omega_{\alpha\gamma} \, \Omega_{\beta\delta} \, (\Omega_{\gamma\delta})^2 \, (Q \otimes Q \otimes Q \otimes Q). \qquad (6.38)$$

Thus, every symbolic expression has, potentially, two different interpretations — one as the transform of a differential polynomial and the second as the symbolic form of a function of the coefficients. The two differ only by a numerical factor and some inessential bracket factors. Since the former interpretation has a wider range of applicability, we shall concentrate on it, although the more classically inclined reader might prefer to concentrate on the symbolic version. The common rules embodied in the algebra of bracket polynomials will produce the same basic identities among covariants of binary forms.

Proofs of the Fundamental Theorems

A variety of different proofs of the two fundamental theorems of classical invariant theory exist in the literature. The most direct, and the one to be used here, is based on certain identities satisfied by the polarization and omega processes, due originally to Capelli, [31]. This mode of proof was introduced by van der Waerden, [221], and strongly advocated by Weyl, [231]. Older proofs appear in [49, 89, 92]. Weitzenböck, [229], and Gurevich, [97], present proofs based on tensor analysis. Kung and Rota, [135], base their proof on certain "straightening laws" in the algebra of bracket polynomials; see Richman, [186], for yet another proof and further references.

Actually, the required differential operators are not the "contravariant" polarizations (5.26) and omega processes (5.2) we have been using so far, but, rather, their "covariant" counterparts, which are differential operators acting on the symbolic variables. First, the covariant polarization processes are the first order differential operators

$$\pi_{\alpha\beta} = \xi_\alpha \, \frac{\partial}{\partial \xi_\beta} + \eta_\alpha \, \frac{\partial}{\partial \eta_\beta}, \qquad (6.39)$$

which polarize the symbolic variables. Just as the contravariant polarizations are unaffected by a linear transformation, the same holds for their covariant versions. Consequently, if $T(\mathbf{x}; \boldsymbol{\xi}_\alpha, \ldots, \boldsymbol{\xi}_\varepsilon)$ is the symbolic form of a (joint) covariant, then any polarization, e.g., $\pi_{\alpha\beta} T$, is also a covariant of the same weight. Note in particular that the polarization of any bracket factor is also a bracket factor:

$$
\pi_{\alpha\beta}[\gamma\,\delta] = \begin{cases} [\alpha\,\delta], & \beta = \gamma, \\ [\gamma\,\alpha], & \beta = \delta, \\ 0, & \text{otherwise}, \end{cases} \qquad \pi_{\alpha\beta}(\gamma\,\mathbf{x}) = \begin{cases} (\alpha\,\mathbf{x}), & \beta = \gamma, \\ 0, & \text{otherwise}. \end{cases}
$$

$$(6.40)$$

Since the polarization operators act as derivations, that is, they are linear and obey the Leibniz Rule, this trivially implies the following lemma:

Lemma 6.27. *The polarization of any bracket polynomial is also a bracket polynomial. The polarization of any syzygy is also a syzygy.*

We also introduce the covariant counterpart

$$
\Xi_{\alpha\beta} = \frac{\partial^2}{\partial \xi_\alpha \partial \eta_\beta} - \frac{\partial^2}{\partial \xi_\beta \partial \eta_\alpha} \tag{6.41}
$$

to the omega process, which also maps covariants to covariants, dropping the weight by 1, and, as we shall see, bracket polynomials to bracket polynomials. (This can be proved directly, although it is not as immediate.) The required *Capelli identity* relates the two processes as follows:

$$
\det \begin{vmatrix} \pi_{\alpha\alpha} + 1 & \pi_{\beta\alpha} \\ \pi_{\alpha\beta} & \pi_{\beta\beta} \end{vmatrix} = (\pi_{\alpha\alpha} + 1)\pi_{\beta\beta} - \pi_{\beta\alpha}\pi_{\alpha\beta} = [\alpha\,\beta]\,\Xi_{\alpha\beta}, \tag{6.42}
$$

where $[\alpha\,\beta] = \xi_\alpha \eta_\beta - \xi_\beta \eta_\alpha$ is a bracket factor of the second kind. When expanding such a non-commutative determinant, the operators are ordered according to their row index. The differential operator identity (6.42) is trivially proved by direct computation using the original definitions (6.39) and (6.41).

Remark: The contravariant Capelli identity is obtained by interchanging the roles of co- and contravariant variables, leading to

$$
(\sigma_{\alpha\alpha} + 1)\sigma_{\beta\beta} - \sigma_{\beta\alpha}\sigma_{\alpha\beta} = [\![\alpha\,\beta]\!]\,\Omega_{\alpha\beta}, \tag{6.43}
$$

which relates the original polarization and omega processes. The symbolic version of (6.43) is the particular case $\alpha = \lambda, \beta = \mu$ of the trivial determinantal product identity

$$\det \begin{vmatrix} (\alpha\,\lambda) & (\beta\,\lambda) \\ (\alpha\,\mu) & (\beta\,\mu) \end{vmatrix} = [\,\alpha\,\beta\,]\,[\![\,\lambda\,\mu\,]\!] = (\xi_\alpha\eta_\beta - \xi_\beta\eta_\alpha)(x_\lambda y_\mu - x_\mu y_\lambda), \quad (6.44)$$

relating all three types of bracket factors.

The proof of the First Fundamental Theorem will be simplified if we employ the artifice, introduced in the remark following Definition 6.12, of identifying the contravariant variables **x** with an additional pair of symbolic variables, namely $\boldsymbol{\xi}_0 = (-y, x)$. In this way, bracket factors of the first and second kind are not distinguished, and so we can assume, without loss of generality, that the transformed covariant $T(\boldsymbol{\xi}_\alpha, \ldots, \boldsymbol{\xi}_\varepsilon)$ depends only on symbolic variables. Indeed, one can readily extend this device to the case when T depends on several different contravariant vectors $\mathbf{x}_\lambda$, each of which is identified with a new pair of symbolic variables $\boldsymbol{\xi}_\lambda = (-y_\lambda, x_\lambda)$. The effect is that all three types of bracket factors coincide, and one thereby deduces an extended version of the First Fundamental Theorem: any covariant polynomial depending on several co- and contravariant vectors is necessarily a polynomial in the three types of bracket factors. The reader should note that the proof will only require the action of the contravariant polarization and omega processes on bracket polynomials and does *not* rely on their behavior under linear maps; thus, our use of this artificial identification of co- and contravariant variables is allowed. (On the other hand, since this artifice is particular to dimension 2, it makes the passage to the higher dimensional versions of the Fundamental Theorems less immediate.)

The proof will proceed by induction on the number of symbolic variables in the transform T. We can assume, without loss of generality, that T is homogeneous of degree n_α in each symbolic vector $\boldsymbol{\xi}_\alpha$. To initiate the induction, we first note that there are no nonconstant covariants $T(\boldsymbol{\xi}_\alpha)$ depending on a single symbolic letter. This is evident from the transformation rules — reflecting the transitivity of the action of $\mathrm{GL}(2)$ on $\mathbb{C}^2 \setminus \{0\}$. Proceeding to the induction step, let $\boldsymbol{\xi}_\alpha$ and $\boldsymbol{\xi}_\beta$ be two of the $m \geq 2$ symbolic vectors occurring in T. Let $R = \pi_{\alpha\beta}T$ and $S = \Xi_{\alpha\beta}T$ be obtained by applying the covariant polarization and omega processes to T. Since T is a covariant, so are R and S. Moreover, since $\pi_{\alpha\beta}$ and $\Xi_{\alpha\beta}$ both involve differentiation with respect to $\boldsymbol{\xi}_\beta$, the degree of both R and S in $\boldsymbol{\xi}_\beta$ is strictly less than that of T. We can therefore use an induction on the degree of covariants in the symbolic vector $\boldsymbol{\xi}_\beta$ to assume that R and S are bracket polynomials. (The induction begins with the case of degree $n_\beta = 0$, which means that the covariant depends on only $m - 1$ symbolic letters.) We now appeal to the Capelli identity (6.42). Applying both sides to T, and using homogeneity, yields

$$(n_\alpha + 1)n_\beta \, T = \pi_{\beta\alpha} R + [\,\alpha\,\beta\,]\, S. \tag{6.45}$$

Induction and Lemma 6.27 imply that the right-hand side of (6.45) is a bracket polynomial, and hence so is T. This justifies the induction step and so completes the proof of the First Fundamental Theorem 6.14 (and its symbolic counterpart, Theorem 6.25).   *Q.E.D.*

A similar inductive argument can be used to prove the Second Fundamental Theorem. Again we identify co- and contravariant variables, and so only need consider a bracket polynomial syzygy

$$S([\,\alpha\,\beta\,],[\,\alpha\,\gamma\,],[\,\beta\,\gamma\,],\dots) \equiv 0, \tag{6.46}$$

depending on bracket factors of the second kind. The goal is to prove that S is necessarily a linear combination of the basic syzygies (6.32):

$$S = \sum_{\alpha,\beta,\gamma,\delta} W_{\alpha\beta\gamma\delta} \, [\![\,\alpha\,\beta\,\gamma\,\delta\,]\!], \tag{6.47}$$

where the coefficients $W_{\alpha\beta\gamma\delta}$ are bracket polynomials. To state this more precisely, we should regard the bracket arguments $[\,\alpha\,\beta\,]$, $[\,\alpha\,\gamma\,]$, ... of S as independent variables, subject to the evident identifications

$$[\,\alpha\,\beta\,] = -[\,\beta\,\alpha\,], \qquad [\,\alpha\,\alpha\,] = 0, \tag{6.48}$$

in order to take care of the trivial first kind of syzygy (6.25). The fact that S defines a syzygy among the brackets means that when we substitute the actual formulae (6.20) for the bracket "variables" in S, we obtain the zero symbolic polynomial, i.e., (6.46) holds. In this manner, the syzygy identity (6.47) is a polynomial identity in the bracket variables, which becomes trivial, $0 = 0$, when the brackets are replaced by their explicit formulae (6.20).

We denote the partial derivatives of S with respect to to the bracket variables by $\partial S/\partial [\,\alpha\,\beta\,]$; the identifications (6.48) imply the conventions

$$\frac{\partial S}{\partial [\,\beta\,\alpha\,]} = -\frac{\partial S}{\partial [\,\alpha\,\beta\,]}, \qquad \frac{\partial S}{\partial [\,\alpha\,\alpha\,]} = 0.$$

The covariant polarization operators (6.39) will act on bracket polynomials according to the basic rules (6.40) and so, in terms of the bracket variables, can be identified with the homogeneous first order differential operator

$$\pi_{\alpha\beta} = \sum_\gamma [\,\alpha\,\gamma\,] \frac{\partial}{\partial [\,\beta\,\gamma\,]}. \tag{6.49}$$

The key identity that plays the role of the Capelli identity (6.42) here is the following:

Lemma 6.28. *Let S be a bracket polynomial. Then the bracket polarizations (6.49) satisfy the following identity:*

$$
\det \begin{vmatrix} \pi_{\alpha\alpha} + 2 & \pi_{\beta\alpha} & \pi_{\gamma\alpha} \\ \pi_{\alpha\beta} & \pi_{\beta\beta} + 1 & \pi_{\gamma\beta} \\ \pi_{\alpha\gamma} & \pi_{\beta\gamma} & \pi_{\gamma\gamma} \end{vmatrix} S =
$$

$$
= \sum_{\lambda,\mu,\nu} \left[\!\!\left[\begin{matrix} \alpha & \beta & \gamma \\ \lambda & \mu & \nu \end{matrix} \right]\!\!\right] \frac{\partial^3 S}{\partial [\alpha\lambda] \partial [\beta\mu] \partial [\gamma\nu]} + \qquad (6.50)
$$

$$
+ 2 \sum_{\nu} [\![\alpha\beta\gamma\nu]\!] \left\{ \frac{\partial^2 S}{\partial [\alpha\beta] \partial [\gamma\nu]} + \frac{\partial^2 S}{\partial [\gamma\alpha] \partial [\beta\nu]} + \frac{\partial^2 S}{\partial [\beta\gamma] \partial [\gamma\nu]} \right\}.
$$

The second summation depends linearly on the basic syzygies (6.32), while the first depends on the syzygy studied in Example 6.20, and so, by (6.33), also depends linearly on them. Therefore, for *any* bracket polynomial S, the left-hand side of (6.50) is a linear combination of syzygies and hence automatically vanishes once the bracket variables are replaced by their actual symbolic expressions. Lemma 6.28 is proved by a tedious direct computation, based on (6.49).

Now, assume that S is a bracket syzygy depending on $m \geq 2$ symbolic letters. We can assume, without loss of generality, that S is homogeneous of degree n_α in each symbolic letter α. More specifically, n_α indicates the degree of S in all bracket variables of the form $[\alpha\varepsilon]$ for all ε. Euler's formula (2.13) combined with (6.49) implies that $\pi_{\alpha\alpha} S = n_\alpha S$.

We shall prove the theorem by induction on m, the number of symbolic letters in S. The initial case $m = 2$ is trivial, since there is only one possible bracket factor, $[\alpha\beta]$, and hence there are no nontrivial syzygies. Turning to the inductive step, let α, β, γ be three of the symbolic letters occurring in S. Define $P = \pi_{\alpha\gamma} S$, $Q = \pi_{\beta\gamma} S$, $R = \pi_{\alpha\beta} S$. Now, the bracket polarization process $\pi_{\alpha\beta}$ reduces the degree n_β, while raising the degree n_α. Therefore, P and Q have degree strictly less than n_γ in the symbolic letter γ, while R has the same degree n_γ but has degree strictly less than n_β in the symbolic letter β. We order the bracket polynomials based on their degrees in these two symbolic letters, so that a polynomial having degrees $\tilde{n}_\beta, \tilde{n}_\gamma$ comes before one with degrees n_β, n_γ if either $\tilde{n}_\gamma < n_\gamma$, or $\tilde{n}_\gamma = n_\gamma$, but $\tilde{n}_\beta < n_\beta$. In this manner, P, Q, R all appear before S in this ordering. Therefore, we can use an inductive argument, relying on the fact that if n_β or $n_\gamma = 0$, the polynomial depends on fewer than m symbolic letters, to assume that P, Q, R are all linear combinations of the basic syzygies (6.32).

We appeal finally to our key identity (6.50). Expanding the determinant along the bottom row and using homogeneity, we find the left-hand side equals

$$(n_\alpha + 2)(n_\beta + 1)n_\gamma S +$$

$$+ \det \begin{vmatrix} \pi_{\beta\alpha} & \pi_{\gamma\alpha} \\ \pi_{\beta\beta} + 1 & \pi_{\gamma\beta} \end{vmatrix} P - \det \begin{vmatrix} \pi_{\alpha\alpha} + 2 & \pi_{\gamma\alpha} \\ \pi_{\alpha\beta} & \pi_{\gamma\beta} \end{vmatrix} Q - n_\gamma \pi_{\beta\alpha} R.$$

If P, Q, R are linear combinations of basic syzygies, then so are any polarizations thereof. Moreover, as we remarked earlier, the right-hand side of (6.50) is also a linear combination of basic syzygies. We conclude that S itself must also be a linear combination of the syzygies (6.32). This completes the proof of the Second Fundamental Theorem. *Q.E.D.*

Remark: The preceding proof of the Second Fundamental Theorem is new. The original proof is due to Pascal, [**177**]; see also [**92**; p. 211] and [**186, 221, 231**] for alternative proofs.

Exercise 6.29. Prove the 3×3 version of the Capelli identity

$$\det \begin{vmatrix} \pi_{\alpha\alpha} + 2 & \pi_{\beta\alpha} & \pi_{\gamma\alpha} \\ \pi_{\alpha\beta} & \pi_{\beta\beta} + 1 & \pi_{\gamma\beta} \\ \pi_{\alpha\gamma} & \pi_{\beta\gamma} & \pi_{\gamma\gamma} \end{vmatrix} = 0, \qquad (6.51)$$

which is valid for the basic covariant polarization operators (6.39). Note that our fundamental syzygy identity (6.50) uses the differential operator (6.51), but where the polarization operators are replaced by their bracket versions (6.49). See Weyl, [**231**; p. 42], for multidimensional generalizations. However, I do not know the corresponding generalization of the syzygy identity (6.50).

Reciprocity

By admitting differentiation with respect to both symbolic and ordinary variables, we can now construct a wide range of additional invariant differentiation processes, of which the omega and scaling processes are but the simplest examples. The basic idea is due to Sylvester, [**207**], although special cases already appear in Boole's original paper, [**24**].

The key idea is that if $\mathbf{a} = (a, b)$ is any covariant vector, the corresponding gradient operator $\nabla_{\mathbf{a}} = (\partial_a, \partial_b)$ transforms contravariantly. We can use our trick of identifying co- and contravariant vectors to then construct an associated covariant differential operator $\widetilde{\nabla}_{\mathbf{a}} = (\partial_b, -\partial_a)$. As a consequence, if $J(\boldsymbol{\xi}_\alpha, \ldots, \boldsymbol{\xi}_\varepsilon)$ is any symbolic representation of a covariant, and we replace one or more of the symbolic variables by their modified gradients

$$\boldsymbol{\xi}_\alpha = (\xi_\alpha, \eta_\alpha) \longmapsto \widetilde{\nabla}_\alpha = \left(\frac{\partial}{\partial \eta_\alpha}, \ -\frac{\partial}{\partial \xi_\alpha} \right), \qquad (6.52)$$

then the result is a covariant differentiation process. As usual, we place all differentiations to the right of multiplications. For example, if we start with a single bracket factor of the second kind $[\alpha\,\beta] = \xi_\alpha \eta_\beta - \xi_\beta \eta_\alpha$, and replace the symbolic variables $\boldsymbol{\xi}_\beta$ by their modified gradient $\widetilde{\nabla}_\beta$, the result is the covariant polarization process (6.39). If we replace both $\boldsymbol{\xi}_\alpha$ and $\boldsymbol{\xi}_\beta$, we obtain the covariant omega process (6.41).

Exercise 6.30. Determine how such a covariant differential operator affects the weights of covariants.

A particularly important case is the following construction. Let Q be a binary form of degree n and P be a binary form of degree m. Let $R = \mathbf{R}[P, Q]$ denote their resultant, which has degree m in the coefficients of Q and degree n in the coefficients of P. Consider the symbolic form $\widehat{R}(\boldsymbol{\xi}_\alpha, \ldots, \boldsymbol{\xi}_\varepsilon; \widetilde{\boldsymbol{\xi}}_\lambda, \ldots, \widetilde{\boldsymbol{\xi}}_\nu)$ of the resultant, where $\alpha, \ldots, \varepsilon$ are m symbolic letters corresponding to the coefficients of Q, while $\lambda, \ldots, \nu$ are symbolic letters corresponding to the coefficients of P. If we replace the symbolic variables for Q by differentiation operators, we obtain a covariant differentiation process

$$\mathcal{D}^* = \widehat{R}(\nabla_\alpha, \ldots, \nabla_\varepsilon; \boldsymbol{\xi}_\lambda, \ldots, \boldsymbol{\xi}_\nu) \qquad (6.53)$$

that maps covariants of order m in Q to covariants of order n in P. It can be shown, cf. [**114, 229**], that this maps nonzero covariants to nonzero covariants, and hence can be used to justify Hermite's famous reciprocity relation, [**102**].

Theorem 6.31. *There is a one-to-one correspondence between covariants of order m and degree j of a binary form Q of degree n and covariants of order n and degree j of a binary form R of degree m.*

Most modern proofs, cf. [**135, 204**], rely on combinatorial formulae counting the number of covariants and do not provide any insight into the actual mechanism discovered by Hermite relating the two types of covariants.

Fundamental Systems of Covariants

Although a constructive algorithm was discovered by Gordan — and will be presented in the following chapter — the determination of a complete Hilbert basis for the covariants of a binary form has been found to be an extremely difficult matter. However, if we are willing to forgo our insistence on preservation of the polynomial structure of the covariants, matters considerably simplify. Indeed, a result due to Stroh, [**203**], and Hilbert, [**107**; p. 64], states that the covariants of orders 1, 2, and 3 provide a general, explicit rational basis for the covariants of a binary form of any order. (See also [**107**; p. 88] for a similar result for the joint covariants of two binary forms.)

Let Q be a binary form of degree $n \geq 4$. Define integers $m+l = n-1$ so that if $n = 2m$ is even, then $l = m - 1$, whereas if $n = 2m + 1$ is odd, then $l = m$. Up to multiple, the only linear (order 1) covariant is the form Q itself. There are m independent quadratic covariants, namely, the even order transvectants

$$S_k = \tfrac{1}{2} (Q, Q)^{(2k)}, \qquad k = 1, \ldots, m. \tag{6.54}$$

Note that S_k has degree $2n - 4k$ and weight $2k$. In particular, $S_1 = H$ is the Hessian, and, if $n = 2m$ is even, S_m is an invariant. There are many possible cubic covariants; for our purposes, the most important of these are the Jacobians of Q with its quadratic covariants, which we denote by

$$T_k = (Q, S_k)^{(1)} = \tfrac{1}{2} (Q, (Q, Q)^{(2k)})^{(1)}, \qquad k = 1, \ldots, l. \tag{6.55}$$

(If $n = 2m$ is even, then $T_m = 0$, since S_m is an invariant.) Note that T_k has degree $3n - 4k - 2$ and weight $2k + 1$. The Stroh–Hilbert Theorem states that the n fundamental covariants $Q, S_1, \ldots, S_m, T_1, \ldots, T_l$ form a rational basis for *all* the covariants; in fact, the only denominators required are powers of Q.

Theorem 6.32. *Let J be any polynomial covariant of the binary form Q of degree n. Then, for some power $N \geq 0$, the covariant $Q^N J$ can be written as a polynomial in the n fundamental covariants $Q, S_1, \ldots, S_m, T_1, \ldots, T_l$.*

Proof: The proof proceeds by a combined induction on the order of the covariant and its weight. According to the First Fundamental Theorem 6.16, every polynomial covariant has a transform given by a linear combination of monomials in the bracket factors of the second kind. The order of a monomial equals the number of different symbolic letters, while the weight equals the number of brackets.

A bracket monomial M represents a reducible covariant if and only if we can split its symbolic letters into two disjoint groups, labeled $\alpha_1, \ldots, \alpha_{m'}$ and $\beta_1, \ldots, \beta_{m''}$, where $m' + m'' = m$, and $m', m'' \geq 1$, such that

$$M(\alpha_1, \ldots, \alpha_{m'}, \beta_1, \ldots, \beta_{m''}) = M'(\alpha_1, \ldots, \alpha_{m'})\, M''(\beta_1, \ldots, \beta_{m''})$$
(6.56)

is a product of brackets depending only on letters in one of the groups. If (6.56) holds, then $J = J' \cdot J''$, where J', J'' have respective transforms M', M''. In particular, if M does not actually depend on the symbolic letters $\beta_1, \ldots, \beta_{m''}$, so that $M'' = 1$ is a constant, then $J = Q^{m''} J'$ is merely a power of Q times another covariant. If our induction has proved the result for lower order covariants, then it clearly holds for reducible covariants of the given order.

Let us now explain the induction step at order m. (We delay initiating the induction since there is a slight twist.) Suppose M is a bracket monomial of order $m \geq 4$ and weight $k \geq 3$ representing a nonzero covariant J. (If $m \geq 4$ and $k < 3$, then it is not hard to see that M is automatically reducible.) We can split the symbolic letters occurring in M into $\alpha_1, \ldots, \alpha_{m'}$ and $\beta_1, \ldots, \beta_{m''}$, where $m' + m'' = m$, and $m', m'' \geq 2$. We write

$$M(\alpha_1, \ldots, \alpha_{m'}, \beta_1, \ldots, \beta_{m''}) =$$
$$= \left\{ \prod_{i=1}^{j} [\alpha_{\mu_i} \beta_{\nu_i}] \right\} M'(\alpha_1, \ldots, \alpha_{m'})\, M''(\beta_1, \ldots, \beta_{m''}),$$
(6.57)

where the initial factor is a product of j "mixed" brackets involving one letter from each of the two groups, while $M'(\alpha_1, \ldots, \alpha_{m'})$ is a bracket monomial of order $m' \geq 2$ and weight $k' \geq 1$, and $M''(\beta_1, \ldots, \beta_{m''})$ is a bracket monomial of order $m'' \geq 2$ and weight $k'' \geq 1$. (We may need to choose a different splitting of symbolic letters if the weights k', k'' are not both at least 1. If no such splitting is possible, the covariant is necessarily reducible.) Note also that $k' + k'' + j = k$. If $j = 0$, then (6.57) reduces to (6.56), and so M is reducible. Otherwise, we proceed as follows. Let us introduce two additional symbolic letters $\alpha_{m'+1}$ and $\beta_{m''+1}$ and treat M as a function of all $m + 2$ letters; this is equivalent to replacing the original covariant J by the reducible covariant $Q^2 J$.

Consider the elementary bracket identity

$$[\alpha\,\beta](\gamma\,\mathbf{x})(\delta\,\mathbf{x}) - [\gamma\,\delta](\alpha\,\mathbf{x})(\beta\,\mathbf{x}) = [\alpha\,\gamma](\beta\,\mathbf{x})(\delta\,\mathbf{x}) - [\beta\,\delta](\alpha\,\mathbf{x})(\gamma\,\mathbf{x}),$$
$$(6.58)$$

which is a simple consequence of the basic syzygy (6.26). Let us form an appropriate j-fold product of this identity:

$$\prod_{i=1}^{j}\left\{ [\alpha_{\mu_i}\,\beta_{\nu_i}](\alpha_{m'+1}\,\mathbf{x})(\beta_{m''+1}\,\mathbf{x}) - [\alpha_{m'+1}\,\beta_{m''+1}](\alpha_{\mu_i}\,\mathbf{x})(\beta_{\nu_i}\,\mathbf{x}) \right\} =$$

$$= \prod_{i=1}^{j}\left\{ [\alpha_{\mu_i}\,\alpha_{m'+1}](\beta_{\nu_i}\,\mathbf{x})(\beta_{m''+1}\,\mathbf{x}) - [\beta_{\nu_i}\,\beta_{m''+1}](\alpha_{\mu_i}\,\mathbf{x})(\alpha_{m'+1}\,\mathbf{x}) \right\}.$$

We multiply both sides of this equality by the product

$$M'(\alpha_1,\ldots,\alpha_{m'})\,M''(\beta_1,\ldots,\beta_{m''})$$

and then expand the results into sums of bracket monomials. Homogeneity of Q implies that each bracket factor of the first kind can be replaced by a numerical factor, which is nonzero since each symbolic letter occurs at most $n = \deg Q$ times. The first term on the left-hand side of the resulting identity is a nonzero multiple of our original bracket monomial (6.57). Every other left-hand term is a reducible bracket monomial of the form

$$\widetilde{M}(\alpha_1,\ldots,\alpha_{m'},\beta_1,\ldots,\beta_{m''})\,[\alpha_{m'+1}\,\beta_{m''+1}]^l, \qquad (6.59)$$

where $l \geq 1$, and $\widetilde{M}$ has the same form (6.57), but the initial product contains l fewer "mixed" bracket factors. Note that (6.59) is the symbolic form of a reducible covariant of the form $\widetilde{J}\cdot(Q,\,Q)^{(l)}$, which equals $2\widetilde{J}S_i$ if $l = 2i$ is even but is 0 if $l = 2i + 1$ is odd. An induction on the number j of mixed brackets proves that each $\widetilde{J}$ is of the required form. On the right-hand side, each term represents a reducible covariant, being a product of bracket monomials $N'(\alpha_1,\ldots,\alpha_{m'+1})$ of order $m' + 1 < m$ and weight l', where $k' \leq l' \leq k' + j < k$, and $N''(\beta_1,\ldots,\beta_{m''+1})$ of order $m'' + 1 < m$ and weight l'' where $k'' \leq l'' \leq k'' + j < k$. Therefore, an induction on the order and weight of the bracket monomial will take care of these two terms.

It remains for us to start the induction. First, as remarked at the outset, every covariant of order 2 is a multiple of the basic quadratic covariants (6.54). We have not yet considered any covariants of order 3. (The preceding implementation of the induction step required order at least 4.) Consider the general bracket monomial

$$M = [\alpha\beta]^p [\alpha\gamma]^q [\beta\gamma]^r, \qquad (6.60)$$

representing a covariant J of order 3 and weight $k = p + q + r$. We assume, without loss of generality, that $p \geq q \geq r$. If $q = r = 0$, then J is reducible, being either 0 if p is odd or $2QS_i$ if $p = 2i$ is even. If $q = 1$ and $r = 0$, then (6.60) represents the cubic covariant T_i if $p = 2i$ is even, whereas Example 6.21 implies that it represents a multiple of QS_{i+1} if $p = 2i + 1$ is odd. In all other cases, we introduce an additional symbolic letter δ, effectively replacing J by QJ, and apply the syzygy (6.26), using homogeneity to replace the bracket factors of the first kind by the explicit numerical factors; this leads to the identity

$$\begin{aligned} nM = (n - p - q + 1)[\alpha\beta]^p[\alpha\gamma]^{q-1}[\beta\gamma]^r[\gamma\delta] + \\ + (n - q - r + 1)[\alpha\beta]^p[\alpha\gamma]^{q-1}[\beta\gamma]^r[\alpha\delta]. \end{aligned} \qquad (6.61)$$

The two monomials on the right-hand side of (6.61) represent covariants of order 4 and weight k, and we can apply the reduction procedure discussed earlier. Specifically, for the first summand, we split the four symbolic letters into the two groups α, β and γ, δ, and so it has the form (6.57), with $M' = [\alpha\beta]^p$ and $M'' = [\gamma\delta]$, while the initial product of mixed brackets is $[\alpha\gamma]^{q-1}[\beta\gamma]^r$. For the second summand, we use the splitting α, δ and β, γ. The preceding reduction algorithm for order 4 monomials will ultimately lead to a linear combination of products of cubic bracket monomials of the same basic form (6.60), but each of *lower* overall weight. Therefore, an induction on the weight $k = p + q + r$ of the cubic covariant will complete the proof. *Q.E.D.*

Exercise 6.33. Suppose J is the cubic covariant represented by the bracket monomial $M = [\alpha\beta]^2[\alpha\gamma]^2[\beta\gamma]^2$. Find an explicit formula of the form

$$Q^3 J = F(Q, S_1, \dots, S_m, T_1, \dots, T_l)$$

by implementing the construction in the proof. If Q is a quartic, then J represents a multiple of its fundamental covariant j, (2.30), and this result provides a syzygy of the form $Q^3 j = F(Q, H, i, T)$. Use this to classify quartics with $j = 0$.

Remark: Hilbert, [**107**; p. 137], states a result that implies the existence of a similar rational basis for the invariants of a binary form. However, this result is non-constructive, and, as far as I know, there is no known explicit rational basis for just the invariants of binary forms of high degree.

Chapter 7

Graphical Methods

One of the barriers awaiting any serious student of the classical theory is the sheer algebraic complexity of many of the constructions. With a view to rendering these complicated manipulations more manageable, not to mention better visualized, Clifford, [51], began developing a graphical method for the description of the invariants and covariants of binary forms, although he died before publishing his findings to any significant extent. Contemporaneously, Sylvester, [206], unveiled his "algebro-chemical theory", the aim of which was to apply the methods of classical invariant theory to the then rapidly developing science of molecular chemistry. Sylvester's theory was never taken very seriously by chemists, and not developed any further by mathematicians, and soon succumbed to a perhaps well-deserved death.[†] The present graphical treatment of invariant theory, based on results of C. Shakiban and the author, [171], is closest to the version developed by Kempe, [125], which builds on Clifford's posthumous notes.

Although Sylvester envisioned his theory as the future of chemistry, it is Clifford's graph theory that, with one slight but important modification, could have become an important tool in computational invariant theory. The correct framework for the algebro-chemical theory is to use "linear combinations of" digraphs or "directed molecules" as the fundamental objects. One can ascribe both a graph theoretical as well as a chemical interpretation to these objects; both are useful for motivating the method. The fundamental syzygies then translate into certain operations which can be performed on digraphs or, equivalently, certain allowable reactions which can occur among directed molecules. The determination of a basis of irreducible digraphs or "atomic molecules" is

---

[†] Sylvester's speculations may not have been totally misguided, however. Modern books on atomic and molecular physics, e.g., [46, 115, 238], are largely treatises in representation theory, a subject that, as we have seen, is not so far removed from classical invariant theory. See also Sartori, [189], for recent applications of invariant theory to atomic physics.

the same problem as the determination of the Hilbert basis for the co-variants of a binary form. Indeed, all the computations in the symbolic calculus of invariant theory have elementary pictorial analogs, justify-ing our guiding principle that "a picture is worth a thousand algebraic manipulations".

Digraphs, Molecules, and Covariants

The starting point is the symbolic representation of an invariant or co-variant depending on one or more binary forms or, more generally, func-tions $Q_\alpha, Q_\beta, \ldots$. According to the First Fundamental Theorem 6.14, the transform T of any covariant J can, without loss of generality, be taken to be a bracket polynomial, meaning that it can be written as a linear combination $T = \sum c_i M_i$ of bracket monomials M_i, each of which is a product of bracket factors of the first and second kind. Of course, owing to the syzygies among bracket factors, not to mention that we have not required that T be symmetric, there are many different such representations of the given covariant J.

The graphical approach deals directly with the bracket monomials, and we show how to associate a *molecule* or *digraph* (short for "directed graph") to each constituent monomial. (See [**44**] for a modern treat-ment of basic graph theory.) Each function Q_α in our system will be represented by an *atom* or, in graph-theoretic terms, a *vertex*. Atoms corresponding to identical Q's are considered to be identical; this reflects the invariance of the symbolic form under permutations of the symbolic letters. The number of each type of atom in the molecule equals the order of the covariant in the corresponding function. For example, a bracket monomial which is cubic in Q and quadratic in R will be repre-sented by a certain pentatomic molecule, three of whose atoms will be of type Q and two of type R. In particular, for the covariants of a single function, Q, all atoms are interchangeable, and the number of atoms equals the degree of the covariant.

Second, the interatomic bonds in our molecule will correspond to the bracket factors of the second kind occurring in the bracket monomial M. We will use *arcs* or *directed bonds* so that we can distinguish between the bracket factors $[\alpha\beta]$ and $[\beta\alpha]$. (For the moment let us ignore the fact that these differ just by a sign.) Thus, $[\alpha\beta]$ will be represented by a bond *from* atom α *to* atom β, whereas $[\beta\alpha]$ will be represented by a bond *from* atom β *to* atom α. If a bracket occurs to the kth power in

M, that is, $[\alpha\beta]^k$ is a factor of M, then there will be k distinct bonds from atom α to atom β.

For example, the Jacobian covariant $(Q, R)^{(1)} = Q_x R_y - Q_y R_x$ has transform $[\alpha\beta]$ and is represented by the diatomic molecule O———●, in which the white atom represents Q and the black atom represents R. Similarly, the Hessian covariant $2H = (Q, Q)^{(2)} = 2(Q_{xx}Q_{yy} - Q_{xy}^2)$ has transform $[\alpha\beta]^2$ and is therefore represented by the diatomic molecule O═══O consisting of two bonds between two identical atoms, each of type Q. The corresponding molecule with two different atoms would represent the general second transvectant $(Q, R)^{(2)}$. The covariant $T = (Q, H)^{(1)}$, with transform $[\alpha\beta][\beta\gamma]^2$, cf. (6.14), has graphical form O———O═══O. As with the Hessian, all the atoms in the molecule representing T are interchangeable since they all correspond to the single function Q; the same molecule with different atoms would correspond to one of the partial transvectants in (5.23).

Bracket factors of the first kind, $(\alpha \mathbf{x})$, will be represented by (undirected) unattached bonds at the atom α. Thus, the scaling covariant $\sigma(Q) = xQ_x + yQ_y$ has transform $(\alpha \mathbf{x})$ and hence is represented by the monatomic ion O— . More generally, a single atom with k such unattached bonds represents the transform $(\alpha \mathbf{x})^k$, which, according to Lemma 6.1, represents the covariant $\sigma^{\underline{k}} Q$; for instance,

$$\underset{\mid}{\overset{\mid}{+}}\!\!-\quad \text{represents}\quad \sigma^{\underline{4}} Q = \sigma(\sigma - 1)(\sigma - 2)(\sigma - 3) Q. \quad (7.1)$$

If Q is a binary form, or, more generally, a homogeneous function of degree n, then (7.1) reduces to $n^{\underline{4}} Q = n(n - 1)(n - 2)(n - 3)Q$. If Q has degree $3, 2, 1$, or 0, then this particular covariant vanishes.

Exercise 7.1. If Q is homogeneous of degree n, then the digraph —O═══O— represents a multiple of the Hessian of Q. Prove this and determine the multiple.

More general covariants have transforms given by linear combinations of bracket monomials. In the chemical formulation, then, a corresponding "linear combination of molecules" will represent the covariant. More specifically, given a bracket polynomial $T = \sum c_i M_i$, where the M_i are its monic bracket monomial constituents, then we define the molecular representation to be the formal sum of digraphs $D = \sum c_i D_i$, where each D_i is the molecular representative of the monomial M_i. For example, the bracket polynomial $2[\alpha\beta]^2[\alpha\gamma] - \frac{1}{3}[\alpha\beta][\beta\gamma][\gamma\alpha]$ has molecular representation

$$2 \; \overset{\text{(digraph)}}{\bigvee} \; - \; \frac{1}{3} \; \overset{\text{(digraph)}}{\bigvee} \; ,$$

which can be interpreted to mean twice the first digraph minus $\frac{1}{3}$ the second. (Chemically, these linear combinations of molecules might be interpreted as "mixtures" of molecular substances, although the admission of negative coefficients stretches this analogy rather thin.) Finally, since covariants must be suitably homogeneous, we are only allowed to sum digraphs having the same number (and types) of atoms and the same number of directed bonds.

Exercise 7.2. Determine the graphical representation of the second transvectant $(Q, H)^{(2)}$ between a form and its Hessian.

A digraph is *reducible* if it is the disjoint union of two subdigraphs or *components*, which we write as $D = D_1 \vee D_2$, meaning there are no bonds in D connecting an atom of D_1 to an atom of D_2. It is not difficult to see that a reducible digraph $D = D_1 \vee D_2$ corresponds to a reducible covariant, $J = J_1 J_2$, which can be written as the product of two simpler covariants; see also (6.56). For example, the reducible digraph on four vertices $\text{O}\!\!=\!\!\!\blacktriangleright\!\text{O} \quad \text{O}\!\!=\!\!\!\blacktriangleright\!\text{O}$ represents 4 times the square of the Hessian, $4H^2$.

Important: The digraph $\text{O}\!\!=\!\!\!\blacktriangleright\!\text{O} \quad \text{O}\!\!=\!\!\!\blacktriangleright\!\text{O}$ is *not* the same as multiple 2 $\text{O}\!\!=\!\!\!\blacktriangleright\!\text{O}$, which represents the covariant $4H$.

Given a molecule representing a bracket monomial, we define the *molecular covalence* of any constituent atom to be the total number of bonds (both directed and unattached) terminating at the atom. For instance, in the molecule $\text{O}\!\!\longrightarrow\!\!\text{O}\!\!=\!\!\!\blacktriangleright\!\text{O}$, the three atoms have respective covalences $1, 3, 2$. For an atom representing a general function $Q(\mathbf{x})$, one can attach as many bonds as desired. However, for covariants of homogeneous polynomials, if we attach too many bonds, then the covariant represented by the molecule will automatically vanish. Indeed, since each bond represents a differentiation process (either scaling or omega), if Q is a polynomial of degree n, it can be differentiated at most n times before it becomes zero, and hence only those molecules in which any atom representing Q has n or fewer bonds can possibly correspond to nonzero covariants. Therefore, if Q is a binary form of degree n, its associated atoms will be said to have (free) *valence* n. Each time we attach a bond to an atom representing such a Q, the valence of the associated atom decreases by 1, so the *valence* of an atom in a molecule is just its free valence minus its molecular covalence. An unattached bond, corresponding to a bracket factor of the first kind, has a valence of 1 since one of its ends is not connected to an atom. The *valence* of the entire molecule M is just the sum of the valences of the constitutive atoms plus the number of unattached bonds. Thus, for a binary form of degree n, the atoms in the molecule O——▸O══▸O have respective valences $n-1$, $n-3$, and $n-2$, and the entire molecule has valence $3n-6$. Note that, in particular, attaching a free bond to a molecule does not alter its total valence, whereas bonding two atoms by an arc reduces the total valence by two.

Remark: In Sylvester's fanciful theory, the atoms corresponding to binary forms can be regarded as correspondingly valenced physical elements — a linear form would correspond to a hydrogen atom, a quadratic form to an oxygen atom, a quartic to carbon, and so on. In this way, each physical molecule or ion would have an invariant-theoretic counterpart.

Consider first a molecule in which there are no unattached bonds, so that the corresponding bracket monomial only contains bracket factors of the second kind. In this case, the valence equals the degree of the covariant in the homogeneous coordinate **x**. In particular, a classical invariant does not depend on **x** at all and so is represented by a *saturated* digraph or *neutral molecule*, meaning one that has valence 0. Covariants correspond to "ions", that is, molecules of positive valence equal to the degree of the covariant in **x**. For example, the Hessian O══▸O is an invariant for a quadratic, $n=2$, but a covariant of degree $2n-4$ for a form of degree $n>2$, since the two atoms each still have valence $n-2$. Similarly, the discriminant of a cubic, as in Exercise 5.5, is represented

(up to a multiple) by the neutral four-atom molecule $\begin{array}{c} O═══▸O \\ \downarrow \qquad \downarrow \\ O═══▸O \end{array}$.

Exercise 7.3. Determine the valence of the atoms in the discriminant digraph when Q is a quartic polynomial. What is the valence of the molecule itself? What does that tell you about the corresponding covariant?

Next consider the unattached bonds. Since these represent bracket factors of the first kind or scaling processes, and we are assuming that the functions are homogeneous polynomials, appending a single unattached bond at an atom of valence v is the same as multiplying the molecule (covariant) by the factor v. For example, the molecule O═══►O─ , which represents the bracket polynomial $[\alpha\beta]^2(\alpha\mathbf{x})$, is equal to $n - 2$ times the Hessian: O═══►O─ $= (n - 2)$ O═══►O . Attaching further such bonds scales the molecule similarly, although one must keep in mind that each unattached bond at an atom decreases its valence by one, so the scaling factors will change, in accordance with Lemma 6.1. For instance, O═══►O─ $= (n - 2)(n - 3)(n - 4)$O═══►O when Q is homogeneous of degree n.

We will retain the term *valence* in the case that the atoms in the digraph represent more general homogeneous functions. Thus, a homogeneous function of degree n will be represented by an atom with valence n. Note that, however, in contrast to the case of binary forms, general valences need not be positive integers but can even be negative or zero without necessarily implying that the corresponding differential polynomial covariant vanishes.

As we discussed in Chapter 6, each bracket polynomial has two possible invariant-theoretic interpretations — either as a transform or as a classical symbolic form. As specified in Theorem 6.23, the two differ only by an overall numerical factor and additional bracket factors of the first kind. Thus, each digraph can be assigned either a transform or a symbolic interpretation. Assume, for simplicity, that all the atoms represent a single binary form of degree n and hence have free valence n. In the symbolic version, each symbolic letter must appear exactly n times in any monomial, and hence each of the atoms must have exactly n attached bonds, either directed (representing bracket factors of the second kind) or unattached at the other end (representing bracket factors of the first kind). If we know the number of directed bonds attached to a given atom, we can reconstruct the number of unattached bonds, and so in this context a slightly modified version of the concept of valence appears. In fact, if we replace all the unattached bonds, by suitable numerical scaling factors as before, the result is precisely the transform representative since these factors are the ones appearing in the equation (6.36) connecting the two versions.

Syzygies and the Algebra of Digraphs

Although there is a one-to-one correspondence between linear combinations of digraphs and bracket polynomials, there is not a one-to-one correspondence between digraphs and covariants owing to the syzygies (6.25), (6.26), (6.27) among the bracket factors. These induce certain equivalence relations among digraphs or, from the chemical point of view, allowable "reactions" among molecules. The implementation leads to an "algebra of digraphs" that directly mirrors the algebra of bracket polynomials. There are three basic rules in the algebra of digraphs:

Rule #1. The first syzygy (6.25) implies that reversing any directed bond in a digraph changes its sign. We represent this rule pictorially by

$$\text{O}\longrightarrow\bullet \;=\; -\,\text{O}\longleftarrow\bullet \;, \tag{7.2}$$

where we are displaying only the affected atoms and bonds; all others are left unchanged.

For instance, if the two atoms are the same, Rule #1 shows that the elementary digraph $\text{O}\longrightarrow\text{O}$ equals its own negative, $-\,\text{O}\longleftarrow\text{O}$, and hence represents the trivial covariant 0. As another application of Rule #1, consider the digraph , which represents the bracket monomial $[\alpha\beta][\beta\gamma][\gamma\alpha]$. This monomial is a symbolic form of the trivial (zero) covariant. Indeed, if we reverse the direction of all three arcs in the digraph, we see that

The two digraphs on each side of this equality are the same, and so our digraph represents a trivial covariant: $2\ $$\ =\ 0$.

Note that Rule #1 implies that the directions of single bonds make a difference in the sign of a digraph, but double (and quadruple) bonds can be simultaneously reversed without changing the covariant. To simplify the pictures, we will often denote double bonds which point in the same direction by plain line segments, so $\text{O}=\!=\!=\text{O}$ will be an alternative way of representing the Hessian $\text{O}=\!=\!\blacktriangleright\text{O}$, which is the negative of the digraph $\text{O}\blacktriangleleft\!=\!=\!\blacktriangleright\text{O}$

Rule #2. The syzygy (6.26) translates into the digraph rule

$$\text{(digraph rule)} \qquad (7.3)$$

which effectively switches directed and unattached bonds. If the atoms represent homogeneous polynomials (or functions), then each of the unattached bonds can be replaced by a multiplicative factor, leading to the alternative version of this rule:

$$v_{\bullet}\left(\text{digraph}\right) = v_{\bullet}\left(\text{digraph}\right) + v_{\circ}\left(\text{digraph}\right), \qquad (7.4)$$

which is valid for homogeneous functions. Here $v_{\circ}$ is the valence of the white atom in the full molecule, etc. Particular attention must be paid to this rule if one or more of the atoms has valence 0, since in this case the associated summand will not contribute to the syzygy. Indeed, one might realistically argue that it is this very feature that is the underlying cause of all of the complications in the algebraic theory of covariants for binary forms!

For example, consider the digraph ⟨digraph⟩ which represents a co-variant of a single binary form of degree $n \geq 3$. (Note that if the degree of the form is 2 or less, the covariant is trivial since two of its atoms have three attached bonds.) If we use the homogeneous version of Rule #2 on one of the upper arcs, we find that

$$(n-2)\,⟨digraph⟩ = (n-2)\,⟨digraph⟩ + (n-2)\,⟨digraph⟩.$$

However, if we now use Rule #1 to reverse one of the single arcs in each of the digraphs on the right-hand side of this equation, we find (after dividing by $n - 2 \neq 0$)

$$⟨digraph⟩ = -\,⟨digraph⟩ - ⟨digraph⟩.$$

But all three of these digraphs are exactly the same, hence

$$3 \quad \triangledown \quad = 3 \quad \triangledown \quad = 0, \qquad (7.5)$$

and our original digraph represents a trivial covariant. The same argument works if the atoms represent a homogeneous function of degree $n \neq 2$, but breaks down for homogeneous functions (not polynomials) of degree 2. For inhomogeneous functions, we cannot conclude that the original digraph represents a trivial differential polynomial covariant. However, if we apply a scaling operator, considering the alternative digraph $\triangledown$, the same argument proves that *this* covariant equals

0. This observation has the interesting consequence that certain differential polynomial covariants can vanish for all homogeneous functions (excluding, in this particular instance, those of degree 2) but be nonzero for general functions.

Rule #3. The remaining syzygy (6.27) says that we can "switch" arcs in digraphs according to the following rule:

$$\circ\!\!-\!\!\bullet \quad = \quad \begin{matrix} \circ & \bullet \\ \downarrow & \downarrow \\ \circ & \bullet \end{matrix} \quad + \quad \times \quad . \qquad (7.6)$$

Note that Rules #2 and #3 are quite similar. Indeed, if we view the unattached bonds in Rule #2 as the ends of directed bonds that are attached to an additional "phantom" atom, they do coincide. Actually, this point of view is a version of our earlier identification of the two types of bracket factors.

For example, we show that the digraph

$$D = \circ\!\!=\!\!\circ\!\!-\!\!\circ\!\!-\!\!\circ\!\!-\!\!\circ \quad = \quad \begin{matrix} \circ\!\!=\!\!\circ\!\!-\!\!\circ \\ \downarrow \\ \circ\!\!\leftarrow\!\!\circ \end{matrix} \quad ,$$

corresponding to the bracket monomial $M = [\alpha\beta]^2 [\beta\gamma][\gamma\delta][\delta\varepsilon]$, is equivalent to a reducible digraph, so M corresponds to a covariant which is the product of two simpler covariants. First applying Rule #1 to the bottom arc, and then Rule #3 to the top and bottom arcs, we find that

On the right-hand side of this equation, the first digraph is reducible, being the product of $T = $ O———▸O══════O and $2H = $ O════▸O . Untangling the second digraph, and reversing the directions of two arcs, we see that it is exactly the same as the original digraph. Thus, the preceding digraph equation takes the form $-D = 2TH + D$, and hence $D = TH$ is also reducible. The reader might find it revealing to compare this elementary "graphic proof" with the more cumbersome algebraic proof it represents.

Remark: The graph theoretic syzygy of Rule #3 coincides with the basic Kauffman bracket relation of modern knot theory, [**124, 137**], upon setting the latter's free parameter equal to 1. The bracket relation underlies the construction of invariant knot polynomials, of importance in recent developments in quantum groups and operator algebras. There are also interesting connections with Penrose's theory of spin networks in quantum mechanics, cf. [**113**; pp. 301–307], [**124**; pp. 443–474]. Our algebra of digraphs might have some as yet unexplored implications for the study of knotted polyhedral skeletons, although it is not clear what invariant-theoretic significance the knot polynomial parameter should have. (Perhaps an invariant theory of quantum groups?)

Graphical Representation of Transvectants

Given a molecular representation of a covariant, we can obtain new, more complicated molecules by having it "react" with other molecules, in particular with free atoms. The basic invariant theoretic mechanism for effecting such reactions is the transvection process.

Proposition 7.4. *The rth transvectant between a given pair of digraphs equals a linear combination of all the different digraphs which can be obtained by connecting the first to the second by r different arcs.*

Proposition 7.4 is a direct consequence of our omega calculus rule (5.22): the rth transvectant of digraphs D, E is the sum

$$(D, E)^{(r)} = \sum F_\nu, \tag{7.7}$$

where the F_ν represent all possible digraphs obtained by connecting all free bond sites in the molecule D to those in the molecule E by r *different* arcs. Here, by "different" we mean that the r arcs should be regarded as having distinct labels, but the labels are dropped after the different summands are constructed. The precise implementation of this scheme will become clear after we consider two easy examples.

Example 7.5. Consider the covariant $T = (Q, H)^{(1)}$ of a binary form Q of degree n, with representative digraph ⭕━━▶⭕═══⭕ . The first transvectant $U = (Q, T)^{(1)}$ will be just the sum of all possible digraphs which can be obtained by joining a single atom ⭕ , representing the form Q itself, to the digraph ⭕━━▶⭕═══⭕ with a single arc. Therefore

$$(Q, T)^{(1)} = \text{[digraph 1]} + \text{[digraph 2]} + \text{[digraph 3]} .$$

(7.8)

If Q is a cubic form, so each atom has valence 3, then the second digraph is 0 since there are four bonds terminating at the middle atom.

We shall prove that all three digraphs can be reduced to simpler form by an application of our syzygy rules. First, applying the homogeneous version of Rule #2 to move the single horizontal arc,

$$(n-1) \text{[digraph]} = (n-2) \text{[digraph]} + (n-1) \text{[digraph]} .$$

(7.9)

The middle digraph is reducible, being just $-4H^2$, while reversing the diagonal arc turns the last digraph into the first, which therefore equals $-2[(n-2)/(n-1)] H^2$. Next,

$$(n-1) \text{[digraph]} = (n-2) \text{[digraph]} + n \text{[digraph]} .$$

But the latter reducible digraph vanishes according to our earlier computation (7.5). Therefore,

$$\text{[digraph]} = \frac{(n-2)}{(n-1)} \text{[digraph]} = -2 \frac{(n-2)^2}{(n-1)^2} H^2 .$$

Finally,

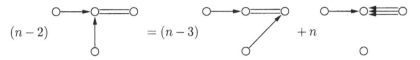

To analyze the latter reducible digraph, we again apply Rule #2,

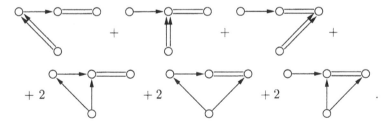

and hence ![O→O⫸O] $= [n/(n-3)]\,Q\,I$, where $I = \frac{1}{2}(Q,\,Q)^{(4)}$ is one-half the fourth transvectant of Q with itself. Combining these, we conclude that

$$U = (Q,\,T)^{(1)} = -6\,\frac{(n-2)^2}{(n-1)^2}\,H^2 + \frac{n^2}{(n-2)(n-3)}\,Q^2\,I. \qquad (7.10)$$

In the cubic case, the second term is not present since the middle digraph in (7.8) vanishes for forms of degree ≤ 3, and one deduces the simpler relation $U = -\frac{3}{2}H^2$. These identities will play a particularly important role in the solution to the equivalence problem for binary forms presented in Chapter 8.

In a similar fashion, the second transvectant $(Q,\,T)^{(2)}$ will be a linear combination of all possible digraphs which can be obtained by joining a single atom by two (distinct) arcs to the digraph for T:

The factors of 2 arise because there are two different ways to attach two labeled arcs to two atoms, and so the last three digraphs each appear twice in the transvectant formula. Again, for forms of low degree, a number of these digraphs are automatically zero. Moreover, if the function is homogeneous, they can all be replaced by simpler digraphs due to further applications of our graph-theoretic syzygies; the details are left as an exercise for the reader.

Exercise 7.6. Analyze the third transvectant $(Q,\,T)^{(3)}$.

Exercise 7.7. Prove that the transvectant $Z = (H, T)^{(1)}$ satisfies the following identity:

$$QZ = \frac{6(n-2)}{n} T^2 + \frac{3(n-2)^3}{n(n-1)^2} H^3 + \frac{n}{n-3} Q^2 HI. \tag{7.11}$$

Exercise 7.8. Let J be represented by the digraph ,

which, for a quartic, is a multiple of its cubic invariant (2.30). Prove the following identity, which includes the fundamental quartic syzygy (2.48) as a special case:

$$Q^3 J = \frac{6(n-2)}{n} T^2 + \frac{3(n-2)^3}{n(n-1)^2} H^3 + \frac{n}{n-3} Q^2 IH. \tag{7.12}$$

See also Exercise 6.33.

If there are unattached bonds on any of the atoms in either D or E, then the new arcs in the rth transvectant $(D, E)^{(r)}$ can be attached at either the atoms in D or E, or to the unattached bonds. For example, if $K = \;$ ⊶ , then

$(Q, K)^{(1)} =$ ⟜⟞ $+$ ⟜⟞ $+$ ⟜⟞ ,

the final summand arising when the arc connects to the unattached bond in K. The justification of the modified transvection rule rests on the commutation formulae (5.28) for the omega and scaling processes.

Transvectants of Homogeneous Functions

We shall now apply our graph theoretic methods to the implementation of Gordan's constructive procedure for determining the Hilbert basis of systems of binary forms. Before beginning, we must first present an important but technical result on transvectants that applies to binary forms, as well as general homogeneous functions.

Definition 7.9. Let $Q_1, \ldots, Q_m$ be a system of homogeneous functions which have respective degrees $n_1, \ldots, n_m$. We will call the system *resonant* if there exist non-negative integers $k_1, \ldots, k_m \in \mathbb{N}$ such that $0 \leq \sum k_\alpha n_\alpha \in \mathbb{N}$ is a non-negative integer.

Note that any system containing a homogeneous polynomial is automatically resonant, whereas if all the functions have either negative and/or irrational degrees of homogeneity, then the system is automatically non-resonant. The key result is the following:

Theorem 7.10. *Let $Q_1, \ldots, Q_m$ be homogeneous functions. Suppose either (a) all the Q_α's are homogeneous polynomials, or (b) the system is non-resonant. Then every partial transvectant of $Q_1, \ldots, Q_m$ can be written as a linear combination of ordinary transvectants.*

Proof: Let us begin with a simple example. Consider the transvectant

$$(P, (Q, R)^{(s)})^{(1)} = \operatorname{tr}(\Omega_{\alpha\beta} + \Omega_{\alpha\gamma})(\Omega_{\beta\gamma})^s (P \otimes Q \otimes R),$$

which is a linear combination of two partial transvectants. If P, Q, R are homogeneous of respective degrees d, e, f, then, according to (6.29),

$$(f - s)\,\Omega_{\alpha\beta}(\Omega_{\beta\gamma})^s(P \otimes Q \otimes R) =$$
$$= (e - s)\,\Omega_{\alpha\gamma}(\Omega_{\beta\gamma})^s(P \otimes Q \otimes R) + d\,(\Omega_{\beta\gamma})^{s+1}(P \otimes Q \otimes R).$$

Therefore, setting the variables equal,

$$\operatorname{tr} \Omega_{\alpha\beta}(\Omega_{\beta\gamma})^s(P \otimes Q \otimes R) =$$
$$= \frac{e - s}{e + f - 2s}\,(P, (Q, R)^{(s)})^{(1)} + \frac{d}{e + f - 2s}\,P \cdot (Q, R)^{(s+1)},$$

and similarly for the other partial transvectant. Therefore, as long as $e + f \neq 2s$, each partial transvectant can be written as a combination of two ordinary transvectants. If Q and R are homogeneous polynomials, and $e + f = 2s$, then $(Q, R)^{(s)}$ is a polynomial of degree 0. In this case, both partial transvectants are automatically zero, so the result holds trivially. Adapting this basic idea, the reader should be able to handle the different partial transvectants in $(P, (Q, R)^{(s)})^{(r)}$ for $r = 2, 3$, etc.

To prove a general result, first note that, by linearity, we need only prove the theorem for covariants represented by bracket monomials, i.e., by single molecules. Let D be a digraph with m atoms $\alpha_1, \ldots, \alpha_m$ of respective valences $\mathbf{v} = (v_1, \ldots, v_m)$. The rth transvectant of D with a single atom α_0 of valence v_0 can be expanded as a sum

$$(Q_\alpha, D)^{(r)} = \sum_I \binom{r}{I} F_I \tag{7.13}$$

over all multi-indices $I = (i_1, \ldots, i_m)$ with $i_1 + \cdots + i_m = r$. In (7.13), F_I denotes the digraph obtained by connecting the single atom α_0 to D by i_ν arcs to atom α_ν, for each $\nu = 1, \ldots, m$. In accordance with the previous simple example, we expect that, in most cases (i.e., provided no resulting denominator vanishes), each of the summands in (7.13) will be equal to a numerical multiple of the transvectant $(Q_\alpha, D)^{(r)}$ modulo lower order transvectants of the form $(Q_\alpha, E)^{(r-s)}$ for $1 \leq s \leq r$ with E

a digraph having the same m atoms as D. If this observation holds, then a straightforward induction on the order of the transvectant suffices to prove the theorem. The requisite lemma relates two of the constituent digraphs in (7.13) and relies on the falling multinomial notation (6.7).

Lemma 7.11. *Let D be a digraph on m atoms with valences* $\mathbf{v} = (v_1, \ldots, v_m)$. *Let F_I and F_J be digraphs appearing in the rth order transvectant* $(Q_\alpha, D)^{(r)}$, *cf.* (7.13). *Then*

$$\mathbf{v}^{\underline{I}} F_J = \mathbf{v}^{\underline{J}} F_I + \mathbf{G}_{IJ}, \tag{7.14}$$

where $\mathbf{G}_{IJ}$ is a linear combination of digraphs occurring in lower order transvectants of the form $(Q, E_\kappa)^{(r-1)}$; each digraph E_κ can obtained by adjoining one additional arc to the original digraph D.

Proof: Using an obvious induction, it suffices to treat the case when F_J can be obtained from F_I by moving a single arc starting at atom α_0 from an atom α_λ in D to a different atom α_μ. In other words, the digraphs have labels $I = (\ldots, k_\lambda + 1, \ldots, k_\mu, \ldots)$ and $J = (\ldots, k_\lambda, \ldots, k_\mu + 1, \ldots)$, where $K = (k_1, \ldots, k_m)$ is some multi-index with $\sum k_\nu = r - 1$. This particular case reduces immediately to the homogeneous version of our Rule #2 for digraphs. Let E denote the digraph obtained from D by attaching one additional arc from atom α_λ to atom α_μ. Equation (7.4) implies that

$$(v_\mu - k_\mu)F_J = (v_\lambda - k_\lambda)F_I + (v_0 - r + 1)\widetilde{F}_K, \tag{7.15}$$

where $\widetilde{F}_K$ is the digraph corresponding to the multi-index K appearing in the transvectant summation (7.13) for $(Q, E)^{(r-1)}$. Multiplying (7.15) by the falling factorial $\mathbf{v}^{\underline{K}}$ reproduces (7.14) in this particular case. The general formula follows by induction.   Q.E.D.

We now apply this lemma to prove a more precise version of Theorem 7.10. We remark that case (b) of the following theorem automatically includes all non-resonant systems.

Theorem 7.12. *Let $Q_1, \ldots, Q_m$ be a system of homogeneous functions. Let D be a digraph of total valence V representing a covariant of the system. If either (a) every function is a binary form, or (b) $V \neq 0, 1, \ldots, 2r - 2$, then every digraph occurring in the rth transvectant $(Q, D)^{(r)}$ can be written as a linear combination of transvectants.*

Proof: For each digraph F_I occurring in (7.13), we apply (7.14) and the falling multinomial identity (6.6) to deduce

$$\mathbf{v}^{\underline{I}} (Q, D)^{(r)} = \sum_J \binom{r}{J} \mathbf{v}^{\underline{I}} F_J$$

$$= \sum_J \binom{r}{J} \left[\mathbf{v}^{\underline{J}} F_I + \mathbf{G}_{IJ} \right] = V^{\underline{r}} F_I + \mathbf{G}_I, \tag{7.16}$$

where $V = v_1 + \cdots + v_m$ is the total valence, and $\mathbf{G}_I$ is a linear combination of digraphs of the form indicated in Lemma 7.11. Therefore, if $V^{\underline{r}} \neq 0$, which requires $V \neq 0, 1, \ldots, r - 1$, then we can write F_I as a multiple (possibly zero) of $(Q, D)^{(r)}$ plus a linear combination of digraphs occurring in transvectants $(Q, E_\kappa)^{(r-1)}$, where the E_κ are obtained by adding in one additional arc to D. Thus, each E_κ has valence $V - 2$. To prove case (b), we use induction on the order r, noting that $V \neq 0, 1, \ldots, 2r - 2$ implies $V - 2 \neq 0, 1, \ldots, 2(r - 1) - 2$. Therefore, each of the digraphs occurring in $\mathbf{G}_I$ is a linear combination of transvectants, and the induction is completed. In case (a), this argument doesn't quite work. Of course, if $V = 0, 1, \ldots, r - 1$, then *any* r^{th} order partial transvectant of Q with D automatically vanishes since at least one atom in the digraph will have more bonds terminating at it than its valence allows, and so, because every atom represents a polynomial, the resulting covariant must vanish. On the other hand, we can still form nonzero r^{th} order partial transvectants when $V = r, \ldots, 2r - 2$, when case (b) does not apply. In this event, the key observation is that if the total valence $V = r$ is the same as the order of the transvectant, then there is only *one* nonzero digraph occurring in the sum for $(Q, D)^{(r)}$ — namely, that one where the r bonds emanating from the atom α_0 terminate at the r remaining free bond sites on D. Any other attachments of these r bonds to D will have more bonds terminating at an atom than its valence allows, and so the covariant will vanish. Therefore, either $V = r$, and the result is immediate, or $V \geq r + 1$, in which case each E_κ has valence $V - 2 \geq r - 1$, and so the induction step can be completed.   *Q.E.D.*

Gordan's Method

A Hilbert basis for the covariants corresponds to the determination of a complete set of "atomic molecules" ("atomicules" in Sylvester's terminology) or *irreducible digraphs*. Indeed, the content of the Hilbert Basis Theorem 2.42 is that any more complicated molecule or digraph is equivalent, under the various digraphical reactions, to some reducible combination (or mixture) of the irreducible digraphs corresponding to the covariants in a Hilbert basis. Gordan devised an efficient, recursive algorithm for generating the Hilbert basis for the covariants of a form of a given degree (and even the joint covariants of several binary forms). In outline, the method begins with the covariant represented by a single atom, i.e., the form Q itself (which we assume to be a homogeneous polynomial). We then successively construct all nontrivial transvectants of it, all transvectants of the nontrivial transvectants, and so on. At each stage, we only need to append one further vertex to the digraphs from the previous stage. The main points to be clarified are (a) which transvectants need to be considered at each stage? and (b) when do we stop the procedure? We begin by stating two elementary lemmas.

Lemma 7.13. *If a digraph D is reducible, and one component of D is an invariant, then each summand in $(Q, D)^{(r)}$ is also reducible, with one component the same invariant.*

Proof: This is more or less obvious. We can't attach any more arcs to a saturated digraph (neutral molecule) representing the invariant, so the same saturated digraph will appear as a component in all summands of the transvectant $(Q, D)^{(r)}$.   *Q.E.D.*

Lemma 7.14. *If D is reducible, and one of the components of D has valence at least r, then $(Q, D)^{(r)}$ is equivalent, modulo lower order transvectants, to a reducible digraph.*

Proof: It suffices to notice that at least one summand in $(Q, D)^{(r)}$ has all the new arcs connected to the high valence component of D, and hence this summand is reducible. Moreover, according to Theorem 7.10 $(Q, D)^{(r)}$ is equivalent, up to a linear combination of lower order transvectants, to a nonzero multiple of this particular summand.   *Q.E.D.*

We can now outline Gordan's method. At each step, we recursively construct a complete set $\mathcal{I}_m$ of "irreducible" digraphs with exactly m vertices whose corresponding covariants appear in a minimal Hilbert basis for the covariants of a binary form of degree n. The method is recursive and terminates in finitely many steps, as guaranteed by the Hilbert Basis Theorem 2.42; see also [**92**; chapter 6] for an alternative justification of its termination. Theorem 6.14 implies that *every* covariant is represented by a digraph and hence appears as a summand in some iterated transvectant. Moreover, Theorem 7.10 implies that it suffices to analyze just one of the summands in each transvectant when constructing the basis.

Step #1: Let $\mathcal{I}_1$ consist of the single monatomic digraph, corresponding to the form itself.

Step #m: To construct $\mathcal{I}_m$ knowing $\mathcal{I}_1, \ldots, \mathcal{I}_{m-1}$, we proceed as follows: Let $\mathcal{B}_{m-1}$ be the set of all digraphs D with $m-1$ vertices constructed by one of the following two rules:

(a) D is an ionic digraph in $\mathcal{I}_{m-1}$, i.e., D has positive valence, and so does not represent an invariant, or

(b) $D = D_1 \vee \cdots \vee D_l$ is a reducible digraph on $m - 1 \geq 4$ vertices, where each constituent $D_\nu \in \mathcal{I}_{j_\nu}$ is an irreducible digraph having $2 \leq j_\nu \leq m - 3$ vertices (so $j_1 + \cdots + j_l = m - 1$) and valence $0 < k_\nu < n$.

From the set $\mathcal{B}_{m-1}$ we construct a set $\mathcal{C}_m$ of digraphs on m vertices by taking one term in each possible transvectant $(Q, D)^{(r)}$, $0 < r \leq n$, that is, just one of all the possible digraphs which can be constructed by attaching a single vertex to D by r arcs. After constructing $\mathcal{C}_m$, one then uses the digraph rules to determine a subset $\mathcal{I}_m \subset \mathcal{C}_m$ of digraphs which do not differ by a reducible digraph. (It is at this stage that complications may arise, since it is sometimes quite tricky to recognize a reducible digraph!) In particular, using Lemma 7.14, one can immediately rule out transvectants $(Q, D)^{(r)}$ of reducible digraphs constructed using Rule (b) if any subcomponent $D_{j_1} \vee \cdots \vee D_{j_\mu}$, $\mu < l$, has valence r or more, i.e., $k_{j_1} + \cdots + k_{j_\mu} > r$. The method terminates when $\mathcal{I}_m$ consists only of saturated digraphs (invariants), and, moreover, Rule (b) does not lead to any irreducible transvectants for any higher m. The Hilbert basis will consist of all the covariants corresponding to all the digraphs appearing in $\mathcal{I}_1 \cup \cdots \cup \mathcal{I}_m$.

To illustrate Gordan's method and demonstrate the power of our "graphical algebra", we show how to construct a complete system of covariants for the binary quadratic, cubic, and quartic.

Example 7.15. For a binary quadratic, we are allowed to attach at most two arcs to any given vertex to obtain a nonzero covariant. We begin with the monatomic digraph $\bigcirc$, which represents the form Q itself. There are only two possible transvectants: $\bigcirc\!\!-\!\!\!\rightarrow\!\!\bigcirc$ and $\bigcirc\!\!=\!\!\!\Rightarrow\!\!\bigcirc$. The first is trivial by Rule #1, and the second is twice the discriminant, which is an invariant. Lemmas 7.13 and 7.14 imply that we cannot get anything further by transvecting again, so we have shown that the only covariants required for a Hilbert basis of a binary quadratic are the form itself and its discriminant.

Example 7.16. Turning to the binary cubic, we begin with Q, i.e., ○, from which we can form three transvectants: ○——▶○ , ○══▶○ , and ○══▶○ . Two of these are trivial by Rule #1, the only nontrivial one being the Hessian $2H =$ ○══○ . Therefore $\mathcal{I}_1 = \{Q\}$, while $\mathcal{I}_2 = \{H\}$. The only digraph in $\mathcal{B}_2$ is the Hessian, which has valence 2, and so we can form the two further representative transvectants:

$$T = (Q, H)^{(1)} = \text{○——▶○══○} \qquad (Q, H)^{(2)} = \text{\vee} \quad ,$$

which are the digraphs in $\mathcal{C}_3$. The second digraph is trivial by Rule #2. Thus, $\mathcal{I}_3 = \{T\}$. Rule (b) in Gordan's method does not apply, so $\mathcal{B}_3 = \{T\}$ also. Now, T has valence 3, so we can form three further transvectants, namely, $(Q, T)^{(1)}$, $(Q, T)^{(2)}$, $(Q, T)^{(3)}$. Representative summands are

○——▶○——▶○══○   ○══○——▶○══○   ⊡ ,

which are the digraphs in $\mathcal{C}_4$. Our earlier computation (7.9) showed that the first is equivalent to $-H^2$. The second is zero by Rule #1, while the third is the discriminant $\Delta = \frac{1}{6}(Q, T)^{(3)}$ given in (2.22). Therefore $\mathcal{I}_4 = \{\Delta\}$. To complete the next step, we note that Rule (a) is no longer applicable since $\mathcal{I}_4$ consists only of an invariant. However, there is one further case from Rule (b) which needs to be taken into account, namely, the reducible digraph ○══○ ○══○ corresponding to $4H^2$. Note that each component has valence 2, so we can form the possibly non-reducible transvectant $(Q, H^2)^{(3)}$. For a cubic, there is only one nonzero summand, which we call D. This digraph is trivial since Rule # 3 shows that

$$D = \quad \boxtimes \quad = \quad \boxtimes \quad + \quad \boxtimes \quad .$$

The two digraphs on the right differ from the one on the left by a single reversed arc, and so the equation has the form $D = -D - D$, which means $D = 0$. There are no more possible irreducible transvectants, and so Gordan's method has terminated. We have therefore proved our claim in Chapter 2: a Hilbert basis for the binary cubic consists of the form Q, the covariants T and H, and the discriminant invariant Δ.

Example 7.17. Finally, we outline how the same method produces the Hilbert basis for the binary quartic. We begin with $\mathcal{I}_1$, which, as always, consists of the only monatomic digraph $\bigcirc$. We can now form four transvectants, two of which are trivial, so $\mathcal{I}_2$ consists of $\bigcirc\!\!=\!\!=\!\!\bigcirc$ and $\bigcirc\!\!\equiv\!\!\equiv\!\!\bigcirc$, which correspond to the Hessian $2H = (Q, Q)^{(2)}$ and the quartic invariant $1152\,i = (Q, Q)^{(4)}$. Since we cannot form nontrivial transvectants from an invariant, we are left with only the Hessian to work with, i.e., $\mathcal{B}_2 = \{H\}$. The possible transvectants in $\mathcal{C}_3$ are represented by

The first digraph represents the covariant $T = (Q, H)^{(1)}$, the second and third are trivial, and the fourth represents the quartic invariant $12^4 j = \frac{1}{12}(Q, H)^{(4)}$. Thus, $\mathcal{I}_3$ consists of the digraphs for T and j. As j is an invariant, we can only get nontrivial transvectants from T. There are four possibilities:

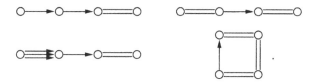

It is not difficult to see that all four are either trivial, or equivalent to reducible digraphs. There are no other possibilities for obtaining irreducible transvectants using either Rule (a) or Rule (b), so Gordan's method is finished, and we have proved that a Hilbert basis for the covariants of the binary quartic consists of Q, H, T, i, and j.

Note that $\Delta = \begin{array}{c}\bigcirc\!\!=\!\!\bigcirc\\ | \quad\;\; |\\ \bigcirc\!\!=\!\!\bigcirc\end{array}$ represents an irreducible invariant for a cubic, that is, when the atoms have free valence 3, but is reducible for a quartic, or, in fact, any higher degree form. This can be seen by the following calculation: Let Q have degree n. We first decompose

$$(n-3)\begin{array}{c}\bigcirc\!\!=\!\!\bigcirc\\ | \quad\;\; |\\ \bigcirc\!\!=\!\!\bigcirc\end{array} = (n-2)\begin{array}{c}\bigcirc\!\!=\!\!\bigcirc\\ |\;\diagdown\; |\\ \bigcirc\!\!=\!\!\bigcirc\end{array} + (n-2)\begin{array}{c}\bigcirc\!\!\equiv\!\!\bigcirc\\ | \quad\;\; |\\ \bigcirc\!\!=\!\!\bigcirc\end{array}$$

or

$$\Delta = \frac{n-2}{n-3}A + \frac{n-2}{n-3}B$$

using Rule #2. (This step is not allowed for a cubic, since $n = 3$.) The second of these digraphs is easily seen to be reducible since

$$(n-3)\ \left[\ \right]\ = (n-3)\ \diagup\ + (n-2)\quad ,$$

hence $A = -[(n-2)/(n-3)]IH$ where $I = \frac{1}{2}(Q, Q)^{(4)}$. As for the first, we begin by applying Rule #3 to the two double bonds:

$$\boxtimes\ =\ \boxtimes\ =\ \boxtimes\ +\ \boxtimes\ =\ 2\ \boxtimes\quad ,$$

which we write as $B = 2C$. On the other hand, using Rule #2, we see that

$$(n-3)\ \boxtimes\ =\ (n-1)\ \boxtimes\ +\ (n-3)\ \boxtimes$$

$$(n-3)\,B = (n-1)\,E - (n-3)\,C,$$

and the digraph E is reducible since

$$(n-3)\ \triangle\ =\ (n-3)\ \triangle\ +\ n\ \triangle$$

$$(n-3)\,E\ =\ -(n-3)\,E + 6n\,QJ,$$

so $E = [3n/(n-3)]QJ$. Since $2C = B = [(n-1)/(n-3)]E - C$, we see that $C = \frac{1}{3}[(n-1)/(n-3)]E = [n(n-1)/(n-3)^2]QJ$, and hence $B = \frac{1}{2}[n(n-1)/(n-3)^2]QJ$. Therefore, provided $n \neq 3$, our original covariant can be written in terms of the basis covariants as

$$D = \frac{n-2}{n-3}A + \frac{n-2}{n-3}B = -\frac{2(n-2)^2}{(n-3)^2}IH + \frac{n(n-1)(n-2)}{2(n-3)^3}JQ.$$

This calculation indicates a general phenomenon. If a digraph D represents a reducible covariant for a binary form of degree n, it also represents a reducible covariant for any binary form of degree $m \geq n$.

However, irreducible digraphs for low degree forms may very well become reducible once the degree of the form is sufficiently large. This motivates one to consider the less intricate problem of constructing a set of irreducible digraphs for forms of infinite (i.e., arbitrarily large) degree, the classical terminology for which is *perpetuant*, [210]. One might also argue that the perpetuants are the irreducible covariants for homogeneous functions of *non-integral* degree, since in this case there are no resonances, and so one will never encounter a division by zero in the implementation of the syzygies. The complete classification of irreducible perpetuants can be found in Stroh, [203] and MacMahon, [146]; see also [65; Chapter 11], [92; pp. 326–331].

Exercise 7.18. Let Q be a quartic. The covariant $U = (Q, T)^{(1)}$ is not in the Hilbert basis, and hence there exists a polynomial identity expressing it in terms of Q, i, j, H, and T. Prove the required identity

$$U = 2^9 3^2 Q \left(108jQ - iH\right). \tag{7.17}$$

Chapter 8

Lie Groups and Moving Frames

So far, the methods that have been introduced for analyzing the invariants of binary forms have been essentially algebraic in nature[†] and, apart from being geared towards the group of linear transformations in the plane (or projective transformations on the line), are also, in principle, applicable to quite general group actions. We have not yet exploited the additional analytical structure of such matrix groups, which has no counterpart in the realm of finite or discrete groups. Namely, the group elements depend analytically on parameters, and hence the group comes equipped with a topology — notions of open and closed sets, convergence, and so on — as well as a differentiable structure, enabling one to apply powerful calculus-based tools that are not available in the purely discrete framework. The resulting combination of group theory and analysis has a remarkable synergy, enabling one to effectively and efficiently deal with a wide range of practical group-theoretic issues.

In this chapter, we begin with a brief survey of the basic theory of matrix Lie groups, emphasizing the general linear group as the prototypical example. We shall assume the reader is familiar with the basics of multivariable calculus, and we will be a bit less systematic in our inclusion of proofs of main results. The interested reader can find all of the relevant details in standard texts on Lie groups, e.g., [**169, 173, 227**]. This chapter includes several innovations that are inspired by new extensions, [**69, 70**], of the Cartan theory of moving frames, cf. [**33, 93, 95**], arising in the basic geometry of curves and surfaces. For sufficiently regular Lie group actions, there is a relatively unexploited, elementary direct method for constructing invariants, based on Cartan's normalization procedure. To underscore the efficacy of the normalization approach,

---

[†] A die-hard algebraist might object that our use of Fourier transforms and differential operators for developing the symbolic method has already placed us well outside the usual algebraic sphere. Nevertheless, since the classical algebraic symbolic method can be reconstructed from this approach, I would still characterize the method as essentially algebraic in nature.

we derive basic results on joint invariants for several geometrically interesting cases. We then present the elementary theory of differential invariants for Lie groups, illustrating with examples from the geometry of curves and from classical invariant theory. Finally, we show how differential invariants can be applied to both solve equivalence problems and classify symmetry groups — both continuous and discrete. We shall apply the method to reveal a novel solution to the basic equivalence problems for binary forms, which implies several striking results concerning the symmetries and canonical forms of classical polynomials.

Lie Groups

The remarkable theory of groups whose elements depend analytically on parameters dates back to the seminal work of the Norwegian mathematician Sophus Lie in the last century. Lie was inspired by Galois' introduction of group theory for solving algebraic equations and attempted to construct a "Galois theory" of differential equations.[†] Through a surprisingly tortuous route — see [**101**] for a fascinating account of the historical details — Lie was led to the continuous groups that now bear his name. The resulting theory has had profound and far-reaching consequences in all areas of mathematics and physics.

In order to keep the exposition as simple as possible, we shall concentrate on the most relevant examples: groups of invertible linear transformations on a finite-dimensional vector space. This will enable the novice to gain familiarity with the basic ideas, while avoiding unnecessary complications at this elementary level. Pedagogically, we find it easier to develop the real theory first; extensions to complex analytic Lie groups are then readily implemented via analytic continuation and are, for the most part, left to the reader to fill in the details.

Let us first briefly review the fundamental concept of a *manifold*, which forms a differential geometric generalization of classical curves and surfaces. Locally, all manifolds look like open subsets of Euclidean space $\mathbb{R}^m$; indeed, the simplest example of a manifold *is* an open subset $X \subset \mathbb{R}^m$ of Euclidean space. We call $m = \dim X$ the *dimension* of X, which indicates the number of local coordinates required to describe

---

[†] Actually, Lie did not achieve his original goal; the more refined Picard–Vessiot theory, cf. [**123**, **148**], is the proper generalization of Galois theory to differential equations. However, it is safe to say that Lie's actual achievements have had the more significant scientific impact.

points thereon. More generally, let $F: \mathbb{R}^n \to \mathbb{R}^k$, $n \geq k$, be any analytic function. A level set $S = \{F(x) = c\}$ is called *regular* if it is nonempty and the Jacobian matrix of F has maximal rank k everywhere on S. Any regular level set S forms an analytic submanifold of $\mathbb{R}^n$ of dimension $m = n - k$. The Implicit Function Theorem allows us to locally solve for k of the variables, say $(x_1, \ldots, x_k)$, in terms of the remaining variables, so $x_i = \varphi_i(x_{k+1}, \ldots, x_n)$, $i = 1, \ldots, k$, thereby providing a local parametrization of the level set by a system of $n-k$ local coordinates. A simple example is the unit sphere $S^1 = \{x^2 + y^2 + z^2 = 1\} \subset \mathbb{R}^3$; the fact that the gradient of $F(x, y, z) = x^2 + y^2 + z^2$ does not vanish on S^1 immediately implies that it is an analytic surface. Most of our examples of manifolds are of one of these two basic types.

Remark: The formal definition of an n-dimensional manifold is a Hausdorff topological space X which is covered by a collection of open subsets $W_\alpha \subset X$, called *coordinate charts*, and one-to-one local coordinate maps $\chi_\alpha: W_\alpha \to V_\alpha \subset \mathbb{R}^n$; the standard coordinates $x^\alpha = (x_1^\alpha, \ldots, x_n^\alpha)$ on V_α provide coordinates for the points on X. The composite "overlap maps" $\chi_{\beta\alpha} = \chi_\beta \circ \chi_\alpha^{-1}: V_\alpha \to V_\beta$, which must at the very least be continuous, define the changes of local coordinates $x_\beta = \chi_{\beta\alpha}(x_\alpha)$ on X. The manifold is *smooth* if the overlap maps are smooth (infinitely differentiable) where defined, and *analytic* if they are analytic. Although for many differential geometric purposes, smooth manifolds are the most natural category in which to work, in this book all manifolds are analytic, and we shall not lose anything by restricting our attention to the (real or complex) analytic category hereafter. An important example is the projective space $\mathbb{RP}^m$, which forms a manifold of dimension m. The coordinate charts are provided by the various canonical homogeneous coordinate representatives; see Chapter 10 for details. Basic manifold theory can be found in any standard text on differential geometry, including [**168, 199, 227**].

A Lie group is a particular type of *topological group*. This means that the group is endowed with a topology in such a manner that both multiplication and inversion are continuous maps. For a Lie group, the topology is a consequence of its underlying structure as a manifold.

Definition 8.1. An analytic *Lie group* G is a group that also carries the structure of an analytic manifold, which is required to be compatible with the group structure so that the group multiplication $(g, h) \mapsto g \cdot h$ and inversion $g \mapsto g^{-1}$ define analytic maps.

A Lie group of dimension r is often referred to as an r-*parameter group*. The "parameters" refer to the local coordinates, $g = (g_1, \ldots, g_r)$, that the group elements depend on.

Example 8.2. The simplest example of an r-parameter Lie group is the r-dimensional abelian vector group $\mathbb{R}^r$. One system of parameters is provided by the usual Cartesian coordinates $x = (x_1, \ldots, x_r) \in \mathbb{R}^r$. (Other choices of local coordinates lead to different, but equally valid, parametrizations.) Note that the group operations, namely vector addition $x + y$ and inversion $-x$, depend analytically on the coordinates.

Example 8.3. The general linear group $GL(2) = GL(2, \mathbb{R})$ consisting of all 2×2 matrices with nonzero determinant forms an analytic, four-parameter Lie group. The (global) parameters are the entries $\alpha, \beta, \gamma, \delta$ of the general nonsingular matrix $A = \left(\begin{smallmatrix} \alpha & \beta \\ \gamma & \delta \end{smallmatrix}\right) \in GL(2)$. We can identify $GL(2) = \{\alpha\delta - \beta\gamma \neq 0\} \subset \mathbb{R}^4$ as an open subset of a four-dimensional Euclidean space, which thereby identifies it as a four-dimensional manifold. Clearly matrix multiplication and inversion are analytic functions of the parameters.

Exercise 8.4. Show that $GL(2, \mathbb{R})$ consists of two connected open subsets of $\mathbb{R}^4$, distinguished by the sign of the determinant. The subset $GL(2, \mathbb{R})^+ = \{\det A > 0\}$ forms a four-parameter subgroup: the group of orientation-preserving planar linear maps.

Example 8.5. The special linear group $SL(2) = SL(2, \mathbb{R})$ forms a three-dimensional submanifold of $\mathbb{R}^4$, defined by the unimodularity constraint $\alpha\delta - \beta\gamma = 1$, which is a regular level set of the determinant function. Since the group operations of matrix multiplication and inversion remain analytic when restricted to $SL(2)$, it has the structure of an analytic three-parameter Lie group. There is no convenient global parametrization, but we can introduce local parameters on various open subsets. For example, on the subset where $\alpha \neq 0$, the matrices in $SL(2)$ can be parametrized by the three matrix entries α, β, γ, so that the general group element is $\begin{pmatrix} \alpha & \beta \\ \gamma & \alpha^{-1}(1 + \beta\gamma) \end{pmatrix}$. Similarly, if $\gamma \neq 0$, we can solve for $\beta = \gamma^{-1}(\alpha\delta - 1)$, and so use α, γ, δ as alternative parameters.

Example 8.6. The planar rotation group $SO(2)$ can be identified with the unit circle $S^1 \subset \mathbb{R}^2$; each rotation through angle θ uniquely corresponds to the point $(\cos\theta, \sin\theta) \in S^1$. A group parameter is the rotation angle θ, and so $SO(2)$ forms a one-parameter Lie group. Similarly, the orthogonal group O(2) consists of two disconnected circles, one containing the rotations and the other the reflections.

Remark: It is not difficult to prove that $\mathbb{R}$ and SO(2) are, up to isomorphism, the only two connected, one-parameter Lie groups; see also Theorem 9.3.

All the preceding examples can be readily generalized to the n-dimensional case. We can identify the space $\mathcal{M}_{n \times n}$ of all real $n \times n$ matrices with the vector space $\mathbb{R}^{n^2}$, the coordinates being provided by the matrix entries. The general linear group GL$(n, \mathbb{R}) = \{\det A \neq 0\}$ forms an open subset of this vector space and so is a manifold of dimension n^2. The group operations are clearly analytic. Similarly, SL$(n, \mathbb{R}) = \{\det A = 1\}$, being a regular level set of the determinant function, forms an $(n^2 - 1)$–dimensional subgroup. It can be shown that the orthogonal group O$(n) = \{A^T A = \mathbb{1}\}$ is a subgroup of dimension $\frac{1}{2}n(n-1)$. These are all particular examples of the most important class of Lie groups:

Definition 8.7. A Lie group $G \subset$ GL$(n, \mathbb{R})$ that forms an analytic submanifold of the general linear group is called a *matrix Lie group*.

Although not every Lie group can be realized as a matrix Lie group, all of the examples we consider in this text are. Moreover, every Lie group agrees locally (in a neighborhood of the identity) with a matrix Lie group. As shown by Cartan, [**34**], the simplest example of a non-matrix Lie group is the simply connected covering group of SL$(2, \mathbb{R})$.

Example 8.8. The Euclidean group SE(2) $=$ SO(2) $\ltimes \mathbb{R}^2$ combines the one-parameter rotation group SO(2) with the two-parameter translation group $\mathbb{R}^2$ as a semi-direct product. As a manifold, SE(2) has the form of a Cartesian product, SO(2) $\times \mathbb{R}^2$, and so can be identified with a solid torus. Similarly, both the five-parameter equi-affine group SA(2) $=$ SL(2) $\ltimes \mathbb{R}^2$, and the six-parameter full affine group A(2) $=$ GL(2) $\ltimes \mathbb{R}^2$ are also Cartesian product manifolds (but *not* Cartesian product groups). All three examples can be realized as matrix Lie groups, contained in GL(3), via the representation $\rho(A, b) = \begin{pmatrix} A & b \\ 0 & 1 \end{pmatrix}$ defined for $A \in$ GL(2), $b \in \mathbb{R}^2$.

Many, but not all — see Theorem 9.3 — matrix Lie groups are topologically closed subsets of GL(n). A general theorem due to Cartan, cf. [**173**; §I.3.5], [**227**; Theorem 3.42], states that the converse holds: any closed subgroup of $GL(n)$ is automatically a matrix Lie group. More generally, one can prove the following result:

Theorem 8.9. *Let* G *be a Lie group, and* $H \subset G$ *a subgroup. If* H *is a topologically closed subset, then* H *is a Lie group.*

Remark: If $H \subset G$ is a normal, closed Lie subgroup, then G/H is itself a Lie group, of dimension $\dim G - \dim H$. If H is not normal, then G/H is called a *homogeneous space*, and forms an analytic manifold of dimension $\dim G - \dim H$.

Lie Transformation Groups

The most important group actions arising in geometry and invariant theory are when a Lie group acts analytically on a manifold. In such cases, the additional structure provides us with new and powerful analytical tools that are not available for more general group actions.

Definition 8.10. A *Lie transformation group* is given by a Lie group G that acts analytically on a manifold X. More precisely, the group action $w(g, x) = g \cdot x$ defines an analytic map $w \colon G \times X \to X$.

Remark: On occasion, we encounter *local transformation groups*, meaning that the image $g \cdot x$ of $x \in X$ may be defined only for group elements g which lie sufficiently near the identity e. In such cases, the transformation laws $g \cdot (h \cdot x) = (g \cdot h) \cdot x$, etc., are imposed only when they make sense. For example, the linear fractional transformations (2.7) give only a local action of $GL(2, \mathbb{R})$ on $\mathbb{R}$ since they are not defined when $\gamma p + \delta = 0$. In this particular case, one can construct a "complete" global transformation group action on the projective line $\mathbb{RP}^1$ by adjoining the point at infinity. See [**176**] for general results on globalizing local group actions and [**168, 169**] for additional details.

A transformation group action is called *connected* if both G and X are topologically connected manifolds. The infinitesimal methods introduced in Chapter 9 will require connectivity, and hence it is important to determine which of the groups we have encountered have this property. If G is any Lie group, then the connected component of G containing the identity element is a normal Lie subgroup $G_0 \subset G$ having the same dimension as G. The quotient group G/G_0 is discrete. Any other connected component $G_* \subset G$ is topologically isomorphic to G_0 and can be obtained by multiplying the elements of G_0 by any fixed element therein, so $G_* = g_* \cdot G_0$ for any $g_* \in G_*$. For example, every matrix $A \in \mathrm{GL}(2, \mathbb{R})$ having negative determinant can be obtained by multiplying a matrix with positive determinant by a fixed reflection, e.g., $\begin{pmatrix} -1 & 0 \\ 0 & 1 \end{pmatrix}$. Thus, $\mathrm{GL}(2, \mathbb{R})/\mathrm{GL}(2, \mathbb{R})^+$ is isomorphic to the discrete two-element group $\mathbb{Z}_2$.

Theorem 8.11. *Suppose G is a Lie group acting globally and transitively on a connected manifold X. If the isotropy subgroup $G_x \subset G$ of any one point $x \in X$ is connected, then G itself is connected.*

The key step in the proof of this theorem is to demonstrate that the orbit(s) of the connected component $G_0 \subset G$ are open submanifolds of X, and hence, since X itself is connected, G_0 also acts transitively on X. Moreover, $G/G_0 \simeq G_x/(G_x \cap G_0) \simeq \{e\}$ by hypothesis. See [173; §I.4] for details. Theorem 8.11 can be immediately applied to prove connectivity of the particular Lie group actions of direct interest in invariant theory.

Remark: If G acts transitively and globally on X, then we can identify $X \simeq G/H$ as a homogeneous space, where $H = G_x$ is the isotropy subgroup of a chosen point $x \in X$.

Proposition 8.12. *The real Lie groups $\mathrm{SO}(n)$, $\mathrm{SL}(n)$, $\mathrm{SE}(n)$, and $\mathrm{SA}(n)$ are all connected. Also, $\mathrm{GL}(n, \mathbb{C})$ is connected. On the other hand, $\mathrm{GL}(n, \mathbb{R})$, $\mathrm{E}(n)$, $\mathrm{A}(n)$, and $\mathrm{O}(n)$ each has two connected components, distinguished by the sign of the determinant.*

Proof: Consider first the rotation group $\mathrm{SO}(n)$. We apply Theorem 8.11 to its standard, transitive action on the unit sphere $S^{n-1} \subset \mathbb{R}^n$. The isotropy subgroup of the point $e_n = (0, \ldots, 0, 1)$, say, is easily seen to be isomorphic to $\mathrm{SO}(n-1)$. Thus, the connectivity of $\mathrm{SO}(n)$ follows from a simple induction starting with the trivial case $\mathrm{SO}(1) = \{1\}$.

In the case of $\mathrm{SL}(n)$, we employ its transitive action on $\mathbb{R}^n \setminus \{0\}$. In this case, the isotropy subgroup $G_{e_n} \subset \mathrm{SL}(n)$ of the basis vector e_n consists of all matrices of the form $\begin{pmatrix} A & b \\ 0 & 1 \end{pmatrix}$ where $\det A = 1$ and can be identified with the equi-affine group $\mathrm{SA}(n-1) = \mathrm{SL}(n-1) \ltimes \mathbb{R}^{n-1}$. Now, $\mathrm{SA}(n-1)$ is topologically isomorphic to the Cartesian product of $\mathrm{SL}(n-1)$ with the vector space $\mathbb{R}^{n-1}$. Therefore, if $\mathrm{SL}(n-1)$ is connected, so is $\mathrm{SA}(n-1)$, and hence, by Theorem 8.11, so is $\mathrm{SL}(n)$. Thus, connectivity of both the unimodular and equi-affine groups follows by induction. The other cases are left as exercises for the reader. *Q.E.D.*

Orbits and Invariance

As we have seen, there is an intimate connection between the invariants of a transformation group and its orbit structure. For sufficiently regular Lie group actions, this correspondence can be made rather precise.

Theorem 8.13. *If G is an r-dimensional Lie group acting analytically on a manifold X, then each orbit is an analytic submanifold of X. Moreover, a point $x \in X$ belongs to an s-dimensional orbit if and only if its isotropy subgroup G_x is a closed Lie subgroup of dimension $r - s$.*

Sketch of Proof: Let $x \in X$ be fixed. Since $G_x \subset G$ is closed, Theorem 8.9 demonstrates that it is a Lie subgroup of G. Therefore, the homogeneous space G/G_x can be endowed with the structure of an analytic manifold of dimension $s = \dim G - \dim G_x$. The group action $g \mapsto g \cdot x$ induces an analytic identification $\psi \colon G/G_x \xrightarrow{\sim} \mathcal{O}_x$ with the orbit through x. See [**173**; §II.1.4] and [**227**] for details.   *Q.E.D.*

In particular, if G acts freely, then its orbits have the same dimension as G itself. The converse is not quite true; if an r-dimensional transformation group G has r-dimensional orbits, then it acts *locally freely*, meaning that the isotropy subgroup of any point is a discrete subgroup of G. For example, the action of $\mathbb{R}$ on $\mathbb{R}^2$ given by rotations $(x, y) \mapsto (x \cos t - y \sin t, x \sin t + y \cos t)$ is locally free on $\mathbb{R}^2 \setminus \{0\}$ since the isotropy subgroup of any nonzero point is the discrete subgroup $2\pi\mathbb{Z} \subset \mathbb{R}$ consisting of all integer multiples of 2π.

In general, orbits can be of differing dimensions, including the fixed points, which have dimension 0. The group is said to act *semi-regularly* on X if all its orbits have the same dimension. A semi-regular group action is *regular* if, in addition, each point $x \in X$ has a system of arbitrarily small neighborhoods U_α whose intersection with each orbit $\mathcal{O}$ is a (pathwise) connected subset $U_\alpha \cap \mathcal{O}$ of the orbit, cf. [**168**, **227**].

Example 8.14. Consider the action (1.12) of the unimodular group SL(2) on the space of binary quadratic forms. As discussed in Chapter 3, there are two types of orbits: the origin $a = b = c = 0$, which has dimension 0, and the level sets of the discriminant $b^2 - 4ac$, each of which has dimension 2. Therefore, SL(2) acts regularly on the open subset consisting of all nonzero binary quadratics. The discriminant is the only independent invariant, and it does distinguish the orbits away from the singular origin. According to Theorem 8.13, the isotropy (or symmetry) subgroup of any nonzero quadratic polynomial is a one-parameter subgroup of SL(2), a fact that is easy to verify directly.

Example 8.15. Let $\alpha \in \mathbb{R}$ be fixed. Consider the action

$$(x, y, z, w) \longmapsto \begin{array}{l} (x \cos t - y \sin t, x \sin t + y \cos t, \\ z \cos a\pi t - w \sin a\pi t, z \sin a\pi t + w \cos a\pi t) \end{array}$$

of the one-parameter group $t \in \mathbb{R}$ on $\mathbb{R}^4$. Note that the functions $x^2 + y^2$ and $z^2 + w^2$ are invariants, and hence the orbits are all contained in the family of tori $\{x^2 + y^2 = c_1, z^2 + w^2 = c_2\}$. If $a \in \mathbb{Q}$ is rational, then the orbits are closed circles (or, more precisely, toroidal knots, [124]), and the action is regular and locally free. On the other hand, if a is irrational, then the orbit through a point (x_0, y_0, z_0, w_0) forms a dense subset of its containing torus, cf. [98]. In this case, the action is only semi-regular, but remains free.

Remark: If G acts on X, then the subset $X_0 \subset X$ consisting of all orbits of maximal dimension — the "generic orbits" — forms an open submanifold, which is dense in X if the action is analytic. The group then acts at least semi-regularly on X_0. The remainder of X contains the singular orbits of lower dimension — which cannot typically be distinguished by continuous invariants.

Not only are the orbits of Lie transformation groups all submanifolds, but in the regular (and even semi-regular) case they locally fit together in a relatively uncomplicated fashion. This is evidenced by the following result, which is a consequence of the powerful Frobenius Theorem, [73], governing the existence of solutions to certain systems of partial differential equations. We refer the reader to [169, 173, 227] for the proof and details.

Theorem 8.16. *Let G be a Lie group acting regularly on an m-dimensional manifold X with s-dimensional orbits. Then, near every point of X, there exists a system of rectifying local coordinates $(y, z) = (y_1, \ldots, y_s, z_1, \ldots, z_{m-s})$ such that any orbit intersects the coordinate chart in at most one slice $\mathcal{S}_c = \{(y, z) \mid z_1 = c_1, \ldots, z_{m-s} = c_{m-s}\}$, for some $c = (c_1, \ldots, c_{m-s}) \in \mathbb{R}^{m-s}$.*

Remark: If the action is semi-regular, the same result holds except that an orbit may intersect the chart in more than one (typically infinitely many) of the slices.

Geometrically, the orbits of a regular or semi-regular Lie transformation group are said to form a *foliation* of the manifold X. Let (y, z)

be the rectifying coordinates as provided by Theorem 8.16. Since the z's are constant on the orbits, they are invariant under the group action and clearly functionally independent. Moreover, since the orbits themselves are completely distinguished by the value of the z's, any other invariant is merely a function $I = I(z)$ of these basic invariants. Reverting back to the original coordinates on X leads to an immediate justification of the Fundamental Theorem on invariants of regular Lie group actions.

Theorem 8.17. *Let G be a Lie group acting regularly on the m-dimensional manifold X with s-dimensional orbits. Then, in a neighborhood N of each point $x_0 \in N \subset X$, there exist $m - s$ functionally independent invariants $I_1(x), \ldots, I_{m-s}(x)$. Any other invariant I defined near x_0 can be locally uniquely expressed as an analytic function of the fundamental invariants: $I = H(I_1, \ldots, I_{m-s})$. The fundamental invariants serve to distinguish the orbits near x_0 in the sense that two points $x, \tilde{x} \in N$ will lie in the same orbit if and only if all the fundamental invariants agree: $I_1(x) = I_1(\tilde{x}), \ldots, I_{m-s}(x) = I_{m-s}(\tilde{x})$.*

Thus, for regular Lie group actions, the fundamental invariants solve the point equivalence problem: two points $x, \tilde{x} \in X$ are equivalent, meaning $\tilde{x} = g \cdot x$ for some $g \in G$ if and only if all their invariants have the same value. The one caveat is that this result is only local, since the points must lie in the same rectifying coordinate chart that defines the invariants. Moreover, it applies only to the regular orbits. The existence of suitable global invariants is a more delicate matter.

Remark: Semi-regular actions are more pathological. For instance, in the semi-regular case $a \notin \mathbb{Q}$ of Example 8.15, there are only two independent continuous invariants, even though the group has dimension 1 and acts on a four-dimensional space. The (continuous) invariants fail to distinguish different orbits lying in the same torus.

Any level set $S = \{I_1(x) = c_1, \ldots, I_k(x) = c_k\}$ of a system of (absolute) invariants is automatically a G–invariant subset. The converse to this result is also true and is easy to prove once the rectifying coordinates of Theorem 8.16 are introduced.

Theorem 8.18. *Suppose G acts regularly on X. A submanifold $S \subset X$ is G–invariant if and only if $S = \{I_1(x) = \cdots = I_k(x) = 0\}$ can be locally expressed by the vanishing of invariants $I_1(x), \ldots, I_k(x)$.*

Remark: It can be shown that if $S = \{F(x) = c\}$ is a nonempty level set of a function $F \colon X \to \mathbb{R}^k$, and the Jacobian matrix of F has maximal rank k on S, then S is a G–invariant subset if and only if F is a relative invariant with respect to some multiplier representation of G on $\mathcal{F}^k(X)$. These results are the differential geometric versions of Gram's Theorem, [231; p. 240], which states that any equations of invariant significance in classical invariant theory can be expressed by the vanishing of classical invariants and covariants. See [68; Proposition 3.6] for details, including a generalization of Theorem 8.17 that determines the number of relative invariants for any "regular" multiplier representation.

Example 8.19. Consider the action of GL(2) on the space $\mathcal{P}^{(n)}$ of binary forms of degree n. If $n \geq 3$, the generic orbits of GL(2) are four-dimensional; see Exercise 9.39. A binary form is *regular* if it belongs to a generic orbit and hence, by Theorem 8.13, has at most a discrete symmetry group. (In Theorem 8.62 we shall prove that a binary form is regular if and only if it is not equivalent to a monomial.) Since $\mathcal{P}^{(n)}$ has dimension $n + 1$, Theorem 8.17 implies that, on the open, dense subset consisting of all regular binary forms, there are precisely $n - 3$ fundamental (absolute) invariants such that any other invariant can be written as a function thereof. Moreover, two regular binary forms are equivalent if and only if their invariants have the same values. For example, binary quartics have a single independent absolute invariant, which is the ratio $I = j^2/i^3$ between their fundamental relative invariants (2.29), (2.30). For regular, complex quartics, the value of I completely determines its equivalence class and canonical form. In the real case, this is true locally, but there are additional global conditions required to distinguish real quartics which are complex- but not real-equivalent; these will be considered shortly.

Exercise 8.20. Consider the action of the special linear group SL(2) on the space of binary cubics. Prove that the discriminant (2.22) is the unique independent invariant. Prove that a cubic is regular if and only if its discriminant does not vanish, and so the discriminant distinguishes regular cubics up to unimodular equivalence.

The preceding discussion shows that if we relax the algebraic condition of polynomial independence and permit functional dependencies, the number of independent invariants is given by a straightforward dimension count. A related result appeared in Theorem 6.32, which provides an explicit system of n rationally independent covariants for a binary form of degree n. However, as far as I know, an explicit system of $n - 3$ functionally independent invariants is not known in general.

Normalization

In favorable situations, one can implement a direct computational algorithm to produce all the invariants of Lie transformation groups. The method of "normalization" lies at the heart of the Cartan approach to moving frames and equivalence problems, [**33, 35, 95, 169**]; an early version can be found in Killing, [**127**]. The normalization procedure presented here forms the foundation of a powerful new "regularized" method of moving frames, recently introduced by M. Fels and the author, [**69, 70**], that can be applied to general Lie group actions. It is surprising that this simple method has not been discussed or exploited in the standard invariant theoretic or geometric literature.

Recall that canonical forms are defined as representative points in each group orbit. When the group acts regularly, the canonical forms can be chosen to depend continuously on the orbits. The most important systems of canonical forms are provided by "cross-sections" to the group orbits; these are defined as submanifolds of the requisite dimension, with the additional property of "transversality", meaning that they do not share any common tangent directions with the orbits.

Remark: The tangent space $TX|_x$ to an m-dimensional manifold X at a point $x \in X$ is the m-dimensional vector space spanned by the tangent vectors to all analytic curves $C \subset X$ at the point $x \in C$.

Definition 8.21. Two submanifolds $N, P \subset X$ are said to intersect *transversally* at a common point $x_0 \in N \cap P$ if they have no nonzero tangent vectors in common: $TN|_{x_0} \cap TP|_{x_0} = \{0\}$.

Definition 8.22. Let G be a Lie transformation group that acts regularly on the m-dimensional manifold X with s-dimensional orbits. A (local) *cross-section* is an $(m - s)$-dimensional submanifold $K \subset X$ such that K intersects each orbit transversally and at most once.

Proposition 8.23. *If a Lie group G acts regularly on a manifold X, then one can construct a local cross-section K passing through any point $x \in X$.*

Proof: According to Theorem 8.17, locally, the orbits are the common level sets of the fundamental invariants $I_1, \ldots, I_{m-s}$. A cross-section can itself be represented as a level set

$$K_1(x) = c_1, \quad \ldots \quad K_s(x) = c_s, \tag{8.1}$$

defined by a "complementary" system of analytic functions. The constants $c_1, \ldots, c_s$ must be chosen so that the solution set to (8.1) is nonempty. According to the Implicit Function Theorem, the solutions to (8.1) will locally define a cross-section K if and only if the Jacobian determinant condition

$$\frac{\partial(K_1, \ldots, K_s, I_1, \ldots, I_{m-s})}{\partial(x_1, \ldots, x_m)} \equiv \det \left| \frac{\partial K_\mu}{\partial x_i}, \frac{\partial I_\nu}{\partial x_i} \right| \neq 0 \qquad (8.2)$$

holds. Clearly, then, a wide choice of functions $K_\nu(x)$ will suffice for the required purpose. For example, in terms of the rectifying local coordinates (y, z) of Theorem 8.16, every cross-section is the graph $K = \{y = h(z)\}$ of some function of the fundamental invariants. *Q.E.D.*

A *coordinate cross-section* occurs when the functions $K_\nu(x) = x_{i_\nu}$, $\nu = 1, \ldots, s$, coincide with s of the given local coordinates on X. (Of course, changing local coordinates changes the notion of coordinate cross-section; on the other hand, any cross-section can be made into a coordinate cross-section by a suitable choice of local coordinates.) In particular, the first s coordinates themselves define a coordinate cross-section,

$$K = \{x_1 = c_1, \ldots, x_s = c_s\}, \qquad (8.3)$$

if and only if

$$\frac{\partial(I_1, \ldots, I_{m-s})}{\partial(x_{s+1}, \ldots, x_m)} \neq 0. \qquad (8.4)$$

Thus, one can refine Proposition 8.23 to state that one can always find a local coordinate cross-section through any point.

Example 8.24. Consider the standard action $\mathbf{x} \mapsto R\mathbf{x}$ of the rotation group SO(3), which is regular on $X = \mathbb{R}^3 \setminus \{0\}$. The orbits are the spheres $\| \mathbf{x} \| = r$, and hence any ray $\mathcal{R}_\mathbf{a} = \{\lambda \mathbf{a} \,|\, \lambda > 0\}$, $\mathbf{a} \neq 0$, provides a (global) cross-section. An open line segment parallel to one of the coordinate axes will be a coordinate cross-section provided it is not tangent to any orbit sphere. For example, the vertical half-lines $\mathcal{L} = \{x = c_1, y = c_2, z > 0\}$ are local cross-sections. The most general local cross-section is given by a curve that transversally intersects each orbit sphere at most once.

Since the elements $k \in K$ of the cross-section can be viewed as canonical forms for general points $x \in X$, their coordinates provide the invariant moduli for the group action. More explicitly, let x be any point whose orbit $\mathcal{O}_x$ intersects the cross-section at the unique point

$k = \xi(x) \in K \cap \mathcal{O}_x$. The components $k_1 = \xi_1(x), \ldots, k_m = \xi_m(x)$ of $k = \xi(x)$ are invariant functions. Indeed, by uniqueness, if $y = g \cdot x$ is any other point in the orbit through x, then $\xi(y) = \xi(g \cdot x) = k = \xi(x)$, which immediately proves invariance. Since K has dimension $m - s$, precisely $m - s$ of these invariant functions will be functionally independent and therefore provide us with a generating system of invariants, in accordance with Theorem 8.16. In the simplest case, for the basic coordinate cross-section (8.3), the first s coordinates of k will be constant, and are hence trivial invariants, while the latter $m - s$ coordinates will provide the fundamental system of invariants. These simple observations will now be formalized into an extremely useful direct method, known as *normalization*, for practical computation of invariants.

For simplicity, let us assume that G acts (locally) freely on X, and so its orbits have the same dimension r as G. We shall also choose the coordinate cross-section $K = \{x_1 = c_1, \ldots, x_r = c_r\}$ defined by the first r coordinates. (This can always be arranged by relabeling the coordinates if necessary.) Let us also introduce local coordinates $g = (g_1, \ldots, g_r)$ on G in a neighborhood of the identity element. We then write out the explicit formulae for the group transformations $\bar{x} = g \cdot x = w(g, x)$ in the given coordinates. The *normalization equations* for the prescribed coordinate cross-section are then obtained by equating the first r components of the function w to the given constants,

$$\bar{x}_1 = w_1(g, x) = c_1, \qquad \ldots \qquad \bar{x}_r = w_r(g, x) = c_r, \qquad (8.5)$$

which must lie in the range of the w's. Since K was assumed to be a cross-section, the Implicit Function Theorem implies that (8.5) can be locally solved for the group parameters in terms of the coordinates x. Write the solution as $g = \gamma(x)$; the resulting map $\gamma \colon X \to G$ is known as the *moving frame* associated with the given cross-section, [**70, 93, 120**].

Remark: In practice, the solution to the normalization equations (8.5) often proceeds in an iterative fashion. One first solves the equation $w_1(g, x) = c_1$ for one of the group parameters, leading to a preliminary normalization, say, $g_1 = \gamma_1(g_2, \ldots, g_r, x)$. Substituting this expression into the remaining equations produces a reduced system of normalization equations for the remaining group parameters $g_2, \ldots, g_r$, which then can be solved in a similar manner. One can even use this method to construct a proper coordinate cross-section; if, at any stage, the Implicit Function Theorem fails to permit one to solve the next equation for a group parameter, then this component should not be a part of the coordinate cross-section, and one should search for an alternative to continue the normalization procedure. If one becomes completely stuck, then this indicates either that the action is not regular, or that it is not free, the latter possibility being discussed shortly. Barring significant algebraic complications, the normalization procedure thus provides an effective, direct method for determining the invariants of regular Lie group actions.

The group element $g = \gamma(x)$ defined by the moving frame map is uniquely characterized by the condition that it maps the point x to its canonical form $k \in K$, whereby

$$k = \gamma(x) \cdot x = w(\gamma(x), x). \tag{8.6}$$

Therefore, the components of (8.6) *are* the invariants or moduli defined by the given cross-section. For the coordinate cross-section (8.3), the first r components of k are just the normalization constants, and the remaining $m - r$ components provide a complete system of functionally independent invariants.

Theorem 8.25. *Given a free, regular Lie group action and coordinate cross-section as described above, let $g = \gamma(x)$ denote the solution to the normalization equations (8.5). Then the functions*

$$I_1(x) = w_{r+1}(\gamma(x), x), \qquad \cdots \qquad I_{m-r}(x) = w_m(\gamma(x), x), \tag{8.7}$$

obtained by substituting the moving frame formulae into the remaining $m - r$ components of w form a complete system of functionally independent invariants for the action of G.

Exercise 8.26. Prove that the moving frame map $\gamma \colon X \to G$ is equivariant, $\gamma(g \cdot x) = \gamma(x) \cdot g^{-1}$, with respect to the actions of G on X and on itself by right multiplication.

Remark: If G does not act freely, then a variant of this procedure will still produce the required invariants. Assuming that the orbits have dimension s, the normalization equations $w_1(g, x) = c_1, \ldots, w_s(g, x) = c_s$ (relabel the coordinates if necessary) can be solved for s of the group parameters in terms of the coordinates x and the remaining $r - s$ group parameters. Substituting the resulting formulae for the group parameters into the remaining components $w_{s+1}(g, x), \ldots, w_m(g, x)$ will still lead to a complete system of functionally independent invariants. The remaining $r - s$ "unnormalized" group parameters will not explicitly occur in the final formulae.

This completes our brief description of the normalization method for computing invariants. We shall now illustrate with a variety of useful, elementary examples.

Example 8.27. The planar Euclidean group SE(2) acts transitively on $\mathbb{R}^2$, and so there are no ordinary Euclidean invariants. The following extended action, though, plays a key role in the Euclidean geometry of planar curves, as discussed below. Consider the free action[†] of SE(2) on $X = \mathbb{R}^4$ that maps a point $(x, y, v, w) \in X$ to

$$\bar{x} = x \cos \theta - y \sin \theta + a, \qquad \bar{y} = x \sin \theta + y \cos \theta + b,$$
$$\bar{v} = \frac{\sin \theta + v \cos \theta}{\cos \theta - v \sin \theta}, \qquad \bar{w} = \frac{w}{(\cos \theta - v \sin \theta)^3}, \qquad (8.8)$$

where θ, a, b serve to parametrize the group. Consider the simple coordinate cross-section $K = \{x = y = v = 0\}$. The solution to the normalization equations $\bar{x} = \bar{y} = \bar{v} = 0$ is readily found; the resulting formulae

$$\theta = -\tan^{-1} v, \qquad a = -\frac{x + yv}{\sqrt{1 + v^2}}, \qquad b = \frac{xv - y}{\sqrt{1 + v^2}}, \qquad (8.9)$$

define the moving frame map from X to SE(2). Substituting the normalizations (8.9) into the remaining component $\bar{w}$, we recover the fundamental invariant

$$I = \frac{w}{(1 + v^2)^{3/2}}. \qquad (8.10)$$

Every other invariant can be (locally) written as a function of I.

Exercise 8.28. Show that one can also express $I = (1 + \bar{v}^2)^{-3/2} \bar{w}$ using the *same* function of the transformed coordinates (8.8); in other words, the group parameters cancel out. Explain why this is true and formulate a general theorem. The T.Y. Thomas Replacement Theorem from Riemannian geometry, [**214**; p. 109], is a particular case, cf. [**70**].

Remark: Unfortunately, most of the standard actions on binary forms are too algebraically complicated for the normalization method to be an effective tool. Even the action (1.12) on binary quadratics is a challenge. It would be interesting to see to what extent more sophisticated Gröbner basis methods, cf. [**54, 204**], might facilitate the normal-

---

[†] The formulae, in fact, only define a local transformation group action, but this does not affect the algorithm. The action can be globalized by treating v and w as projective coordinates.

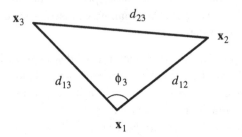

Figure 4.   Law of Cosines.

ization computations in such situations and lead to complete systems of functionally independent invariants.

Joint Invariants

One particularly fruitful application of the normalization method is to the determination of joint invariants for classical group actions. Let us first illustrate the basic idea with the well-known case of the Euclidean group. For simplicity, we restrict our attention to the planar case.

Example 8.29. Although the usual action of the proper Euclidean group SE(2) on $X = \mathbb{R}^2$ has no ordinary invariants, there are joint invariants, which are simply classified using the normalization procedure. We let $(R, \mathbf{a}) \in \mathrm{SE}(2)$ act simultaneously on $m \geq 2$ coordinate planes $X^{(m)} = X \times \cdots \times X \simeq \mathbb{R}^{2m}$ according to $\overline{\mathbf{x}}_k = R\,\mathbf{x}_k + \mathbf{a}$. The joint action is regular and free on $X^{(m)} \setminus D^{(m)}$, where $D^{(m)} = \{\mathbf{x}_1 = \mathbf{x}_2 = \cdots = \mathbf{x}_m\}$ is the diagonal.

We begin by normalizing $\overline{\mathbf{x}}_1 = 0$ by setting $\mathbf{a} = -R\mathbf{x}_1$. We can then substitute the normalized values of the translation parameters into the remaining transformation formulae and continue to normalize the resulting expressions $\overline{\mathbf{x}}_k = R\,(\mathbf{x}_k - \mathbf{x}_1)$, which now depend only on the angular group parameter θ. As long as $\mathbf{x}_1 \neq \mathbf{x}_2$, the second component of $\overline{\mathbf{x}}_2$ can be normalized to zero by specifying the rotation angle θ; we are thus led to the basic moving frame formulae

$$\theta = -\sphericalangle(\mathbf{x}_2 - \mathbf{x}_1, \mathbf{e}_1), \qquad \mathbf{a} = -R_\theta\,\mathbf{x}_1.$$

Here $\sphericalangle(\mathbf{x}, \mathbf{y}) = \tan^{-1}(y_2/y_1) - \tan^{-1}(x_2/x_1)$ denotes the angle from $\mathbf{x}$ to $\mathbf{y}$. The first component of $\overline{\mathbf{x}}_2$ then reduces to the Euclidean interpoint distance $d_{12} = d(\mathbf{x}_1, \mathbf{x}_2) = \|\mathbf{x}_1 - \mathbf{x}_2\|$, which thereby forms the

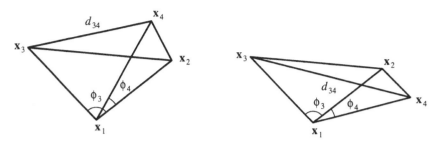

Figure 5. Four-Point Euclidean Configurations.

fundamental two-fold joint invariant for the Euclidean group. We now substitute the normalized value for the rotation angle into the remaining transformation formulae:

$$\bar{\mathbf{x}}_k = (d_{1k} \cos \phi_k, d_{1k} \sin \phi_k), \qquad (8.11)$$

where

$$d_{1k} = d(\mathbf{x}_1, \mathbf{x}_k) = \| \mathbf{x}_1 - \mathbf{x}_k \|, \qquad \phi_k = \sphericalangle(\mathbf{x}_k - \mathbf{x}_1, \mathbf{x}_2 - \mathbf{x}_1). \quad (8.12)$$

According to Theorem 8.25, both components of (8.11) are (proper) Euclidean invariants. We have therefore proved the following result.

Theorem 8.30. *Every joint invariant of the planar, orientation-preserving Euclidean group* SE(2) *is a function of the interpoint distances* d_{1k}, $k \geq 2$, *and angles* ϕ_k, $k \geq 3$, *as given in* (8.12).

In particular, other distances d_{jk} and angles $\sphericalangle(\mathbf{x}_k - \mathbf{x}_l, \mathbf{x}_m - \mathbf{x}_n)$ can all be re-expressed as functions of the particular ones given in (8.12). For instance, referring to Figure 4, the Law of Cosines allows us to express $\cos \phi_3$ in terms of the lengths d_{12}, d_{13}, d_{23} of the sides of the triangle formed by $\mathbf{x}_1, \mathbf{x}_2, \mathbf{x}_3$. Therefore, we can express the positive angle $|\phi_3|$ in terms of the indicated distances. Note that the sign of ϕ_3 is an orientation-preserving Euclidean invariant but can be reversed if we extend our action to the full Euclidean group E(2). However, this trick can only be played on three-fold joint invariants; once we include a fourth point, the signs of the angles ϕ_3 and ϕ_4 can be simultaneously changed, but not each individual sign. On the other hand, unless $\mathbf{x}_1, \mathbf{x}_2$, and $\mathbf{x}_3$ are collinear, the distance d_{34} serves to distinguish between the two possible configurations, as shown in Figure 5, having different relative signs for ϕ_3 and ϕ_4. Thus, we have demonstrated a version of the classical theorem in Weyl, [**231**; Theorem 2.9.A], that the only joint

Euclidean invariants are functions of the distances between points. In the orientation-preserving case, the result holds provided the points do not all lie on a single triangle, otherwise the signs of the angle invariants cannot be determined by the interpoint distances alone.

Example 8.31. Next consider the equi-affine group SA(2) acting on multiple copies of $\mathbb{R}^2$ via $\overline{\mathbf{x}}_k = A\mathbf{x}_k + \mathbf{a}$, where $\det A = 1$. We begin by setting $\overline{\mathbf{x}}_1 = 0$, which leads to the normalization $\mathbf{a} = -A\mathbf{x}_1$ of the translation component. The remaining transformation formulae now become $\overline{\mathbf{x}}_k = A(\mathbf{x}_k - \mathbf{x}_1)$. Assuming $\mathbf{x}_2 \neq \mathbf{x}_1$, we can normalize $\overline{\mathbf{x}}_2 = \mathbf{e}_1$ to be the first basis vector. Furthermore, if $\mathbf{x}_3 - \mathbf{x}_1$ is not parallel to $\mathbf{x}_2 - \mathbf{x}_1$, we can normalize $\overline{\mathbf{x}}_3 = 2a_{123}\mathbf{e}_2$, where a_{ijk} denotes the (signed) area of the triangle with vertices $\mathbf{x}_i, \mathbf{x}_j, \mathbf{x}_k$, cf. (4.14). The associated moving frame map $\rho \colon X = \{a_{123} \neq 0\} \to \mathrm{SA}(2)$ is given by

$$A = \begin{pmatrix} 1 & 0 \\ 0 & \det B \end{pmatrix} B^{-1}, \qquad b = -A\mathbf{x}_1,$$

where the matrix $B = (\mathbf{x}_2 - \mathbf{x}_1, \mathbf{x}_3 - \mathbf{x}_1)$ has the indicated columns, and $\det B = 2a_{123}$. The additional joint invariants are the components of $\overline{\mathbf{x}}_k = A(\mathbf{x}_k - \mathbf{x}_1)$ for $k \geq 4$, which are $-a_{13k}/a_{123}$ and $2a_{12k}$, and hence combinations of area invariants.

Theorem 8.32. *Every joint invariant of the equi-affine group is a function of the signed triangular areas* $a_{ijk} = \frac{1}{2}(\mathbf{x}_j - \mathbf{x}_i) \wedge (\mathbf{x}_j - \mathbf{x}_k)$. *In fact, a generating system is provided by the areas* a_{12k}, a_{13k}.

Exercise 8.33. Use the normalization method to analyze the joint invariants of the Euclidean group SE(n), the equi-affine group SA(n), and the full affine group A(n) acting on $\mathbb{R}^n$.

Example 8.34. Finally, we treat the projective action of GL($2, \mathbb{C}$) on $\mathbb{CP}^1$. The group transformations map p_k to $\bar{p}_k = (\alpha p_k + \beta)/(\gamma p_k + \delta)$. According to Example 4.30, we can normalize the first three points to take on any prescribed values, say, $1, \infty, 0$. We begin with $\bar{p}_3 = 0$ by setting $\beta = -\alpha p_3$. Next, normalize $\bar{p}_2 = \infty$ by setting $\delta = -\gamma p_2$. Third, normalize $\bar{p}_1 = 1$ by setting $\alpha = \gamma(p_1 - p_2)/(p_1 - p_3)$. The remaining components will then form invariants:

$$\bar{p}_k = \frac{\alpha p_k + \beta}{\gamma p_k + \delta} = \frac{(p_1 - p_2)(p_k - p_3)}{(p_1 - p_3)(p_k - p_2)} \equiv [\, p_1, p_2, p_3, p_k \,],$$

reconfirming our earlier observation that the four-fold cross-ratios (4.15) generate the joint projective invariants. In fact, only $n - 3$ of these cross-ratios are functionally independent, so we can take $[\,p_1,p_2,p_3,p_k\,]$, $k = 4,\ldots,n$, as our fundamental invariants. Note that we have not normalized the remaining group parameter γ, which reflects the non-effectiveness of the projective action of GL(2). It is easier to leave the result in this form; explicitly imposing the unimodularity constraint introduces unnecessary square roots in the final formulae.

Exercise 8.35. Determine how to express other cross-ratios, e.g., $[\,p_3,p_2,p_1,p_4\,]$, $[\,p_1,p_2,p_4,p_5\,]$, in terms of the fundamental cross-ratios. *Hint*: Use the Replacement Theorem from Exercise 8.28.

Exercise 8.36. Determine the joint invariants of the projective action of GL($3,\mathbb{C}$) on $\mathbb{CP}^2$. Generalize to higher dimensions.

Prolongation of Group Actions

Whenever a group acts on functions, there is an induced action on their derivatives. The process of extending the group action to derivatives is known as "prolongation". If we identify the graph of a function $f\colon\mathbb{R}\to\mathbb{R}$ with a curve, say, then the first prolongation of a group transformation can be identified with its action on the tangent lines to the curve. The prolongation procedure lies at the foundation of Lie's theory of symmetry groups of differential equations, [**140**, **168**], as well as Cartan's geometry of moving frames, [**33**, **93**, **120**, **70**].

Let us look first at a simple example, motivated by the geometry of curves in the (oriented) Euclidean plane. The underlying group is the proper Euclidean group SE(2). Suppose $\mathcal{C}$ is a curve that (locally) coincides with the graph of a function $y = f(x)$. Given a Euclidean transformation $\bar{\mathbf{x}} = R\mathbf{x} + \mathbf{a}$, the transformed curve $\overline{\mathcal{C}} = R\cdot\mathcal{C} + \mathbf{a}$ will, generically, also be a graph $\bar{y} = \bar{f}(\bar{x})$; the transformed function $\bar{f}$ can be obtained by eliminating x from the two parametric equations

$$\bar{x} = x\cos\theta - f(x)\sin\theta + a, \qquad \bar{y} = x\sin\theta + f(x)\cos\theta + b. \qquad (8.13)$$

The first prolongation of the Euclidean transformation (8.13) will map the tangent line to the curve $\mathcal{C}$ at a point $\mathbf{x}$ to the tangent line to $\overline{\mathcal{C}}$ at the image point $\bar{\mathbf{x}}$; similarly the second prolongation will map the osculating circle at $\mathbf{x}$ to that at $\bar{\mathbf{x}}$, and so on. The actual formulae for how the derivatives of the function transform are found by implicit differentiation of (8.13), so

$$\bar{f}'(\bar{x}) = \frac{\sin\theta + f'(x)\cos\theta}{\cos\theta - f'(x)\sin\theta}, \qquad \bar{f}''(\bar{x}) = \frac{f''(x)}{(\cos\theta - f'(x)\sin\theta)^3}, \qquad (8.14)$$

and so on. Note that if $f'(x) = \cot\theta$, then the denominators vanish; this happens when the Euclidean transformation produces a vertical tangent at the point $\bar{x}$, and hence the transformed curve is no longer the graph of a function.

The transformation rules (8.14) serve to define an action of the Euclidean group. This will become more transparent if we introduce new coordinates to represent the derivatives of the function $y = f(x)$. Let us set $v = f'(x)$, $w = f''(x)$, and let $\bar{v} = \bar{f}'(\bar{x})$, $\bar{w} = \bar{f}''(\bar{x})$ be the corresponding transformed derivative coordinates. Then the transformation laws (8.13), (8.14) coincide with the local action[†] of SE(2) given in (8.8). In this manner, we can regard the process of prolongation in a purely geometric manner, allowing us to apply our general theory of transformation groups to this novel situation.

This process of introducing new coordinates to represent the derivatives of a function on a space serves to define the so-called *jet space*, a concept first formalized by Ehresmann, [64], around 1950. The essential ideas, though, appear in numerous nineteenth-century works, particularly those of Lie, such as [140]. We shall not attempt to define jet spaces formally, but content ourselves with a few self-evident examples; the interested reader can refer to [27, 168, 169] for theoretical and practical details.

For example, the second order jet space corresponding to scalar functions $y = f(x)$ is the four-dimensional Euclidean space $J^2 \simeq \mathbb{R}^4$ having the preceding coordinates x, y, v, w. The second order *prolongation* or *jet*, of a function $y = f(x)$ is the function $f^{[2]} \colon \mathbb{R} \to \mathbb{R}^3$ given by the additional formulae $v = f'(x)$ and $w = f''(x)$. Note that the graph of f defines a curve in $\mathbb{R}^2$, while the graph of its second prolongation defines a curve in J^2. In general, the nth jet space for curves in $\mathbb{R}^2$ is the Euclidean space $J^n \simeq \mathbb{R}^{n+2}$. The coordinates $\mathbf{z}^{[n]} = (x, y^{[n]}) \equiv (x, y, y_1, \dots, y_n)$ represent the derivatives up to order n and thereby serve to define the nth order prolongation $y^{[n]} = f^{[n]}(x) \equiv (f(x), f'(x), \dots, f^{(n)}(x))$ of a function $y = f(x)$. Note that the graph of $f^{[n]}$ traces out a curve in J^n, which is the nth prolongation of the curve defined by the graph of f. We let $g^{[n]}$ denote the nth prolongation of a group transformation g. By definition, if $y = f(x)$ is a curve with jet $\mathbf{z}^{[n]} = (x, f^{[n]}(x))$, then the

[†] To define a global action, we can either switch to more general parametrized curves or complete the space of derivatives (or "jet space"), in the same manner that projective space completes Euclidean space, by including elements corresponding to curves with vertical tangents, [168; §3.5].

transformed point $\bar{\mathbf{z}}^{[n]} = g^{[n]} \cdot \mathbf{z}^{[n]} = (\bar{x}, \bar{f}^{[n]}(\bar{x}))$ equals the jet of the transformed function $\bar{y} = \bar{f}(\bar{x})$.

As we showed earlier, the formulae for the prolonged group transformations are found by implicit differentiation. For simplicity we deal just with the case of curves in the plane, although the general ideas (suitably adapted) will work in higher dimensional situations. If the group acts on the independent variable x according to $\bar{x} = \xi(g, x, y)$, then we obtain the prolonged transformations by successively differentiating the corresponding dependent variable transformation formula $\bar{y} = \eta(g, x, y)$ using the implicit differentiation operator (the subscripts on ξ indicating partial derivatives)

$$\frac{d}{d\bar{x}} = \frac{1}{\xi_x(x, f(x)) + \xi_y(x, f(x))f'(x)}\frac{d}{dx} = \frac{1}{\xi_x(x, y) + v\,\xi_y(x, y)}\frac{d}{dx}.$$
(8.15)

For example, the first derivatives transform according to the (x, y)-dependent linear fractional transformations

$$\bar{v} = \frac{d\bar{y}}{d\bar{x}} = \frac{\eta_x + \eta_y f'(x)}{\xi_x + \xi_y f'(x)} = \frac{\eta_x + v\,\eta_y}{\xi_x + v\,\xi_y}.$$

The second order derivative formula is

$$\bar{w} = \frac{d\bar{v}}{d\bar{x}} = \frac{d^2\bar{y}}{d\bar{x}^2} = \frac{\begin{bmatrix} (\eta_{xx} + 2v\eta_{xy} + v^2\eta_{yy} + w\eta_y)(\xi_x + v\xi_y) - \\ -(\xi_{xx} + 2v\xi_{xy} + v^2\xi_{yy} + w\xi_y)(\eta_x + v\eta_y) \end{bmatrix}}{(\xi_x + v\xi_y)^3},$$

and so on. In particular, the third order Euclidean prolongation formulae are given by

$$\bar{x} = x\cos\theta - y\sin\theta + a, \qquad \bar{y} = x\sin\theta + y\cos\theta + b,$$

$$\bar{v} = \frac{d\bar{y}}{d\bar{x}} = \frac{\sin\theta + v\cos\theta}{\cos\theta - v\sin\theta}, \qquad \bar{w} = \frac{d^2\bar{y}}{d\bar{x}^2} = \frac{w}{(\cos\theta - v\sin\theta)^3}, \qquad (8.16)$$

$$\bar{z} = \frac{d^3\bar{y}}{d\bar{x}^3} = \frac{z(\cos\theta - v\sin\theta) + 3w^2\sin\theta}{(\cos\theta - v\sin\theta)^5}.$$

Exercise 8.37. Determine the fourth order prolongation of the proper Euclidean group SE(2). Do the same for the equi-affine group SA(2) acting on curves in the plane.

Differential Invariants

The invariants of prolonged transformation group actions play an important role in the geometry of submanifolds and, in fact, provide a complete solution to the associated equivalence problem. The original definition dates back to the work of Halphen, [**99**], and Lie, [**139**]; see [**169**] for a modern treatment of the subject.

Definition 8.38. A *differential invariant* is an invariant function for a prolonged transformation group.

The *order* of a differential invariant is that of the highest order derivative occurring therein. In particular, ordinary invariants are differential invariants of order 0. For example, a second order differential invariant for the planar Euclidean group will be a function $I(x, y, v, w)$ which is unchanged under the second prolonged action (8.8). In other words, if $y = f(x)$ is any analytic function, and $\bar{y} = \bar{f}(\bar{x})$ its image under a Euclidean transformation, then we require

$$I(\bar{x}, \bar{f}(\bar{x}), \bar{f}'(\bar{x}), \bar{f}''(\bar{x})) = I(x, f(x), f'(x), f''(x)). \qquad (8.17)$$

According to Example 8.27, the second order prolonged action admits a single fundamental differential invariant, namely

$$I = \frac{w}{(1 + v^2)^{3/2}} = \frac{f''(x)}{(1 + f'(x)^2)^{3/2}} = \kappa. \qquad (8.18)$$

We recognize this as the classical Euclidean curvature of the curve $y = f(x)$, whose invariance is well known. We conclude that every second order Euclidean differential invariant is a function of the curvature invariant. Furthermore, we may apply the normalizations given in Example 8.27 to the third order derivative coordinate $\bar{z}$ in (8.16), and so recover the fundamental third order differential invariant

$$J = \frac{(1 + v^2)z - 3vw^2}{(1 + v^2)^3} = \frac{(1 + f'(x)^2)f'''(x) - 3f'(x)f''(x)^2}{(1 + f'(x)^2)^3}. \qquad (8.19)$$

The geometric interpretation of J will appear shortly. Every third order differential invariant can be (locally) written as a function of I and J.

Exercise 8.39. Use the formulae found in Exercise 8.37 to determine a fourth order Euclidean differential invariant.

Exercise 8.40. More generally, we can consider curves given in parametrized form $x = g(t)$, $y = h(t)$. Determine the second prolonged action of the Euclidean group on parametrized curves. (*Hint*: Introduce four new coordinates representing the first and second order derivatives of the parametric functions.) Use the normalization procedure to determine differential invariants for curves under a fixed parametrization. The parametric curvature invariant is characterized by its additional invariance under all reparametrizations $t \mapsto \phi(t)$.

The formulae for the higher order prolonged group actions rapidly become rather complicated. Fortunately, it is not necessary to compute all higher order differential invariants in this explicit, computationally intensive manner. A more enlightened approach is to introduce certain invariant differential operators that generate higher order differential invariants from simpler, lower order invariants. These differential operators arise from certain geometrically significant invariant one-forms, of which the simplest is the classical Euclidean arc length element.

A pedestrian way of motivating one-forms is that they are the objects that one integrates when evaluating line integrals over a manifold. They are the simplest examples of differential forms, which play a crucial role in the geometry and topology of manifolds; we refer the reader to [**25**, **161**, **168**, **198**] for a more complete development.

In the simple case $X = \mathbb{R}$, a *one-form* is given by an expression $\omega = h(x)\,dx$, where $h(x)$ is an analytic function and dx denotes the usual infinitesimal element. Of particular importance are the "exact" one-forms given by the differentials $df(x) = f'(x)\,dx$ of analytic functions. Under an analytic map $\bar{x} = \xi(x)$, we have the standard covariant change of variables formula $d\bar{x} = \xi'(x)\,dx$. Therefore, the one-form $\bar{\omega} = \bar{h}(\bar{x})\,d\bar{x}$ will be transformed, or "pulled back", to the one-form $\omega = \bar{h}(\xi(x))\,\xi'(x)\,dx$ under the map ξ. In particular, if two functions correspond, $\bar{f}(\bar{x}) = f(x)$, then so do their differentials: $d\bar{f} = df$.

Definition 8.41. Let G act analytically on X. A one-form is called G–*invariant* if it is unchanged by all the group transformations.

More explicitly, $\omega = h(x)\,dx$ is G–invariant if and only if

$$h(x) = h(w(g,x)) \, \frac{\partial w}{\partial x} \, (g,x) \qquad (8.20)$$

for every group transformation $\bar{x} = g \cdot x = w(g,x)$. Equation (8.20) says that the coefficient $h(x)$ of a one-form gives a relative invariant for the Jacobian multiplier, as described in Exercise 4.39. In particular, the differential dI of any G–invariant function I is a G–invariant one-form.

For example, in the case of a Euclidean group transformation (8.13) restricted to a curve $y = f(x)$, we find

$$d\bar{x} = (\cos\theta - f'(x)\sin\theta)\,dx = (\cos\theta - v\sin\theta)\,dx, \qquad (8.21)$$

where we use the jet notation for derivatives introduced earlier. Equation (8.21) can be interpreted as saying that dx is a relative invariant for the (prolonged) Euclidean group with weight $\mu = \cos\theta - v\sin\theta$. On the other hand, a direct computation shows that the function $1 + v^2 = 1 + f'(x)^2$ transforms according to

$$1 + \bar{v}^2 = \frac{1 + v^2}{(\cos\theta - v\sin\theta)^2}, \qquad (8.22)$$

under the prolonged Euclidean group (8.8), and so $\sqrt{1 + v^2}$ forms a relative invariant of the reciprocal weight. Therefore, their product, which is the classical Euclidean arc length element

$$ds = \sqrt{1 + v^2}\, dx = \sqrt{1 + f'(x)^2}\, dx, \qquad (8.23)$$

is a Euclidean-invariant one-form. The invariant one-form (8.23) can, in fact, be obtained directly via our normalization procedure. We merely substitute the moving frame formulae (8.9) for the group parameter θ into (8.21). This procedure works in general, cf. [**70**].

Once we have determined an invariant one-form ds, Proposition 4.41 implies that every other invariant one-form is obtained by multiplying it by a differential invariant.

Proposition 8.42. *Let* $\omega = P\, dx$ *be a nonzero* G–*invariant one-form. A second one-form* $\tilde{\omega} = \tilde{P}\, dx$ *is* G–*invariant if and only if the ratio* $\tilde{\omega}/\omega = \tilde{P}/P$ *is a differential invariant.*

If $ds = P\, dx$ is any one-form and f a function, then we define the derivative $df/ds = (1/P)\, df/dx$ to be the ratio between the differential of f and the one-form ds. Since differentials of invariant functions are invariant one-forms, Proposition 8.42 has the following important immediate consequence.

Proposition 8.43. *If* $ds = P\, dx$ *is an invariant one-form for a prolonged transformation group, then the associated invariant differential operator*

$$\frac{d}{ds} = \frac{1}{P}\frac{d}{dx} \qquad (8.24)$$

maps differential invariants to differential invariants.

Consequently, differentiation with respect to arc length defines a Euclidean-invariant process: if I is any differential invariant, so is

$$\frac{dI}{ds} = \frac{1}{\sqrt{1 + f'(x)^2}}\frac{dI}{dx}.$$

For example, the derivative of curvature with respect to arc length,

$$\frac{d\kappa}{ds} = \frac{1}{\sqrt{1 + f'(x)^2}}\frac{d\kappa}{dx} = \frac{(1 + f'(x)^2)f'''(x) - 3f'(x)f''(x)^2}{(1 + f'(x)^2)^3}, \qquad (8.25)$$

is the same as the third order differential invariant (8.19) obtained by normalization.

Since the dimension of the nth order jet space is $n+2$, and the three-parameter Euclidean group SE(2) acts freely and regularly thereon for all $n \geq 1$, Theorem 8.17 implies that there are precisely $n - 1$ functionally independent differential invariants of order $\leq n$. However, $n - 2$ of them already appear as differential invariants of order $n - 1$ or less. Therefore, precisely one new n^{th} order differential invariant appears at each order $n \geq 2$. On the other hand, the kth derivative of the curvature invariant with respect to arc length is a differential invariant of order $k + 2$. We conclude that a complete system of differential invariants for the planar Euclidean group is provided by the curvature κ, and its successive derivatives with respect to arc length $\kappa_s, \kappa_{ss}, \ldots$.

Exercise 8.44. Prove that the fourth order differential invariant K that was obtained by normalizing the fourth order prolongation of SE(2) in Exercise 8.39 can be written in terms of the curvature invariants as $K = \kappa_{ss} + 3\kappa^3$. The explicit, general formulae relating the normalized higher order differential invariants and the differentiated lower order invariants can be found in [**70**].

The remarkable fact is that this general structure for the differential invariants of curves in the Euclidean plane extends, in a natural fashion, to the geometry of curves (scalar functions) under *almost all* continuous transformation groups acting on $\mathbb{R}^2$. There is a single (up to functions thereof) lowest order differential invariant, which we identify with the group-invariant curvature, and a single (up to constant multiples) invariant one-form of lowest order, known as the group-invariant arc length. Higher order differential invariants are obtained by differentiating the curvature with respect to the group-invariant arc length. The required conditions are embodied in the following definition.

Definition 8.45. An r-dimensional Lie group acting on $X \subset \mathbb{R}^2$ is said to be *ordinary* if its $(r - 2)^{\text{nd}}$ prolongation acts transitively on an open subset of the jet space J^{r-2}.

Note that the jet space of order $r - 2$ has dimension r, and so this is the maximal order at which the group can act transitively. Almost all transformation groups are ordinary; the exceptions are those that do not act transitively on X itself, and a handful that are said to "pseudo-stabilize", [**169**]. (Non-effective actions are also, technically speaking, not ordinary, but they can almost always be replaced by an ordinary, effectively acting version, as outlined in Proposition 3.40.) Differential invariants for non-ordinary groups are also well understood, but the results are slightly different. We note that Lie, [**138**], completely classified all possible finite-dimensional transformation groups on the plane, up to change of coordinates, and their associated differential invariants; see [**169**] for complete details.

Exercise 8.46. Prove that the elementary similarity groups

$$(x, y) \longmapsto (\lambda x + a, \lambda^k y + b)$$

are ordinary except when $k = 1$. Find the simplest differential invariant and an invariant differential operator in each case. Explain why the case $k = 1$ is exceptional.

The Fundamental Theorem governing differential invariants for ordinary planar transformation groups can now be stated. We refer the reader to [**169**] for the complete proof and generalizations to higher dimensions.

Theorem 8.47. *Let G be an ordinary r-dimensional transformation group acting on $\mathbb{R}^2$. Then G admits a unique (up to functions thereof) differential invariant κ, of order equal to $r - 1$, called the G-invariant curvature, and a unique (up to constant multiple) G-invariant one-form, $ds = P \, dx$, of order at most $r - 2$, called the G-invariant arc length element. Moreover, every other differential invariant of G is a function $I = I(\kappa, \kappa_s, \kappa_{ss}, \ldots)$ of the G-invariant curvature and its derivatives with respect to the G-invariant arc length.*

Remark: The curvature invariant κ is defined locally on an open subset $\mathcal{V}^{r-1} \subset J^{r-1}$ where G acts regularly. The Lie classification of planar transformation groups, [**138, 169**], demonstrates that, at least in the neighborhood of a point (x_0, y_0), κ can be chosen to be globally defined on the entire regular subset $\mathcal{V}^{r-1} \subset J^{r-1}$ consisting of *regular jets*, meaning those whose prolonged group orbit has dimension equal to that of G. Moreover, the level sets of κ can be chosen to coincide with the prolonged orbits, and so two regular jets $\mathbf{z}^{[r-1]}, \overline{\mathbf{z}}^{[r-1]} \in \mathcal{V}^{r-1}$ lie in the same orbit if and only if $\kappa(\mathbf{z}^{[r-1]}) = \kappa(\overline{\mathbf{z}}^{[r-1]})$. We shall avoid the more pathological singular subset where the orbits have less than maximal dimension; see [**70, 170**] for a discussion of what can be done in such cases.

Exercise 8.48. Use the results of Exercise 8.37 to prove that the fundamental differential invariant of the five-parameter equi-affine group

SA(2) acting on curves $y = f(x)$ in the plane is

$$I = \frac{3f''f'''' - 5(f''')^2}{9(f'')^{8/3}}, \qquad (8.26)$$

known as the (equi-)affine curvature. Show that the equi-affine arc length element is $\sqrt[3]{f''(x)}\, dx$, and hence all higher order equi-affine differential invariants are obtained by differentiating the curvature with respect to arc length. These quantities form the foundation of the equi-affine geometry of planar curves, cf. [**95, 30**].

Differential Invariants for Binary Forms

Let us now apply these constructions to study the fundamental GL(2) multiplier representation $\rho_{n,0}$, (4.19), that underlies the invariant theory of binary forms of degree n and weight 0. (The restriction to weight 0 is not essential but does keep the subsequent formulae simpler.) Our goal is to obtain explicit formulae for the fundamental differential invariants as certain classical absolute rational covariants, as well as the invariant differentiation process, which is closely related to the first transvection operation. Subsequently, we shall apply these results to the determination of symmetries and resolution of the fundamental equivalence problem for binary forms.

We shall retain our original notation p instead of x to denote the projective coordinate and use $u = Q(p)$ as the dependent variable. Thus, we consider the planar action of GL(2) given by

$$\bar{p} = \frac{\alpha p + \beta}{\gamma p + \delta}, \qquad \bar{u} = (\gamma p + \delta)^{-n} u = \sigma^{-n} u, \qquad (8.27)$$

where we abbreviate

$$\sigma = \gamma p + \delta, \qquad \Delta = \alpha\delta - \beta\gamma \neq 0.$$

Since

$$d\bar{p} = \frac{\alpha\delta - \beta\gamma}{(\gamma p + \delta)^2}\, dp = \frac{\Delta}{\sigma^2}\, dp, \qquad (8.28)$$

we can recursively apply the implicit differentiation operator

$$\frac{d}{d\bar{p}} = \frac{\sigma^2}{\Delta}\frac{d}{dp} = \frac{(\gamma p + \delta)^2}{\alpha\delta - \beta\gamma}\frac{d}{dp} \qquad (8.29)$$

to $\bar{u}$, as in (8.27), to determine the prolonged transformation rules for functions $u = Q(p)$. According to Theorem 8.47, since GL(2) has dimension 4, we must prolong to third order to recover the fundamental differential invariant. Let us use the abbreviations

$$u = Q(p), \qquad v = Q'(p), \qquad w = Q''(p),$$
$$z = Q'''(p), \qquad y = Q''''(p), \tag{8.30}$$

for the derivative coordinates in the third order jet space $\mathbf{J}^3$. We find

$$\bar{p} = \frac{\alpha p + \beta}{\gamma p + \delta}, \qquad\qquad \bar{u} = \frac{u}{\sigma^n}, \tag{8.31}$$

$$\bar{v} = \frac{\sigma v - n\gamma u}{\Delta \sigma^{n-1}}, \qquad\qquad \bar{w} = \frac{\sigma^2 w - 2(n-1)\gamma\sigma v + n(n-1)\gamma^2 u}{\Delta^2 \sigma^{n-2}},$$

$$\bar{z} = \frac{\sigma^3 z - 3(n-2)\gamma\sigma^2 w + 3(n-1)(n-2)\gamma^2\sigma v - n(n-1)(n-2)\gamma^3 u}{\Delta^3 \sigma^{n-3}},$$

$$\bar{y} = \frac{\left(\begin{array}{c} \sigma^4 y - 4(n-3)\gamma\sigma^3 z + 6(n-2)(n-3)\gamma^2\sigma^2 w - \\ -4(n-1)(n-2)(n-3)\gamma^3\sigma v + n(n-1)(n-2)(n-3)\gamma^4 u \end{array}\right)}{\Delta^4 \sigma^{n-4}}.$$

The general formula is readily established by induction:

$$\bar{Q}^{(m)} = \frac{1}{\Delta^m \sigma^{n-m}} \sum_{j=0}^{m} (-1)^{m-j} \binom{m}{j} \frac{(n-j)!}{(m-j)!} \gamma^{m-j} \sigma^j Q^{(j)}. \tag{8.32}$$

Let us now apply our normalization method. We leave aside the exceptional cases $n = 0$, which is not ordinary since it is not transitive on $\mathbb{R}^2$, and $n = 1$, for reasons to appear shortly. The p-axis $\{u = 0\}$ consists of singular orbits, and so we restrict our attention to the upper half plane[†] $X = \{u > 0\}$, corresponding to positive $Q(p) > 0$.

Remark: If n is even, then the upper and lower half planes are distinct orbits, whereas for n odd, the reflection $u \mapsto -u$, which takes one to the other, can be obtained as a group transformation. Thus, for $Q(p) < 0$, if $n = \deg Q$ is odd, we can use a group element to switch to the positive case, whereas if n is even, a second "branch" of negative polynomials must be analyzed separately. Throughout the discussion, we will avoid roots of Q; in particular, we are assuming $Q \not\equiv 0$ is nontrivial.

We begin by normalizing $\bar{p} = 0$, which gives

$$\beta = -\alpha p, \qquad \text{whereby} \qquad \Delta = \alpha\sigma. \tag{8.33}$$

The second normalization $\bar{u} = 1$ results in

$$\delta = \sqrt[n]{u} - \gamma p, \qquad \text{whereby} \qquad \sigma = \gamma p + \delta = \sqrt[n]{u}. \tag{8.34}$$

[†] In the complex case, there are no sign restrictions since $\mathbb{C} \setminus \{0\}$ remains connected.

Here we choose the positive branch of the nth root. Next we normalize $\bar{v} = 0$; this leads to

$$\gamma = \frac{1}{n}\, u^{(1-n)/n}\, v. \tag{8.35}$$

Substituting (8.33), (8.34), (8.35) into the formula for $\bar{w}$, we find that

$$\bar{w} = \frac{u^{2(1-n)/n}}{\alpha^2}\left(uw - \frac{n-1}{n}v^2\right) \tag{8.36}$$

$$= \frac{Q^{2(1-n)/n}}{\alpha^2}\left(QQ'' - \frac{n-1}{n}(Q')^2\right) = \frac{Q^{2(1-n)/n}}{\alpha^2}\frac{H}{n(n-1)},$$

where, according to (5.12), $H = \frac{1}{2}(Q, Q)^{(2)}$ is the classical Hessian of Q. (This identification breaks down when $n = 1$, although the actual moving frame construction continues to be valid, as the reader can check.) The points where the Hessian vanishes, $H = 0$, are singularities of the second prolonged action. We exclude the completely singular case when $H \equiv 0$, which, according to Proposition 2.23, requires that $Q \neq (ap+b)^n$ is not the nth power of a linear form. For real functions, the computation again splits into two branches, depending on the sign of the Hessian H. To be specific, let us assume that $H > 0$ on the domain of interest. We can then normalize $\bar{w} = 1/[n(n-1)]$ by setting $\alpha = Q^{(1-n)/n}\sqrt{H}$. Combining this with the previous normalizations leads to the basic moving frame formulae for classical invariant theory:

$$\alpha = Q^{(1-n)/n}\sqrt{H}, \qquad \beta = -p\,Q^{(1-n)/n}\sqrt{H},$$
$$\gamma = \frac{1}{n}Q^{(1-n)/n}, \qquad \delta = Q^{1/n} - \frac{1}{n}p\,Q^{(1-n)/n}. \tag{8.37}$$

Substituting the normalizations (8.37) into the higher order transformation rules gives us the differential invariants. The fundamental "curvature" differential invariant is obtained from $\bar{z}$ in (8.31); after multiplying by $n^2(n-1)$, we find

$$J = -\frac{n^2(n-1)}{H^{3/2}}\left[Q^2Q''' - \frac{3(n-2)}{n}QQ'Q'' + \frac{2(n-1)(n-2)}{n^2}(Q')^3\right]$$

$$= \frac{T}{H^{3/2}}, \tag{8.38}$$

where, referring back to (5.14), $T = (Q, H)^{(1)}$ is a familiar covariant!

The quantity J is a differential invariant for the orientation-preserving subgroup $\mathrm{GL}(2,\mathbb{R})^+$. However, for the full group $\mathrm{GL}(2,\mathbb{R})$, a matrix with negative determinant will reverse the sign of T, and so J is only a relative differential invariant whose weight is given by the sign multiplier

$\mu_s(A) = \text{sign}(\det A)$. (This is due to our restricting the moving frame (8.37) to matrices with positive determinant.) The sign ambiguity can be easily remedied by adopting its square

$$J^2 = \frac{T^2}{H^3}, \qquad \text{where} \qquad \begin{aligned} H &= \tfrac{1}{2}(Q, Q)^{(2)}, \\ T &= (Q, H)^{(1)}, \end{aligned} \qquad (8.39)$$

as the fundamental differential invariant. In the complex-valued case, although we did not need to fix the sign of Q or H, there is still an unspecified sign in J due to the ambiguity in the branching of the square root of H. In this case too, the ambiguity can be resolved by using (8.39) as the fundamental invariant.

As we have seen, higher order differential invariants can be determined in two different ways: direct substitution or invariant differentiation. The methods do not lead to the same higher order differential invariants, although according to Theorem 8.47, one can re-express one set of differential invariants in terms of the other. A fundamental fourth order differential invariant is given by substituting the moving frame formulae (8.37) into the final prolongation formula in (8.31); we compute

$$\bar{y} = \frac{1}{n^3(n-1)} K + \frac{3(n-2)}{n^3(n-1)},$$

where, referring to (5.15),

$$K = \frac{U}{H^2}, \qquad \text{where} \qquad U = (Q, T)^{(1)}. \qquad (8.40)$$

Therefore, we can take the fundamental fourth order differential invariant to be the absolute rational covariant K. Alternatively, we can construct the invariant differential operator by normalizing the one-form

$$d\bar{p} = \Delta \, \sigma^{-2} \, dx,$$

which leads to the invariant "arc length element" and associated invariant differential operator

$$ds = \frac{\sqrt{H}}{Q} \, dp, \qquad \text{and} \qquad \frac{d}{ds} = \frac{Q}{\sqrt{H}} \frac{d}{dp}. \qquad (8.41)$$

Higher order differential invariants are then obtained by invariantly differentiating the fundamental curvature invariant (8.38) with respect to the arc length form. In particular, since dJ/ds has fourth order, we can express it in terms of the normalized differential invariants K and J; the actual formula is

$$\frac{dJ}{ds} = \frac{U}{nH^2} - \frac{3T^2}{2nH^3} = \frac{K - \tfrac{3}{2}J^2}{n}. \qquad (8.42)$$

Although the original differential invariant J had a sign ambiguity, its derivative (8.42) does not, and so is a differential invariant under the full group GL(2).

One can prove (8.42) by direct computation, or using the algorithm described in [**70**]; an alternative approach is the following. Let $n > 2$. If Z is a polynomial covariant of weight m, then $N = H^{-m/(2n-4)}Z$ is an absolute rational covariant. Using (5.11), we find

$$
\frac{dN}{ds} = \frac{Q}{\sqrt{H}} \frac{d}{dp}\left(\frac{Z}{H^{m/(2n-4)}}\right) = \frac{Q\left[(2n-4)HZ' - mZH'\right]}{(2n-4)H^{(m+3n-6)/(2n-4)}}
$$
$$
= \frac{Q\,(H,\,Z)^{(1)}}{(2n-4)H^{(m+3n-6)/(2n-4)}}\,,
\tag{8.43}
$$

which is a multiple of the product of the absolute covariants corresponding to Q and to $(H,\,Z)^{(1)}$. On the other hand, using the equation for

$$
T = (Q,\,H)^{(1)} = (2n-4)Q'H - nQH',
$$

we find

$$
Q\,(H,\,Z)^{(1)} = m\,QZH' - (2n-4)\,QHZ'
$$
$$
= \frac{m(2n-4)}{n}\,ZQ'H - (2n-4)\,QZ'H - \frac{m}{n}\,Z\,T
\tag{8.44}
$$
$$
= \frac{2n-4}{n}\,H\cdot(Q,\,Z)^{(1)} - \frac{m}{n}\,Z\,T.
$$

Therefore, substituting into (8.43),

$$
\frac{dN}{ds} = \frac{(Q,\,Z)^{(1)}}{nH^{(m+n-2)/(2n-4)}} + \frac{m}{n(2n-4)}\,JN,
\tag{8.45}
$$

where J is given in (8.38). The second summand is a lower order differential invariant. We have therefore proved the following classification.

Theorem 8.49. *A complete system of differential invariants for the multiplier representation $\rho_{n,0}$ is provided by $J_m = H^{-m/2}Z_m$, for $m = 3, 4, \ldots$. The covariants $Z_2 = H$, $Z_3 = T = (Q, H)^{(1)}$, $\ldots$, $Z_{m+1} = (Q, Z_m)^{(1)}$, are obtained by successive transvection of Q with its Hessian $H = \frac{1}{2}(Q, Q)^{(2)}$.*

Exercise 8.50. Use the graph theoretic methods of Chapter 7 to verify the syzygy (8.44).

Equivalence and Signature Curves

Recall that two subsets $S, \bar{S} \subset X$ of a manifold are said to be equivalent under a transformation group G acting on X if there is a group transformation $g \in G$ mapping one to the other: $\overline{S} = g \cdot S$. Theorem 8.17 provides a solution to the equivalence problem for points on X. The equivalence of (finite) collections of points are similarly treated using the joint invariants corresponding to a Cartesian product action of G on $X \times \cdots \times X$. The equivalence problem for submanifolds $S \subset X$ is the primary focus of Cartan's theory of moving frames and is based on the appropriate differential invariants. We shall describe only the simplest case of planar curves, although general results of Cartan, cf. [33, 70], have far wider applicability.

Definition 8.51. Let G be a planar transformation group acting on $X \subset \mathbb{R}^2$. Two planar curves $\mathcal{C}$ and $\overline{\mathcal{C}}$ in X are *equivalent* under G if there is a group transformation $g \in G$ such that $\overline{\mathcal{C}} = g \cdot \mathcal{C}$.

For example, if $G = \mathrm{SE}(2)$ is the Euclidean group, then Euclidean equivalent curves can be obtained from each other by a combination of translations and rotations. Similarly, if $G = \mathrm{GL}(2)$ acts via the multiplier representation (8.27), then two curves given by the graphs of functions $u = Q(p)$ and $\bar{u} = \overline{Q}(\bar{p})$ are equivalent if and only if the functions can be mapped to each other via a transformation of the form (8.27).

The solution to the general equivalence problem for curves under a planar transformation group relies on the functional relationships between their differential invariants. Assuming G is an ordinary transformation group, usually we need only the two lowest order differential invariants: namely, the group-invariant curvature κ and its derivative κ_s with respect to the group-invariant arc length element.

Definition 8.52. A plane curve $\mathcal{C} \subset \mathbb{R}^2$ is *G–regular* if the differential invariants κ, κ_s are defined and analytic on $\mathcal{C}$. The *G–invariant signature set* associated with a regular planar curve is the set $\mathcal{S} \subset \mathbb{R}^2$ parametrized by the curvature κ and its first derivative with respect to the arc length: $\mathcal{S} = \{ (\kappa(\mathbf{x}), \kappa_s(\mathbf{x})) \mid \mathbf{x} \in \mathcal{C} \}$. The curve $\mathcal{C}$ is *nonsingular* if its signature set $\mathcal{S}$ is a nondegenerate curve, called the *signature curve*.

In other words, if we parametrize $\mathcal{C}$ by $\mathbf{x}(t)$, then nonsingularity requires that the tangent $d(\kappa, \kappa_s)/dt$ to the signature curve be nowhere zero, and so, at least locally, the signature curve is a regular plane curve. In particular, the signature set is not allowed to degenerate to a point. (This important case will be discussed shortly.) However, it may (and often does) have self-intersections. Since derivatives with respect to arc length are scalar multiples of t derivatives, nonsingularity is guaranteed by requiring that the first and second derivatives of curvature with respect to arc length do not simultaneously vanish: $(\kappa_s, \kappa_{ss}) \neq 0$.

The importance of the signature curve lies in the fact that it characterizes the original curve up to a group transformation. This fact has recently been applied to the basic computer vision problem of object recognition, [**29**].

Theorem 8.53. *Let G be an ordinary transformation group acting on $\mathbb{R}^2$. Two nonsingular analytic curves $\mathcal{C}$ and $\overline{\mathcal{C}}$ are equivalent under G, so $\overline{\mathcal{C}} = g \cdot \mathcal{C}$ for some $g \in G$, if and only if their signature curves are identical: $\overline{\mathcal{S}} = \mathcal{S}$.*

Proof: First note that since κ and κ_s are differential invariants, their values are identical for any two curves related by a group transformation. Therefore, equivalent curves automatically have identical signature curves. To prove the converse, recall that $\kappa(x, y^{[r-1]})$ depends on $(r-1)^{\text{st}}$ order derivatives of $y = f(x)$, and hence its derivative $\kappa_s(x, y^{[r]})$ has order $r = \dim G$. Let us work locally, and assume that the common signature curve is given by specifying κ_s as a function of κ, so that

$$\kappa_s\left(x, y, \frac{dy}{dx}, \ldots, \frac{d^r y}{dx^r}\right) = H\left(\kappa\left(x, y, \frac{dy}{dx}, \ldots, \frac{d^{r-1} y}{dx^{r-1}}\right)\right). \quad (8.46)$$

(This will be the case provided $\kappa_s \neq 0$, which, by nonsingularity, holds almost everywhere on the curve.) Therefore, any curve $y = f(x)$ having the given signature curve will be a solution to the rth order ordinary differential equation (8.46). The curve can be uniquely recovered from its initial conditions

$$f(x_0) = y_0, \ f'(x_0) = y_1, \ \ldots \ f^{(r-2)}(x_0) = y_{r-2}, \ f^{(r-1)}(x_0) = y_{r-1}. \quad (8.47)$$

Now, suppose we are given two regular curves with the same signature curve. Let $\mathbf{x}_0 \in \mathcal{C}$ and $\overline{\mathbf{x}}_0 \in \overline{\mathcal{C}}$ be two points mapping to the *same* point on the common signature curve; in particular

$$\kappa(x_0, f(x_0), \ldots, f^{(r-1)}(x_0)) = \kappa(\bar{x}_0, \bar{f}(\bar{x}_0), \ldots, \bar{f}^{(r-1)}(\bar{x}_0)).$$

Let

$$\mathbf{z}^{[r-2]} = (x_0, y_0, y_1, \ldots, y_{r-2}) = (x_0, f(x_0), f'(x_0), \ldots, f^{(r-2)}(x_0)),$$
$$\overline{\mathbf{z}}^{[r-2]} = (\bar{x}_0, \bar{y}_0, \bar{y}_1, \ldots, \bar{y}_{r-2}) = (\bar{x}_0, \bar{f}(\bar{x}_0), \ldots, \bar{f}^{(r-2)}(\bar{x}_0)),$$

be the corresponding points in J^{r-2}. Since G is ordinary, it acts transitively on the open subset of the $(r-2)^{\text{nd}}$ order jet space defined by the regular curves. Therefore, we can find a group transformation $g \in G$ whose order $r - 2$ prolongation maps the point $\mathbf{z}^{[r-2]}$ to $\overline{\mathbf{z}}^{[r-2]} = g^{[r-2]} \cdot \mathbf{z}^{[r-2]}$. Let $\widetilde{C} = g \cdot C$. Then, at the point $\widetilde{\mathbf{x}}_0 = g \cdot \mathbf{x}_0$, the derivatives of $\widetilde{C}$ agree with those of $\overline{C}$, that is,

$$\widetilde{\mathbf{z}}^{[r-2]} = \overline{\mathbf{z}}^{[r-2]} = g^{[r-2]} \cdot \mathbf{z}^{[r-2]}. \tag{8.48}$$

Moreover, since κ is an invariant, its value is unaffected by the group transformation g, so

$$\kappa(x_0, y_0, \ldots, y_{r-1}) = \kappa(\widetilde{x}_0, \widetilde{y}_0, \ldots, \widetilde{y}_{r-1}) = \kappa(\overline{x}_0, \overline{y}_0, \ldots, \overline{y}_{r-1}). \tag{8.49}$$

Equations (8.48), (8.49) imply that the order $r - 1$ derivatives for $\widetilde{C}$ and $\overline{C}$ also coincide: $\widetilde{y}_{r-1} = \overline{y}_{r-1}$. Therefore, both $\widetilde{C}$ and $\overline{C}$ satisfy the same initial value problem (8.46), (8.47). By the basic uniqueness theorem for solutions to ordinary differential equations, $\widetilde{C} = \overline{C}$ coincide locally. Analytic continuation implies that they agree globally.   *Q.E.D.*

Example 8.54. Let $G = \mathrm{SE}(2)$ act on $\mathbb{R}^2$. Consider an ellipse

$$\frac{x^2}{a^2} + \frac{y^2}{b^2} = 1, \qquad \text{or} \qquad y = \pm\sqrt{b^2 - r^2 x^2}, \qquad r = \frac{b}{a}. \tag{8.50}$$

Substituting into (8.18), (8.25), we find, after some algebra,

$$\kappa = \frac{-ab}{(a^2 + (r^2 - 1)x^2)^{3/2}}, \qquad \frac{d\kappa}{ds} = \frac{3(r^2 - 1)xy}{r(a^2 + (r^2 - 1)x^2)^3}. \tag{8.51}$$

In particular, if the ellipse is circular, then $a = \pm b$, and hence $\kappa = -1/a$ is constant.[†]

The Euclidean signature curve $\mathcal{S}$ for the ellipse, then, is parametrized by (8.51). Eliminating x, we find the elegant explicit formula

$$\left(\frac{d\kappa}{ds}\right)^2 = -9\kappa^{8/3}\left(\kappa^{2/3} - \frac{a^{2/3}}{b^{4/3}}\right)\left(\kappa^{2/3} - \frac{b^{2/3}}{a^{4/3}}\right). \tag{8.52}$$

The sign of κ_s is uniquely determined by that of the function xy. The signature curve (8.52) is graphed in Figure 6. According to Theorem 8.53, the only curves with signature curve given by (8.52) are the ellipses obtained from (8.50) by a Euclidean transformation. In the circular case, the signature curve $\mathcal{S} = \{(-1/a, 0)\}$ degenerates to a point.

[†] Our choice of sign gives the circle a negative curvature. But since κ is a signed curvature, the negative value is indicative of the fact that the upper semi-circle is parametrized by x in the *clockwise* direction in (8.50).

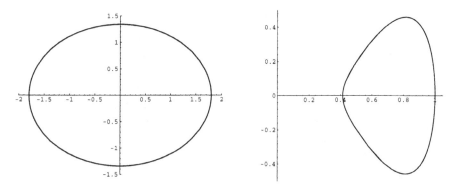

Figure 6. Ellipse and Its Signature Curve.

Remark: The more traditional approach is to characterize a curve up to Euclidean motion by its curvature as a function of arc length: $\kappa = \kappa(s)$. A classical result, cf. [**95**], states that two plane curves are equivalent if and only if these functions are the same, modulo a phase shift: $\bar{\kappa}(\bar{s}) = \kappa(s)$ where $\bar{s} = s + k$ for some constant k. The signature curve approach has the advantage of (a) entirely relying on local information, since the arc length itself must be determined by integration, and (b) eliminating the final translational ambiguity. In practical computations, the main disadvantage is the introduction of an extra derivative via κ_s. See [**29**] for invariant numerical algorithms that can be used to approximate the signature curve.

Symmetries of Curves

Recall that the *symmetry group* of a curve $\mathcal{C} \subset \mathbb{R}^2$ is the subgroup consisting of all group elements $g \in G$ that map $\mathcal{C}$ to itself: $g \cdot \mathcal{C} = \mathcal{C}$. Symmetries are merely self-equivalences and hence are also embodied in the structure of the signature curve, as we describe now.

Consider first the case when the curve $\mathcal{C}$ is nonsingular, so its signature set $\mathcal{S}$ is a nondegenerate curve. The inverse image of a point on $\mathcal{S}$ will then consist of a discrete number of points in the original curve $\mathcal{C}$. Let us define the *index* of the curve to be the minimal such number. Under our nondegeneracy assumption, it can be proved that, generically, the inverse image of any point in $\mathcal{S}$ has cardinality equal to the curve's index; indeed, the only exceptions are points of self-intersection of the signature curve. The following result is an immediate consequence of our basic equivalence Theorem 8.53.

Theorem 8.55. *If a curve is nonsingular, then it admits a discrete symmetry group, whose cardinality is the index of the curve.*

Example 8.56. In view of Example 8.54, any noncircular ellipse forms a nondegenerate curve under the Euclidean group. As we move once around the ellipse, its signature curve $\mathcal{S}$, as given in (8.52), is traced twice, and hence the ellipse has index 2. This mirrors the fact that an ellipse admits a single proper Euclidean symmetry — rotation through 180°. There are, of course, reflectional symmetries too; since the invariant $\kappa_s \mapsto -\kappa_s$ changes sign under Euclidean reflections, they induce reflectional symmetries (about the κ axis) of the signature curve. Alternatively, one can use the square κ_s^2 as the full Euclidean differential invariant, and then the revised (or folded) Euclidean signature curve parametrized by (κ, κ_s^2) is traced 4 times, indicating the fact that an ellipse has a four-fold discrete Euclidean symmetry group.

Of particular importance are degenerate curves whose G–invariant curvature is constant; this implies that the associated signature set degenerates to a single point. These are the curves of maximal symmetry and thereby play a distinguished role. We shall consider only those that are *regular* in the sense that their $(r-1)^{\text{st}}$ order jet lies in the regular subset $\mathcal{V}^{r-1} \subset \mathrm{J}^{r-1}$, as in the remark following Theorem 8.47, and hence they admit a well-defined curvature invariant, whose level sets are the prolonged group orbits in $\mathcal{V}^{r-1}$.

Theorem 8.57. *Let G be an ordinary transformation group acting on $X \subset \mathbb{R}^2$. Let $\mathcal{C} \subset X$ be an analytic curve with G–invariant signature set $\mathcal{S}$. Then the following conditions are equivalent:*
(i) $\mathcal{C}$ has constant G–invariant curvature κ.
(ii) $\mathcal{S}$ degenerates to a point.
(iii) $\mathcal{C}$ is the orbit of a one-parameter subgroup of G.
(iv) $\mathcal{C}$ admits a one-parameter symmetry group.

Proof: The equivalence of (i) and (ii) is clear, as is that of (iii) and (iv). Suppose first that the curve $\mathcal{C} = H \cdot \mathbf{x}_0$ is the orbit of a one-parameter subgroup $H \subset G$, which is (or at least is contained in) the curve's symmetry group. Note that H acts transitively on its orbit $\mathcal{C}$. But the curvature is a differential invariant and hence unchanged by the transformations in H, which immediately implies that κ must remain constant on the curve. Conversely, any curve having constant curvature $\kappa = c$ will be recovered as a solution to the order $r - 1$ ordinary differential equation $\kappa(x, y^{[r-1]}) = c$, uniquely determined by its initial data $\mathbf{z}_0^{[r-2]} = (x_0, f(x_0), \ldots, f^{(r-2)}(x_0))$. By transitivity of the prolonged action on J^{r-2}, for any two such jets $\mathbf{z}_0^{[r-2]}, \mathbf{z}^{[r-2]}$ corresponding to two points $\mathbf{z}_0, \mathbf{z} \in \mathcal{C}$ there is a group transformation $g \in G$ mapping $\mathbf{z}_0^{[r-2]}$ to $\mathbf{z}^{[r-2]}$. Using the uniqueness of solutions to the initial value problem, we conclude that g is a symmetry of the curve. Therefore, the curve admits at least a one-parameter family of symmetries — parametrized by points on the curve. In fact, with a little more work, the symmetries so constructed can be shown to form a one-parameter group.   *Q.E.D.*

Remark: Assuming that the curvature κ distinguishes regular orbits in J^{r-1}, two maximally symmetric curves are equivalent if and only if they have the same (constant) curvature.

In Euclidean geometry, the curves having constant Euclidean curvature are the circles and straight lines. Each is the orbit of a particular one-parameter subgroup of SE(2), which also forms the symmetry group of the curve. Circles of the same radius have the same curvature, and clearly only these are equivalent under a Euclidean transformation.

Exercise 8.58. The maximally symmetric curves for the equi-affine group are those whose equi-affine curvature (8.26) is constant. Prove that these are the conic sections. Produce the fundamental differential equation for conic sections, meaning an ordinary differential equation whose solutions are all conic sections in the plane. (*Hint*: Differentiate the equation $\kappa = $ const.) Further, prove that two ellipses are equi-affine equivalent if and only if they enclose the same area.

Remark: Another interesting example is the action of SL(3, $\mathbb{R}$) on the projective plane $\mathbb{RP}^2$. See [**33, 69**] for the explicit formula for the projective curvature and arc length. The maximally symmetric projective curves are the so-called "W–curves" studied by Lie and Klein, [**129, 141**; pp. 68–82]. For the one-parameter subgroup e^{tJ} given by a matrix exponential, the corresponding W-curve is obtained by projecting a solution $x(t) = e^{tJ}x_0$, where $x_0 \in \mathbb{R}^3$, down to $\mathbb{RP}^2$.

Remark: In higher dimensions, there are corresponding, but deeper, results due to Cartan that characterize the submanifolds of a homogeneous space up to group transformations via the associated signature set parametrized by the fundamental differential invariants. For example, for a nondegenerate surface in three-dimensional space, the Euclidean signature set is a surface in a six-dimensional space parametrized by the mean curvature, the Gaussian curvature, and their derivatives with respect to the Euclidean-invariant Frenet frame on the surface. The maximally symmetric submanifolds are those all of whose differential invariants are constant, and so the signature set degenerates to a single point. They are realized as the orbits of higher dimensional subgroups of G. See [**33**, **93**, **95**, **120**, **70**], for precise statements and a variety of geometric examples.

Equivalence and Symmetry of Binary Forms

Let us now apply these results to binary forms. Recall that two binary forms $Q(p)$ and $\overline{Q}(\bar{p})$ are equivalent under the multiplier representation $\rho_{n,0}$ if and only if

$$\overline{Q}(\bar{p})(\gamma p + \delta)^n = Q(p), \qquad (8.53)$$

for some projective transformation $\bar{p} = (\alpha p + \beta)/(\gamma p + \delta)$. If we regard the graph of $Q(p)$ as defining a curve $\mathcal{C} = \mathcal{C}_Q$ in the plane, condition (8.53) means that the corresponding curves are equivalent under a group transformation (8.27). In this manner, the equivalence problem for binary forms can be regarded as a very particular case of our general treatment of equivalence and symmetry of planar curves under a Lie transformation group.

The solution to the equivalence problem is effected by analyzing the associated signature curve, which is parametrized by the absolute rational covariants J^2 and dJ/ds, as given in (8.39), (8.42). We assume that Q is nondegenerate, meaning that its Hessian H does not vanish identically, and so Q is *not* the nth power of a linear form. In view of (8.42), we will find it more convenient to use the simpler rational covariant K, cf. (8.40), in place of the differentiated invariant dJ/ds.

Definition 8.59. The *signature set* $\mathcal{S} = \mathcal{S}_Q \subset \mathbb{C}^2$ of a nondegenerate complex-valued binary form $Q(p)$ is parametrized by the two fundamental absolute rational covariants,

$$\mathcal{S}_Q = \left\{ (J(p)^2, K(p)) = \left(\frac{T(p)^2}{H(p)^3}, \frac{U(p)}{H(p)^2} \right) \,\middle|\, H(p) \neq 0 \right\}. \qquad (8.54)$$

The binary form is *nonsingular* at a point p provided $H(p) \neq 0$ and $(J'(p), K'(p)) \neq 0$, and so $\mathcal{S}_Q$ is (locally) a nondegenerate curve.

A binary form is singular at all points if and only if it is maximally symmetric; such forms will be classified later in this chapter. Otherwise, by analyticity, $Q(p)$ will be nonsingular at all but a finite number of values of p.

Exercise 8.60. Reformulate the singularity condition in terms of the vanishing of certain covariants of the form. An interesting open problem is to characterize these singular values geometrically.

A direct application of our general equivalence Theorem 8.53 produces the following Fundamental Equivalence Theorem for binary forms.

Theorem 8.61. *Two nondegenerate binary forms* $Q(p)$ *and* $\overline{Q}(\bar{p})$ *of degree* $n \geq 2$ *are equivalent under the multiplier representation of weight* n*, as in* (8.53)*, if and only if their signature curves are identical:* $\mathcal{S}_Q = \mathcal{S}_{\overline{Q}}.$

Consequently, and rather surprisingly, a complete solution to the equivalence problem for complex binary forms depends on merely two absolute rational covariants — J and K.

In the real case, the signs of the Hessian and, possibly, of the form itself come into play. However, when $T \neq 0$, the sign of H equals the sign of the invariant $J^2 = T^2/H^3$ and hence can be directly determined from the signature curve. Therefore, Theorem 8.61 holds as stated for real forms of odd degree. If the degree $n = \deg Q$ is even, then there are, in fact, two distinct signature curves,

$$
\begin{aligned}
\mathcal{S}_Q^+ &= \{(J(p), K(p)) \mid Q(p) > 0\}, \\
\mathcal{S}_Q^- &= \{(J(p), K(p)) \mid Q(p) < 0\},
\end{aligned}
\tag{8.55}
$$

indexed by the sign of the form. (If Q is of one sign, then one of these will be empty.) In this case, both signature curves must agree, $\mathcal{S}_Q^+ = \mathcal{S}_{\overline{Q}}^+$ and $\mathcal{S}_Q^- = \mathcal{S}_{\overline{Q}}^-$, in order that the forms be equivalent.

Remark: Theorem 8.61 also applies as stated to homogeneous functions of any degree $n \neq 0, 1$. The case $n = 1$ can be handled by the same moving frame computation, but, as remarked above, one must use an alternative to the Hessian covariant (which vanishes identically when the degree is 1) to construct the required differential invariants; see (8.36). The case $n = 0$ needs to be handled slightly differently owing to the intransitivity of the action (8.27) on the (p, u)–plane.

Remark: Theorem 8.61 was first proved in [**167**] by an alternative method based on the solution to an equivalence problem arising in the calculus of variations. The direct approach provided by the method of moving frames is based on general results in [70]. Surprisingly, there is *no* classical counterpart to this result, although Clebsch, [49; pp. 365, 421] and Hilbert, [106; §10], do discuss the equivalence of binary forms in some depth.

If Q and $\overline{Q}$ are nonsingular and have identical signature curves, then one can explicitly determine all the transformations mapping Q to $\overline{Q}$ by solving the two rational equations

$$J(p)^2 = \overline{J}(\bar{p})^2, \qquad K(p) = \overline{K}(\bar{p}). \tag{8.56}$$

The second of these two equations merely serves to delineate the appropriate branch of the signature curve. This is important, since the first equation, which takes the form $T(p)^2\,\overline{H}(\bar{p})^3 = \overline{T}(\bar{p})^2\,H(p)^3$, is, in general, a polynomial equation of degree $6n - 12$ for p in terms of $\bar{p}$; as such, many of its roots will often be spurious. At a generic point p — meaning at nonsingular points where the common signature curve does not cross itself — each solution $\bar{p} = \varphi(p)$ to (8.56) will define an equivalence between the two binary forms; in particular, the theory guarantees φ is necessarily a linear fractional transformation!

As evidenced by Theorems 8.55 and 8.57, the signature curve also encodes the symmetries, meaning self-equivalences of a binary form. Let us first consider the maximally symmetric binary forms[†] whose fundamental invariant J is constant, and hence the signature set reduces to a single point.

Theorem 8.62. *Let $Q(p)$ be a binary form of degree $n \geq 2$. The following conditions are equivalent:*

(a)   *Q is complex-equivalent to a monomial p^k, with $k \neq 0, n$.*

(b)   *The covariant T^2 is a constant multiple of $H^3 \not\equiv 0$.*

(c)   *The signature set is a single point.*

(d)   *Q admits a one-parameter symmetry group.*

(e)   *The graph of Q coincides with the orbit of a one-parameter subgroup of* GL(2).

Proof: Most of these follow immediately from Theorem 8.57. We explicitly compute the relevant covariants for the monomial $Q = p^k$; the

---

[†] However, note that the forms with vanishing Hessian admit a larger symmetry group; see Theorem 8.66.

result is
$$H = -(n-1)k(n-k)\,p^{2k-2},$$
$$T = -2(n-1)k(n-k)(2k-n)\,p^{3k-3}, \qquad (8.57)$$
$$U = -6(n-1)k(n-k)(2k-n)^2\,p^{4k-4}.$$
Therefore, the absolute covariants
$$J^2 = \frac{T^2}{H^3} = -\frac{4(2k-n)^2}{(n-1)k(n-k)}, \qquad K = \frac{U}{H^2} = -\frac{6(n-2k)^2}{(n-1)k(n-k)},$$
$$(8.58)$$
are both constant, and so the signature set is a point. On the other hand, if $J^2 = -c^2$ is constant, then we can choose the monomial $Q(p) = p^k$ with exponent
$$k = \frac{n}{2}\left(1 \pm \sqrt{\frac{\lambda}{\lambda+4}}\right), \qquad \text{where} \qquad \lambda = \frac{(n-1)c^2}{4}, \qquad (8.59)$$
to represent it; the only other monomials giving the same value for J are constant multiples of either p^k or p^{n-k}. Theorem 8.61 implies that any other binary form giving the same constant value for J is equivalent to such a monomial. In particular, only integral values of k are allowed if Q is a polynomial of degree n. $\qquad$ Q.E.D.

Exercise 8.63. What are the real canonical forms for maximally symmetric binary forms? In other words, which additional real forms are complex- but not real-equivalent to a monomial?

Exercise 8.64. Note that the proof of Theorem 8.62 does not require that $Q(p)$ be a polynomial and hence k need not necessarily be integral. Analyze the particular case $J^2 = 16/(n-1)$ in detail.

Exercise 8.65. Give a direct proof that a form admits a one-parameter symmetry group if and only if it is equivalent to a monomial. *Hint*: This uses the fact that every one-parameter subgroup has the form e^{tA} for some 2×2 matrix A. The key step is to diagonalize the matrices in the one-parameter subgroup. What happens if A is not diagonalizable?

Theorem 8.66. Let $Q(p)$ be a nonzero binary form of degree n. Then the symmetry group of Q is:
(a) A two-parameter group if and only if $H \equiv 0$ if and only if Q is equivalent to a constant.
(b) A one-parameter group if and only if $H \not\equiv 0$ and T^2 is a constant multiple of H^3 if and only if Q is complex-equivalent to a monomial p^k, with $k \neq 0, n$.

(c) A finite group in all other cases. The cardinality of the group
     equals the index of the signature curve $\mathcal{S}_Q$.

The first case is proved by direct computation, using the fact that
Q is the nth power of a linear form, and hence equivalent to $\pm p^n$. The
second and third cases are direct consequences of Theorem 8.61. If we
replace the binary form by a more general analytic function, then the
final, nonsingular case can admit an infinite, discrete symmetry group.
For example, a periodic function $Q(p)$ admits an infinite discrete group
of translational symmetries.

The number of symmetries of a nonsingular complex binary form
can, in fact, be determined explicitly! Fixing a nonsingular point p_0, let

$$h_0 = H(p_0) \neq 0, \qquad t_0 = T(p_0), \qquad u_0 = U(p_0), \qquad (8.60)$$

denote the values of the fundamental covariants. We assume that $j_0^2 =
h_0^{-3} t_0^2$ and $k_0 = h_0^{-2} u_0$ defines a generic point $(j_0^2, k_0) \in \mathcal{S}_Q$ on the
signature curve, meaning that it is not a point of self-intersection. Then
the index of Q is equal to the number of values of p, including p_0 and,
possibly, ∞, which map to (j_0, k_0), i.e., the number of common solutions
to the pair of equations $J(p)^2 = j_0^2$, $K(p) = k_0$. Clearing denominators,
we see that the index equals the number of common roots to the two
polynomial equations

$$\begin{aligned}
A(p) &\equiv h_0^3 \, T(p)^2 - t_0^2 \, H(p)^3 = 0, \\
B(p) &\equiv h_0^2 \, U(p) - u_0 \, H(p)^2 = 0,
\end{aligned} \qquad (8.61)$$

of respective degrees $6(n-2), 4(n-2)$. In Exercise 2.38, we learned
an algorithm for explicitly determining the number of common roots to
a pair of polynomial equations. Let $R_j = \mathbf{R}_j[A, B]$, $j = 0, 1, 2, \ldots$,
denote the jth *subresultant* of the polynomials (8.61). The index k is
then characterized by the conditions $R_0 = R_1 = \cdots = R_{k-1} = 0$, while
$R_k \neq 0$. Now, the only non-constructive part of this algorithm is the
determination of the initial generic point p_0. Fortunately, one does not
need to inspect the signature curve for possible self-intersections to fix
an appropriate value. The subresultants are polynomial functions of
p_0. If $Q(p)$ has index k, then the polynomial equations (8.61) will have
at least k common roots for *all* values of p_0, and hence the first k of
them will vanish for all p_0, while the kth one will not be identically zero.
The roots of $R_k(p_0) = 0$ are, in fact, precisely the non-generic points of
self-intersection of the signature curve!

Theorem 8.67. *Let $Q(p)$ be a binary form of degree n which is not* *complex equivalent to a monomial. Let* $H = \frac{1}{2}(Q, Q)^{(2)}$, $T = (Q, H)^{(1)}$, $U = (Q, T)^{(1)}$, *and define the bivariate polynomials*

$$A(p,q) = H(q)^3 T(p)^2 - T(q)^2 H(p)^3,$$
$$B(p,q) = H(q)^2 U(p) - U(q) H(p)^2. \tag{8.62}$$

Let $R_k(q) = \mathbf{R}_k[A, B]$ *be their* k^{th} *subresultant, taken with respect to* p. *Then the index of* $Q(p)$, *meaning the cardinality of its symmetry group, is equal to the first natural number* $k = 0, 1, 2, \dots$ *for which* $R_k(q) \not\equiv 0$.

The fact that Q is not equivalent to a monomial implies that T^2 is a not a constant multiple of H^3, and hence $A(p, q) \not\equiv 0$ is a nontrivial polynomial. Therefore, for generic $q = p_0$, the first symmetry equation (8.61) will have degree $6n - 12$, and hence the index is automatically bounded by this number. If U is a constant multiple of H^2, then $B(p, q) \equiv 0$, and so every root of A provides a symmetry; in this case the index exactly equals $6n - 12$. We shall call this the case of *maximal discrete symmetry*. For example, if

$$(Q, Q)^{(4)} \equiv 0, \tag{8.63}$$

then the syzygy (7.10) automatically implies that U is a constant multiple of H^2 and the form has maximal discrete symmetry. I do not know a good geometric interpretation of the maximal symmetry condition $U = cH^2$. In the non-maximal case, the index is bounded by the (generic) degree $4n - 8$ of $B(p, q) \not\equiv 0$.

Corollary 8.68. *Let* k *denote the index of a binary form* Q *of degree* n *which is not complex-equivalent to a monomial. Then*
(a) $k = 6n - 12$ *if and only if* $U = cH^2$ *for some constant* c, *or*
(b) $k \leq 4n - 8$ *if* U *is not a constant multiple of* H^2.

This theorem is, as far as I know, original. I do not know if the bound on the index is optimal. The real case is a little trickier, since one must determine the number of common *real* solutions to (8.61), and, in the case of even degree, whether the sign of Q is the same at each solution. An explicit algorithm for effecting this restricted count remains to be worked out.

Remark: According to (5.10), the transvectant equation (8.63) has the explicit form

$$n(n - 1)QQ'''' - 4(n - 1)^2 Q'Q''' + 3(n - 2)(n - 3)(Q'')^2 = 0. \tag{8.64}$$

Motivated by our integration of the Hessian equation $(Q, Q)^{(2)} = 0$ in (5.13), let us introduce the logarithmic derivative $R = -Q'/(nQ)$, which results in the remarkable third order ordinary differential equation

$$R''' - 12RR'' + 18R'^2 = \alpha\,(R' - R^2)^2, \qquad (8.65)$$

where $\alpha = 6n^2/(n-1)$. Equation (8.65) is a rescaled version of an equation discovered by Chazy, [**45**], in his deep study of third order ordinary differential equations having the "Painlevé property", cf. [**116**; Chapter 14]. Chazy proves that the general solution to (8.65) can be written as

$$p = \frac{\varphi(z)}{\psi(z)}, \qquad R = \frac{1}{\psi(z)}\,\frac{d\psi}{dp}, \qquad (8.66)$$

where $\varphi(z)$ and $\psi(z)$ are two arbitrary linearly independent solutions of the hypergeometric equation

$$z(1-z)\frac{d^2\psi}{dz^2} + \left(\frac{1}{2} - \frac{7}{6}z\right)\frac{d\psi}{dz} + \frac{1}{6(\alpha - 24)}\,\psi = 0, \qquad (8.67)$$

demonstrating a deep, and relatively unexplored, connection between hypergeometric functions and classical invariant theory. Interestingly, equation (8.67) arises in Schwarz's theory of algebraic hypergeometric functions which admit discrete symmetry groups, cf. [**108**; §10.3]. Further details can be found in [**47, 169**; pp. 196–199], including connections with differential invariants and elliptic functions. The Chazy equation (8.65) appears as a similarity reduction of the Yang–Mills equations, [**43**], and the Prandtl boundary layer equations from fluid mechanics, [**47**]. Takhtajan, [**211**], uses Chazy's result to connect the theory of modular forms with the solution of integrable soliton equations.

Example 8.69. Let us apply these results to binary cubics. According to the remark following (7.10), the covariant U can be simply written as $U = (Q,\,T)^{(1)} = -\frac{3}{2}H^2$. Thus, when $H \neq 0$, the second rational covariant (8.40) is a constant, $K = -\frac{3}{2}$, and so, for a nondegenerate case, the signature curve is a horizontal line in $\mathbb{C}^2$. Using this observation, the signature approach of Theorem 8.61 immediately recovers the classification table for complex cubics on p. 28; for $Q \not\equiv 0$ the cases are

(a)   The degenerate case when the Hessian vanishes identically, $H \equiv 0$, in which case the form is the cube of a linear form.

(b)   The maximally symmetric case when the signature curve degenerates to a point, and so T^2 is a constant multiple of H^3, in which case the cubic has a double root.

(c)   The generic case when the signature curve is a horizontal line and the cubic has three simple roots.

At this point, the reader may be wondering how the sign of the discriminant enters into the classification of nondegenerate real cubics, since the signature curve remains a horizontal line $K = -\frac{3}{2}$. The resolution of this apparent paradox is that the signature curve is not, in fact, the entire horizontal line! Recall the basic cubic syzygy (2.47),

$$T^2 + H^3 = 2^4 3^6 \Delta \, Q^2. \tag{8.68}$$

If $\Delta < 0$, the cubic has three real roots and $H(p) < 0$ is negative definite. (This fact follows either by computing H from the canonical form, or by noting that, according to the second identity in (2.46), the discriminant of the quadratic form H is positive, and hence H itself is of one sign, which is determined by its leading coefficient.) Since $J = T^2/H^3$, (8.68) implies that $-1 \leq J(p) \leq 0$, and hence the signature curve in this case is the horizontal line segment $\mathcal{S}_Q = \left\{ \left(a, \frac{3}{2}\right) \mid -1 \leq a \leq 0 \right\}$. On the other hand, if $\Delta > 0$, then (8.68) implies $T^2 + H^3 \geq 0$. In this case, $H(p)$ is indefinite; when $H > 0$, then $J \geq 0$, while when $H < 0$, then $J < -1$. Therefore, the signature curve for a cubic with complex roots consists of two pieces: $\mathcal{S}_Q = \left\{ \left(a, \frac{3}{2}\right) \mid a \geq 0 \right\} \cup \left\{ \left(a, \frac{3}{2}\right) \mid a < -1 \right\}$. We see that the two real signature curves cover *different* portions of the same horizontal line, and so the two cubics cannot be real-equivalent.

Since U is a constant multiple of H^2, all nondegenerate cubics have maximal discrete symmetry groups of index 6, which equals the number of different permutations of the three roots. In the real case, one requires real solutions to (8.62), and hence Q has index 6 if $\Delta < 0$, but index 2 if $\Delta > 0$. The reader may find it instructive to find the explicit symmetries of the two canonical forms.

Example 8.70. Turning to a binary quartic, to find the signature curve we use the fundamental syzygy for the quartic (2.48) combined with the identity (7.10):

$$T^2 = -\tfrac{16}{9} H^3 + 2^{10} 3^2 \, i \, Q^2 H - 2^{14} 3^4 \, j \, Q^3, \qquad U = -\tfrac{8}{3} H^2 + 2^9 3^2 \, i \, Q^2.$$

Here $i = \frac{1}{576} I = \frac{1}{1152} (Q, Q)^{(4)}$ and j are the fundamental quartic invariants (2.29), (2.30). Let us introduce the rational covariant $S = 48 \, Q/H$ in terms of which our absolute covariants assume the parametric form

$$J^2 = -\tfrac{16}{9} + 4 \, i \, S^2 - 12 \, j \, S^3, \qquad K = -\tfrac{8}{3} + 2 \, i \, S^2. \tag{8.69}$$

The last two equations give a simple parametrization of the signature curve, which consequently takes the implicit form

$$9 j^2 \left(K + \tfrac{8}{3} \right)^3 = 2 i^3 \left(K - \tfrac{1}{2} J + \tfrac{16}{9} \right)^2. \tag{8.70}$$

Thus, the different signature curves are distinguished by the absolute invariant $a = j^2/i^3$. In particular, the signature curve is (contained in) a straight line if and only if either $i = 0$ or $j = 0$, cf. [**97**; Ex. 25.6].

Referring back to our table of canonical forms for complex-valued quartics appearing on p. 29, we see that the nondegenerate cases, for which J is not constant, are types I and II. In case I, $Q = p^4 + \mu p^2 + 1$, $\mu \neq \pm 2$, the absolute invariant takes the value

$$a = \frac{j^2}{i^3} = \frac{\mu^2(36 - \mu^2)^2}{27(12 + \mu^2)^3} . \tag{8.71}$$

Notice that the following six values for the quartic parameter,

$$\pm\mu, \qquad \pm\frac{12 - 2\mu}{2 + \mu}, \qquad \pm\frac{12 + 2\mu}{2 - \mu},$$

give the same value for a in (8.71). The associated quartics are all equivalent; the relevant equivalence transformations are supplied by the matrices

$$\begin{pmatrix} 1 & 0 \\ 0 & 1 \end{pmatrix}, \qquad \frac{1}{\sqrt[4]{2 + \mu}}\begin{pmatrix} 1 & 1 \\ 1 & -1 \end{pmatrix}, \qquad \frac{1}{\sqrt[4]{2 - \mu}}\begin{pmatrix} i & 1 \\ i & -1 \end{pmatrix},$$

$$\begin{pmatrix} i & 0 \\ 0 & 1 \end{pmatrix}, \qquad \frac{1}{\sqrt[4]{2 - \mu}}\begin{pmatrix} i & i \\ 1 & -1 \end{pmatrix}, \qquad \frac{1}{\sqrt[4]{2 - \mu}}\begin{pmatrix} 1 & -i \\ 1 & i \end{pmatrix}.$$

A quartic of type II, $Q = p^2 + 1$, also has J not constant. The absolute invariant now takes on the exceptional value $a = \frac{1}{27}$, which is *not* attained by a quartic of type I (except in the degenerate case $\mu = \pm 2$, when type I degenerates to type III, and is equivalent to a monomial, as indicated by the fact that J is now constant and so the signature curve is a single point).

The degenerate cases have point signature sets, and so $S = 48\,Q/H$ must be constant. Therefore, a quartic is complex-equivalent to a monomial if and only if its Hessian is a constant multiple of Q itself. Thus, in accordance with our general Theorem 8.61, the signature curve provides a complete classification of canonical forms and equivalence classes of quartics.

Exercise 8.71. How many symmetries does a nondegenerate quartic have? Does the index depend on μ?

Exercise 8.72. Discuss the corresponding classification of real binary quartics constructed in Exercise 2.25 from the point of view of signature curves. See [**167**] for details.

Exercise 8.73. Prove that a binary form $Q(x,y)$ of degree $n \geq 3$ is complex-equivalent to a sum of two n^{th} powers, that is, to $x^n + y^n$, if and only if its covariants H, T, U are related by the equation

$$HU - \frac{n-3}{n-2} T^2 + \frac{2n(n-2)}{(n-1)^2} H^3 = 0. \tag{8.72}$$

In particular, a quartic is the sum of two fourth powers if and only if $j = 0$. See [**96**, **134**, **183**, **184**], for more general results on expressing a binary form as a sum of powers.

Chapter 9

Infinitesimal Methods

As Lie first discovered, the true power of Lie group theory lies in the ability to work on an infinitesimal level. Each Lie group-theoretic object and property has an infinitesimal counterpart. Passage from the finite to the infinitesimal typically leads to a linearization of the underlying conditions, and hence a significant simplification. Moreover, as long as the group is connected, one can always reconstruct the finite object and its properties from their infinitesimal analogs. The resulting combination of group theory and analysis has proven to be exceptionally potent. A striking example is the Killing–Cartan classification of simple Lie groups, [126, 32, 53], which was established over a century ago and forms a (short) part of many textbooks on the subject, [173, 223]. The corresponding problem for simple finite groups was only recently completed and required the arduous labor of many researchers; moreover, the proof runs to thousands of journal pages, cf. [91].

The infinitesimal counterpart of a Lie group is known as a Lie algebra, an algebraic object of independent interest and wide applicability, ranging from physics to abstract algebra. When the group acts on an open subset of Euclidean space, or, more generally, a smooth manifold, the "infinitesimal generators" in the Lie algebra are realized as first order differential operators acting on smooth functions. Our primary focus will be on the infinitesimal approach to invariants and symmetry, leading to the fundamental differential operators which annihilate the invariants and covariants. Interestingly, Cayley, [41], realized the importance of these operators in the context of binary forms many years before Lie arrived on the scene. However, Cayley did not understand that this device was not particular to the general linear group but forms a particular instance of a vast and far-ranging theory; the honor of making this discovery thus fell to Sophus Lie. The infinitesimal operators also play a key role in Hilbert's proof of his Basis Theorem for classical invariants, [106], where they are combined to form a differential operator that, like the symmetrization operator in finite group theory, produces invariants directly from more general functions.

One-Parameter Subgroups

As we have already seen, each Lie group contains a wide variety of subgroups. Of primary importance are the "one-parameter subgroups", meaning connected subgroups that depend smoothly on a single real parameter.[†] The prototypical one-parameter group is the additive group of real numbers, $\mathbb{R}$, which motivates the basic definition.

Definition 9.1. A *one-parameter subgroup* of a Lie group G is defined as the image of a nonconstant analytic group homomorphism $\rho: \mathbb{R} \to G$.

In Example 3.18, we discovered that any matrix exponential defines a one-parameter subgroup of the general linear group. Let us prove that these form a complete list.

Theorem 9.2. *Each one-parameter subgroup of* $\mathrm{GL}(n, \mathbb{R})$ *is given by a matrix exponential* $\rho_J(t) = \exp tJ$, *where* J *is an arbitrary nonzero* $n \times n$ *matrix. We call* J *the* infinitesimal generator *of the one-parameter subgroup* $H_J = \{\exp tJ\}$.

Proof: Consider the one-parameter subgroup which is parametrized by $\rho: \mathbb{R} \to \mathrm{GL}(n, \mathbb{R})$. The formula

$$J = \frac{d\rho(t)}{dt}\bigg|_{t=0} \tag{9.1}$$

serves to define its infinitesimal generator. Geometrically, J can be identified with the tangent to the curve $\rho(t)$ at the identity matrix $\mathbb{1} = \rho(0)$. Since ρ is a group homomorphism,

$$\rho(s + t) = \rho(s) \cdot \rho(t). \tag{9.2}$$

Let us differentiate the matrix equation (9.2) with respect to s and set $s = 0$. In view of (9.1), we conclude that $\rho(t)$ is a matrix-valued solution to the initial value problem:

$$\frac{d\rho}{dt} = J \cdot \rho, \qquad \rho(0) = \mathbb{1}. \tag{9.3}$$

The unique (and globally defined) solution to this linear system of first order ordinary differential equations is provided by the matrix exponential $\rho(t) = \exp tJ$.   *Q.E.D.*

[†] As in the preceding chapter, for specificity we concentrate on the real, analytic versions here.

Remark: If the matrices $K = \lambda J$, $\lambda \neq 0$, are nonzero scalar multiples of each other, then $\exp tK = \exp \tilde{t}J$, with $\tilde{t} = \lambda t$, and hence the two matrices generate the same one-parameter subgroup, albeit parametrized by different group homomorphisms.

According to Exercise 3.26, every one-parameter subgroup is isomorphic to either the additive group of real numbers, $\mathbb{R}$, or to the circle group $S^1 \simeq \mathrm{SO}(2)$, depending on the (complex) Jordan canonical form of its infinitesimal generator J. The former class divides into two subcategories, depending on whether the subgroup forms a topologically closed subset or not. The complete result is not hard to establish using the general formula for a matrix exponential, cf. [**98**].

Theorem 9.3. *A one-parameter subgroup* $\{\exp tJ\} \subset \mathrm{GL}(n, \mathbb{R})$ *is isomorphic to*

(a) S^1 *if and only if* J *is diagonalizable over* $\mathbb{C}$, *and all its eigenvalues are purely imaginary and rational multiples of each other, or*

(b) $\mathbb{R}$, *but is not a topologically closed subset, if and only if* J *is diagonalizable over* $\mathbb{C}$, *and all its eigenvalues are purely imaginary, but not all rational multiples of each other, or*

(c) $\mathbb{R}$, *and is topologically closed, in all other cases.*

We let $\mathfrak{gl}(n, \mathbb{R})$ denote the space of infinitesimal generators of one-parameter subgroups of $\mathrm{GL}(n, \mathbb{R})$ — including the zero matrix, which generates the trivial subgroup $\{e\}$. Thus, we can identify $\mathfrak{gl}(n, \mathbb{R}) \simeq \mathcal{M}_{n \times n} \simeq \mathbb{R}^{n^2}$ with the n^2-dimensional vector space consisting of *all* $n \times n$ matrices. The choice of notation is convenient for later developments.

Not every matrix lies in a one-parameter subgroup of $\mathrm{GL}(n, \mathbb{R})$. Indeed, since

$$\det(\exp tJ) = \exp(t \operatorname{tr} J), \tag{9.4}$$

matrices with negative determinant are not (real) exponentials.

Definition 9.4. The *exponential map*

$$\exp : \mathfrak{gl}(n, \mathbb{R}) \longrightarrow \mathrm{GL}(n, \mathbb{R})^+ \tag{9.5}$$

is defined by evaluating the matrix exponential $\exp tJ$ at $t = 1$.

The exponential map is still not onto the set of matrices with positive determinant. The following result is not too difficult to establish; see [**173**; §I.3.4] for details.

Proposition 9.5. *A matrix $A \in \mathrm{GL}(n, \mathbb{R})$ lies in the image of the exponential map if and only if each negative eigenvalue of A has an even number of associated Jordan blocks of each size. On the other hand, every complex matrix $A \in \mathrm{GL}(n, \mathbb{C})$ can be written as the exponential, $A = \exp J$, of a complex $n \times n$ matrix $J \in \mathfrak{gl}(n, \mathbb{C})$.*

Exercise 9.6. Prove the two-dimensional version of Proposition 9.5: a 2×2 real matrix can be written as the exponential of a real matrix if and only if either both eigenvalues are positive, or the matrix is diagonalizable and has only one negative eigenvalue.

Matrix Lie Algebras

Let $G \subset \mathrm{GL}(n, \mathbb{R})$ be a matrix Lie group. The subset

$$\mathfrak{g} = \{ J \mid \{\exp tJ\} \subset G \} \subset \mathfrak{gl}(n, \mathbb{R})$$

consisting of all matrices whose one-parameter subgroup is contained in G forms a subspace of $\mathfrak{gl}(n, \mathbb{R})$, known as the space of *infinitesimal generators* of G. If one identifies $\mathfrak{gl}(n, \mathbb{R})$ with the tangent space to $\mathrm{GL}(n, \mathbb{R})$ at the identity matrix $\mathbb{1}$, then $\mathfrak{g} \simeq T G|_e$ will be the subspace tangent to the submanifold $G \subset \mathrm{GL}(n, \mathbb{R})$ at $e = \mathbb{1}$. The question then arises: Does every vector subspace of $\mathfrak{gl}(n, \mathbb{R})$ generate a Lie subgroup? The answer is "no", because we have not taken the commutators of infinitesimal generators into account yet.

Suppose $J \in \mathfrak{g}$ is the infinitesimal generator of a one-parameter subgroup $H_J = \{\exp tJ\} \subset G$. Let $A \in G$ be any other group element. Then the conjugate subgroup $A^{-1} H_J A$ consists of all matrices of the form $A^{-1} e^{tJ} A = \exp t(A^{-1} J A)$ and hence equals the one-parameter subgroup $H_{A^{-1}JA}$ generated by the conjugate matrix $A^{-1} J A$. Thus, if $J \in \mathfrak{g}$ and $A \in G$, then $A^{-1} J A \in \mathfrak{g}$ also, so that the space of infinitesimal generators is closed under conjugation. Now let $A = \exp tK \in G$ itself belong to a one-parameter subgroup generated by some $K \in \mathfrak{g}$. Then $e^{-tK} J e^{tK} \in \mathfrak{g}$ for all t. Expanding the two matrix exponentials in power series and collecting terms, we conclude that

$$e^{-tK} J e^{tK} = J + t(JK - KJ) + \cdots \in \mathfrak{g}, \qquad \text{for all} \quad t \in \mathbb{R}. \quad (9.6)$$

Since $\mathfrak{g}$ forms a subspace of $\mathfrak{gl}(n, \mathbb{R})$, we conclude that the matrix coefficient of each power of t in (9.6) must belong to $\mathfrak{g}$. In particular:

Proposition 9.7. *If J and K are infinitesimal generators of a Lie subgroup $G \subset \mathrm{GL}(n, \mathbb{R})$, then so is their commutator $[J, K] = JK - KJ$.*

Remark: The higher order terms in (9.6) can all be expressed in terms of higher order matrix commutators between J and K. For example, the order t^2 terms are $\frac{1}{2}K^2J - KJK + \frac{1}{2}JK^2 = \frac{1}{2}[[J, K], K]$.

Subspaces of $\mathfrak{gl}(n, \mathbb{R})$ which are closed under matrix commutator play a distinguished role, first recognized by Lie himself.

Definition 9.8. A *matrix Lie algebra* is a subspace $\mathfrak{g} \subset \mathfrak{gl}(n, \mathbb{R})$ with the property that $J, K \in \mathfrak{g}$ implies $[J, K] \in \mathfrak{g}$.

In particular, $\mathfrak{gl}(n, \mathbb{R})$ is itself a matrix Lie algebra.

Exercise 9.9. The matrix commutator is clearly bilinear and skew-symmetric: $[J, K] = -[K, J]$. Prove that it also satisfies the *Jacobi identity*

$$[J, [K, L]] + [L, [J, K]] + [K, [L, J]] = 0. \qquad (9.7)$$

Remark: In general, an abstract *Lie algebra* is defined as a vector space equipped with a bilinear, skew-symmetric bracket operation that satisfies the Jacobi identity (9.7). An example is the Poisson bracket, (5.30), which defines a Lie algebra structure on the space of binary forms.

The fundamental theorem from Lie group theory (in our simplified matrix version) can now be stated.

Theorem 9.10. *There is a one-to-one correspondence between connected r-dimensional Lie subgroups $G \subset \mathrm{GL}(n, \mathbb{R})$ and r-dimensional matrix Lie algebras $\mathfrak{g} \subset \mathfrak{gl}(n, \mathbb{R})$.*

See [**173, 223, 227**] for a proof. Since each infinitesimal generator generates a one-parameter subgroup, the exponential map restricts to $\exp: \mathfrak{g} \to G$ which maps the Lie algebra onto a neighborhood of the identity in G. As with the general linear group, even if G is connected, not every group element lies in the image of exp. (See [**62**] for a recent survey on known results about the image of the exponential map.) However, when G is connected, every group element can be written as a product of exponentials, and this is sufficient for most practical purposes. This result is proved by a standard open and closed topological connectivity argument; see [**173, 227**] for details.

Lemma 9.11. *Let $G \subset \mathrm{GL}(n, \mathbb{R})$ be a connected Lie group. Then we can write any element $A \in G$ as a finite product of exponentials, so that $A = (\exp J_1)(\exp J_2) \cdots (\exp J_N)$ for $J_1, \ldots, J_N \in \mathfrak{g}$.*

Example 9.12. Formula (9.4) implies that the Lie algebra $\mathfrak{sl}(n) = \mathfrak{sl}(n, \mathbb{R})$ of the unimodular subgroup $SL(n) = SL(n, \mathbb{R})$ consists of all matrices having zero trace: $\mathfrak{sl}(n) = \{J \in \mathfrak{gl}(n) \mid \operatorname{tr} J = 0\}$. Note that $\mathfrak{sl}(n)$ has dimension $n^2 - 1$, which implies that $SL(n)$ is a Lie group of the same dimension. The reader should check that the commutator of any two trace zero matrices also has trace zero. The orthogonal groups $O(n)$ and $SO(n)$ have the same Lie algebra, denoted $\mathfrak{so}(n)$, consisting of all skew-symmetric $n \times n$ matrices. Indeed, if $J^T + J = 0$, then the matrix-valued function $C(t) = e^{tJ} \left(e^{tJ}\right)^T$ is constant since

$$\frac{d}{dt} e^{tJ} \left(e^{tJ}\right)^T = e^{tJ} (J + J^T) \left(e^{tJ}\right)^T = 0.$$

Moreover, $C(0) = \mathbb{1}$, and hence $C(t) = \mathbb{1}$ for all t, which implies that $A(t) = e^{tJ}$ is orthogonal, and conversely. Again, since $\mathfrak{so}(n)$ forms a subspace of dimension $\frac{1}{2}n(n - 1)$, its associated matrix Lie groups[†] $SO(n)$ and $O(n)$ also have dimension $\frac{1}{2}n(n - 1)$.

Exercise 9.13. Prove that the one-parameter subgroups generated by matrices J and K commute, so $e^{tJ}e^{sK} = e^{sK}e^{tJ} = e^{tJ+sK}$ for all $s, t \in \mathbb{R}$, if and only if $[J, K] = 0$. Use this to conclude that a matrix Lie group $G \subset GL(n)$ is abelian if and only if its Lie algebra $\mathfrak{g} \subset \mathfrak{gl}(n)$ is *abelian*, meaning that its generators commute: $[J, K] = 0$ for $J, K \in \mathfrak{g}$.

To check that a given subspace $\mathfrak{g} \subset \mathfrak{gl}(n, \mathbb{R})$ forms a Lie algebra, it suffices to verify that the commutator of basis elements $\{J_1, \ldots, J_r\}$ lies in the subspace, meaning that

$$[J_i, J_j] = \sum_{k=1}^{r} c_{ij}^k J_k \tag{9.8}$$

for certain scalars $c_{ij}^k \in \mathbb{R}$. These coefficients are known as the *structure constants* for the Lie subalgebra $\mathfrak{g}$ relative to the given choice of basis.

Exercise 9.14. Determine the structure constants of the Lie algebra $\mathfrak{gl}(n, \mathbb{R})$ relative to the standard basis, which are the matrices $E_{\alpha\beta}$ having a single 1 in the (α, β) position and zeros elsewhere. Prove that the polarization operators $\sigma_{\alpha\beta}$, for $1 \le \alpha, \beta \le n$, as defined in (5.26), span a Lie algebra isomorphic to $\mathfrak{gl}(n)$, whose bracket is given by the commutator $[\sigma, \tau] = \sigma \cdot \tau - \tau \cdot \sigma$.

[†] Recall that $O(n)$ is not connected. A given Lie algebra can have various disconnected Lie subgroups associated with it.

Exercise 9.15. Prove that the structure constants of a Lie algebra satisfy the basic identities

$$c_{ij}^k = -c_{ij}^k, \qquad \sum_{l=1}^{r} \left\{ c_{ij}^l c_{lk}^m + c_{ki}^l c_{lj}^m + c_{jk}^l c_{li}^m \right\} = 0. \qquad (9.9)$$

Remark: Suppose c_{jk}^i is a collection of $\frac{1}{2} r^2 (r - 1)$ constants satisfying (9.9). Ado's Theorem, cf. [7], [117; chap. VI], states that there exists a matrix Lie algebra having these structure constants relative to some basis $J_1, \ldots, J_r$. As a consequence, every Lie group can be locally (but not necessarily globally) identified with a matrix Lie group, constructed via Theorem 9.10.

Definition 9.16. A *Lie algebra homomorphism* is a linear map $\varphi \colon \mathfrak{g} \to \mathfrak{h}$ between Lie algebras that preserves the commutator:

$$[\varphi(J), \varphi(K)] = \varphi([J, K]).$$

Remark: A general result, based on the exponential map, cf. [173], states that any Lie algebra homomorphism corresponds to a local (meaning in a neighborhood of the identity element) group homomorphism.

Exercise 9.17. Any one-dimensional subspace of $\mathfrak{gl}(n, \mathbb{R})$ is, trivially, an (abelian) Lie algebra. Prove that there are, up to isomorphism, only two different types of two-dimensional Lie algebras: an abelian one, and a non-abelian one for which we can choose a basis J_1, J_2 such that $[J_1, J_2] = J_1$.

Let us specialize these general constructions to our favorite group, GL(2). We introduce the following basis[†]

$$J^- = \begin{pmatrix} 0 & 1 \\ 0 & 0 \end{pmatrix}, \quad J^0 = \begin{pmatrix} 1 & 0 \\ 0 & -1 \end{pmatrix}, \quad J^+ = \begin{pmatrix} 0 & 0 \\ 1 & 0 \end{pmatrix}, \quad \mathbb{1} = \begin{pmatrix} 1 & 0 \\ 0 & 1 \end{pmatrix}, \qquad (9.10)$$

of the Lie algebra $\mathfrak{gl}(2)$. The associated one-parameter subgroups are readily found by exponentiating these four matrices:

$$\begin{pmatrix} 1 & t \\ 0 & 1 \end{pmatrix}, \quad \begin{pmatrix} e^t & 0 \\ 0 & e^{-t} \end{pmatrix}, \quad \begin{pmatrix} 1 & 0 \\ t & 1 \end{pmatrix}, \quad \begin{pmatrix} e^t & 0 \\ 0 & e^t \end{pmatrix}. \qquad (9.11)$$

The basis (9.10) was designed so that the Lie algebra $\mathfrak{sl}(2)$ for the three-dimensional unimodular subgroup SL(2) is spanned by the trace-free

---

[†] The notation comes from quantum mechanics, [136].

matrices J^-, J^0, J^+. We note the commutation relations

$$[J^-, J^0] = -2J^-, \qquad [J^+, J^0] = 2J^+, \qquad [J^-, J^+] = J^0,$$

verifying that these three matrices span a Lie algebra. The identity generator commutes with every other generator: $[J^a, \mathbb{1}] = 0$, which is because its associated one-parameter subgroup forms the center of $GL(2)$ — see Exercise 9.13. These relations serve to identify the structure constants for the Lie algebra $\mathfrak{gl}(2)$.

Example 9.18. The two-dimensional matrix Lie algebra $\mathfrak{u} \subset \mathfrak{gl}(2)$ spanned by J^- and J^0 corresponds to the two-parameter subgroup consisting of upper triangular unimodular matrices $\begin{pmatrix} \lambda & \beta \\ 0 & \lambda^{-1} \end{pmatrix}$ with $\lambda > 0$. On the other hand, the two-parameter subgroup given by the matrices $\begin{pmatrix} a & b \\ 0 & 1 \end{pmatrix}$ with $a > 0$ is isomorphic to the affine group $A(1)$ which acts on $\mathbb{R}$ according to $x \mapsto ax + b$. Its Lie algebra $\mathfrak{a}(1)$ has a basis consisting of the matrices $K^0 = \begin{pmatrix} 1 & 0 \\ 0 & 0 \end{pmatrix}$, $K^- = \begin{pmatrix} 0 & 1 \\ 0 & 0 \end{pmatrix}$, which have the commutation relations $[K^-, K^0] = -K^-$. Now, according to Exercise 9.17, there is, up to isomorphism, only one non-abelian two-dimensional Lie algebra. Indeed the linear map that takes $J^- \mapsto K^-$, $J^0 \mapsto 2K^0$ provides the isomorphism $\mathfrak{u} \xrightarrow{\sim} \mathfrak{a}(1)$. In this case, the map $a = \lambda^2$, $b = \lambda\beta$, provides the corresponding (in this case global) group isomorphism.

Vector Fields and Orbits

Lie's infinitesimal method lets us replace rather complicated nonlinear group transformations by simpler infinitesimal counterparts, leading to the linearization and consequent simplification of basic invariance criteria. The infinitesimal invariance conditions form a system of linear partial differential equations, enabling one to apply direct analytical methods to reconstruct the required invariants. The key step is to identify the infinitesimal generators of a transformation group action as certain differential operators that act on functions defined on the space.

Definition 9.19. Let $G \subset GL(n, \mathbb{R})$ be a matrix Lie group, acting as a transformation group on a manifold X. Let $J \in \mathfrak{g}$ be the infinitesimal generator of the one-parameter subgroup $\{\exp tJ\} \subset G$. The associated *infinitesimal generator* of the action of G on X is the first order differential operator $\mathbf{v}_J$, defined so that

$$\mathbf{v}_J(F(x)) = \frac{d}{dt} F\left(e^{-tJ} \cdot x\right)\bigg|_{t=0} \qquad (9.12)$$

for every smooth function $F: X \to \mathbb{R}$.

The quantity $\mathbf{v}_J(F(x))$ represents the infinitesimal change in $F(x)$ under the one-parameter group e^{tJ}. In other words, if $\overline{F} = e^{tJ} \cdot F$ is the transformed function, as in (4.3), then

$$\overline{F}(x) = F(e^{-tJ} \cdot x) = F(x) + t\, \mathbf{v}_J(F(x)) + \cdots .$$

In particular, if $\overline{F} = F$ is invariant under the group transformations, then $\mathbf{v}_J(F(x))$ will vanish; see Theorem 9.28.

Remark: The more typical convention is to use e^{tJ} instead of $e^{-tJ} = (e^{tJ})^{-1}$ in the definition (9.12) of the infinitesimal generator $\mathbf{v}_J$. Our version is dictated by the action (4.3) of the group transformations on functions. The only effect is on the overall sign of $\mathbf{v}_J$.

Example 9.20.   Consider our usual projective action (2.7) of $GL(2, \mathbb{R})$ on the real line $X = \mathbb{R}$. Let us compute the infinitesimal generators for the four fundamental one-parameter subgroups (9.11); we let $\mathbf{v}_a$ denote the infinitesimal generator associated with J^a, as given in (9.10). First, we compute

$$\mathbf{v}_-(F(p)) = \frac{d}{dt} F\left(e^{-tJ^-} \cdot p\right)\bigg|_{t=0} = \frac{d}{dt} F(p - t)\bigg|_{t=0} = -\frac{\partial F}{\partial p} .$$

Similarly,

$$\mathbf{v}_0(F(p)) = \frac{d}{dt} F(e^{-2t}p)\bigg|_{t=0} = -2p\frac{\partial F}{\partial p} ,$$

$$\mathbf{v}_+(F(p)) = \frac{d}{dt} F\left(\frac{p}{1 - tp}\right)\bigg|_{t=0} = p^2\frac{\partial F}{\partial p} ,$$

$$\mathbf{v}_e(F(p)) = \frac{d}{dt} F(p)\bigg|_{t=0} = 0.$$

Therefore, the corresponding infinitesimal generators are the first order differential operators

$$\mathbf{v}_- = -\frac{\partial}{\partial p}, \qquad \mathbf{v}_0 = -2p\frac{\partial}{\partial p}, \qquad \mathbf{v}_+ = p^2\frac{\partial}{\partial p}, \qquad \mathbf{v}_e = 0. \qquad (9.13)$$

The general formula for the infinitesimal generator of a one-parameter group is readily established. Let us write out the group transformations as $e^{tJ} \cdot x = w(t, x)$. Applying the chain rule to differentiate

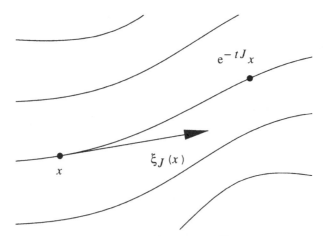

Figure 7.   Flow and Infinitesimal Generator.

$F(e^{-tJ} \cdot x) = F(w(-t, x))$ with respect to t yields the first order partial differential operator

$$\mathbf{v}_J = \sum_{i=1}^{n} \xi_J^i(x) \frac{\partial}{\partial x^i} \,, \quad \text{where} \quad \xi_J(x) = \frac{\partial}{\partial t} w(-t, x)\bigg|_{t=0} = -\frac{\partial w}{\partial t}(0, x).$$

$$(9.14)$$

The coefficient functions $\xi_J(x) = (\xi_J^1(x), \dots, \xi_J^m(x))$ depend analytically on x and linearly on the matrix J. We regard $\xi_J(x)$ as defining a *vector field*[†] on the space X. The associated one-parameter group of transformations $e^{-tJ} \cdot x$ is also known as the *flow* generated by the vector field ξ_J. This terminology comes from fluid mechanics, [**234**], in which context the vector field ξ_J represents the velocity field of a steady fluid flow. The orbit $\mathcal{O}_x = \{e^{-tJ} \cdot x \mid t \in \mathbb{R}\}$ through $x \in X$ is the *streamline* of the fluid flow, and $\xi_J(x)$ is its tangent vector at x. The geometry is illustrated in Figure 7.

Remark: According to Theorem 8.57, the streamlines or orbits of one-parameter subgroups are the curves of constant G-invariant curvature and, hence, are the maximally symmetric curves.

[†] In differential geometry, [**168, 227**], a vector field on a manifold *is* a first order differential operator; we maintain the distinction here only insofar as the manifold X is viewed as a subset of Euclidean space. Readers conversant with general manifold theory can readily make the identification between the two.

The flow can be reconstructed directly from the infinitesimal generator by solving a system of first order ordinary differential equations, whose integral curves are the streamlines. Thus, the theory of one-parameter transformation groups is, in effect, identical to the theory of autonomous first order ordinary differential equations.

Theorem 9.21. *Let* $\mathbf{v}_J$ *be the infinitesimal generator of the one-parameter transformation group* $\{\exp tJ\}$ *acting on* $X \subset \mathbb{R}^m$. *Given a point* $x_0 \in X$, *the group transformations* $x(t) = \exp(tJ) \cdot x_0$ *form the unique solution to the autonomous initial value problem*

$$\frac{dx}{dt} = -\xi_J(x), \qquad x(0) = x_0. \tag{9.15}$$

Proof: We differentiate the group law $e^{(s+t)J} \cdot x = e^{sJ} \cdot e^{tJ} \cdot x$ with respect to s and set $s = 0$. The result is an extended version,

$$\xi_J\left(e^{tJ} \cdot x\right) = -\frac{d\left[e^{tJ} \cdot x\right]}{dt}, \tag{9.16}$$

of equation (9.14). Consequently, the function $x(t) = e^{tJ} \cdot x_0$ satisfies the initial value problem (9.15). The uniqueness theorem for first order systems of ordinary differential equations completes the proof. *Q.E.D.*

Corollary 9.22. *An infinitesimal generator* $J \in \mathfrak{g}$ *belongs to the subalgebra of the isotropy subgroup* $G_x \subset G$ *of a point* $x \in X$ *if and only if the associated vector field vanishes there, so* $\xi_J(x) = 0$.

Example 9.23. For example, we can recover the subgroups of GL(2) corresponding to the infinitesimal generators (9.13) by integrating the associated ordinary differential equations

$$\frac{dp}{dt} = 1, \qquad \frac{dp}{dt} = 2p, \qquad \frac{dp}{dt} = -p^2, \qquad \frac{dp}{dt} = 0.$$

In accordance with (9.15), the right-hand side in each case is merely minus the coefficient of $\partial/\partial p$ in the vector field. Indeed, the corresponding solutions with initial data $p(0) = p_0$ are

$$p = p_0 + t, \qquad p = e^{2t}p_0, \qquad p = \frac{p_0}{tp_0 + 1}, \qquad p(0) = p_0,$$

as promised.

For a transformation group, the *commutator*

$$[\mathbf{v}_J, \mathbf{v}_K] = \mathbf{v}_J \cdot \mathbf{v}_K - \mathbf{v}_K \cdot \mathbf{v}_J \tag{9.17}$$

between two infinitesimal generators is known as their *Lie bracket*. As with the matrix commutator, the Lie bracket indicates commutativity (or lack thereof) of the associated flows. It is not hard to see that the Lie bracket is both bilinear and skew-symmetric, and satisfies the Jacobi identity (9.7), and therefore defines a Lie algebra structure on the infinite-dimensional space of analytic vector fields on the manifold X. See [169] for a proof of the following basic result.

Theorem 9.24. *The map* $\sigma: J \mapsto \mathbf{v}_J$ *that associates a Lie algebra element* $J \in \mathfrak{g}$ *with its infinitesimal generator* $\mathbf{v}_J$ *on* X *forms a Lie algebra homomorphism, so* σ *is linear:* $\mathbf{v}_{cJ+dK} = c\,\mathbf{v}_J + d\,\mathbf{v}_K$, *for* $J, K \in \mathfrak{g}$, *and* $c, d \in \mathbb{R}$, *and preserves commutators:* $\mathbf{v}_{[J,K]} = [\,\mathbf{v}_J, \mathbf{v}_K\,]$.

Example 9.25. Let us apply this result to the projective action of $GL(2)$. Consider a general 2×2 matrix

$$J = \begin{pmatrix} a & b \\ c & d \end{pmatrix} = b\,J^- + \tfrac{1}{2}(a-d)\,J^0 + c\,J^+ + \tfrac{1}{2}(a+d)\,\mathbf{1}.$$

Linearity implies that the corresponding infinitesimal generator is

$$\mathbf{v}_J = b\,\mathbf{v}_- + \frac{a-d}{2}\,\mathbf{v}_0 + c\,\mathbf{v}_+ + \frac{a+d}{2}\,\mathbf{v}_e = \left[-b + (d-a)p + cp^2\right]\frac{\partial}{\partial p}. \quad (9.18)$$

This can easily be checked directly. The infinitesimal generators $\mathbf{v}_-$, $\mathbf{v}_0$, $\mathbf{v}_+$ obey the same commutation relations as the matrices J^-, J^0, J^+:

$$[\,\mathbf{v}_-, \mathbf{v}_0\,] = -2\mathbf{v}_-, \qquad [\,\mathbf{v}_+, \mathbf{v}_0\,] = 2\mathbf{v}_+, \qquad [\,\mathbf{v}_-, \mathbf{v}_+\,] = \mathbf{v}_0, \quad (9.19)$$

while $\mathbf{v}_e \equiv 0$ commutes with every generator (9.18).

The infinitesimal generators are intimately tied to the underlying geometry of the group action.

Proposition 9.26. *Let* G *act on* X, *and let* $x \in X$. *The vector space* $V|_x = \mathrm{Span}\{\xi_J(x) \mid J \in \mathfrak{g}\}$ *spanned by all the vector fields determined by the infinitesimal generators at* x *coincides with the tangent space to the orbit* $\mathcal{O}_x$ *of* G *that passes through* x, *so* $V|_x = T\mathcal{O}_x|_x$. *In particular, the dimension of* $\mathcal{O}_x$ *equals the dimension of* $V|_x$. *Moreover, the isotropy subgroup* $G_x \subset G$ *has dimension* $\dim G - \dim \mathcal{O}_x = r - s$.

Proof: The streamlines $e^{tJ} \cdot x$ of the one-parameter subgroups of G passing through x describe curves in the full group orbit $\mathcal{O}_x$. Hence their tangent vectors, namely $\xi_J(x)$, belong to the tangent space to $\mathcal{O}_x$. Since the exponential maps the Lie algebra onto a neighborhood of the identity in G, every tangent vector to $\mathcal{O}_x$ can be found in this way. The final statement then follows from Corollary 9.22. *Q.E.D.*

Example 9.27. Consider the standard action $\mathbf{x} \mapsto R\mathbf{x}$ of the rotation group $\mathrm{SO}(3)$ on $\mathbb{R}^3$. A basis for the infinitesimal generators is given by $\mathbf{v}_x = z\partial_y - y\partial_z$, $\mathbf{v}_y = x\partial_z - z\partial_x$, and $\mathbf{v}_z = y\partial_x - x\partial_y$, corresponding to the vector fields $\xi_x = (0, z, -y)$, $\xi_y = (-z, 0, x)$, $\xi_z = (y, -x, 0)$. These generate the one-parameter subgroups corresponding to rotations around the x, y, and z axes respectively. For example, the flow of the first vector field, $(x(t), y(t), z(t)) = (x, y\cos t - z\sin t, y\sin t + z\cos t)$, can be obtained by solving the linear system of ordinary differential equations $\dot{x} = 0$, $\dot{y} = -z$, $\dot{z} = y$. Note that, at any point (x, y, z), the three vector fields ξ_x, ξ_y, ξ_z span the tangent space to the sphere centered at the origin, reconfirming the connection between infinitesimal generators and orbits.

Infinitesimal Invariance

We now apply the infinitesimal generators to the computation of group invariants. Recall that if $J \in \mathfrak{g}$ generates a one-parameter subgroup $H_J \subset G$, then $\mathbf{v}_J(F(x))$ represents the infinitesimal change in the function $F(x)$ under H_J. In particular, if $F(x)$ is an invariant, then it does not change value under the action of G, and hence we expect $\mathbf{v}_J(F(x)) \equiv 0$ to vanish. It turns out that, for connected Lie group actions, this condition is both necessary and sufficient.

Theorem 9.28. *Let $G \subset \mathrm{GL}(n)$ be a connected Lie group acting on $X \subset \mathbb{R}^n$. A function $I \colon X \to \mathbb{R}$ is G-invariant if and only if I satisfies the infinitesimal invariance conditions*

$$\mathbf{v}_J(I) = 0 \qquad \text{for all infinitesimal generators} \quad J \in \mathfrak{g}. \qquad (9.20)$$

Remark: Since the map to infinitesimal generators is linear, we only need check the infinitesimal invariance condition (9.20) for a *basis* of $\mathfrak{g}$.

Proof: According to the definition (9.12), if

$$I(e^{-tJ}x) = I(x), \qquad (9.21)$$

then (9.20) holds automatically. To prove the converse, we require a slight extension of (9.12):

$$\mathbf{v}_J(I)\left(e^{-tJ} \cdot x\right) = \frac{d}{dt} I\left(e^{-tJ} \cdot x\right), \qquad (9.22)$$

valid for all t. As in the proof of (9.16), this is proved by differentiating both sides of $I\left(e^{-(s+t)J} \cdot x\right) = I\left(e^{-sJ} \cdot e^{-tJ} \cdot x\right)$ with respect to s and setting $s = 0$. Thus, if I satisfies the infinitesimal invariance criterion (9.20), then (9.22) implies that $I(e^{-tJ}x)$ does not depend on t, and hence (9.21) holds. Therefore, $I(g \cdot x) = I(x)$ for all $g = e^J$ lying in the image of the exponential map $\exp: \mathfrak{g} \to G$. As we have seen, not every group element lies in the image of exp, but straightforward induction based on Lemma 9.11 completes the proof of Theorem 9.28. *Q.E.D.*

Remark: If G is not connected, then the infinitesimal invariance criterion (9.20) only implies invariance of the function under the connected component of G containing the identity. To prove that I is invariant under all group transformations, we need also to check its invariance under a single group element in each of the other connected components. For example, to prove invariance of a function under the full general linear group $\mathrm{GL}(n, \mathbb{R})$, we need to check its infinitesimal invariance under the generators in $\mathfrak{gl}(n, \mathbb{R})$, which proves invariance under $\mathrm{GL}(n, \mathbb{R})^+$, and then its invariance under the action of a single orientation-reversing matrix, e.g., $\mathrm{diag}(-1, 1, 1, \dots, 1)$.

Example 9.29. For the standard action of $\mathrm{SO}(3)$ on $\mathbb{R}^3$ considered in Example 9.27, it is a simple matter to check that the function $r^2 = x^2 + y^2 + z^2$ is annihilated by the infinitesimal generators $\mathbf{v}_x, \mathbf{v}_y$, and $\mathbf{v}_z$. For example, $\mathbf{v}_x(r^2) = z\partial_y(r^2) - y\partial_z(r^2) = 2yz - 2yz = 0$. The rotational invariance of r^2 is then a consequence of Theorem 9.28.

Consider a one-parameter group with infinitesimal generator (9.14). The corresponding infinitesimal invariance condition (9.20) is the linear partial differential equation

$$\xi_J^1(x) \frac{\partial I}{\partial x_1} + \xi_J^2(x) \frac{\partial I}{\partial x_2} + \cdots + \xi_J^m(x) \frac{\partial I}{\partial x_m} = 0. \qquad (9.23)$$

The solutions of (9.23) are effectively found by the classical method of characteristics, cf. [**168**]. We replace the partial differential equation by the characteristic system of ordinary differential equations

$$\frac{dx^1}{\xi_J^1(x)} = \frac{dx^2}{\xi_J^2(x)} = \cdots = \frac{dx^m}{\xi_J^m(x)}. \qquad (9.24)$$

The general solution to (9.24) can be (locally) written in the form $I_1(x) = c_1, \dots, I_{m-1}(x) = c_{m-1}$, where the c_i are constants of integration, and the resulting functions $I_1, \dots, I_{m-1}$ form a complete set of functionally independent invariants of the one-parameter group generated by $\mathbf{v}_J$.

Example 9.30. Consider the (local) one-parameter transformation group

$$(x,y,z) \longmapsto \left(x\cos t - y\sin t, x\sin t + y\cos t, \frac{\sin t + z\cos t}{\cos t - z\sin t} \right). \quad (9.25)$$

(This forms a one-parameter subgroup of the first order prolonged Euclidean action considered in Example 8.27.) Differentiating with respect to t, we find that its infinitesimal generator is the vector field $\mathbf{v} = y\partial_x - x\partial_y - (1+z^2)\partial_z$. Therefore, the infinitesimal invariance condition (9.20) is

$$\mathbf{v}(I) = y\frac{\partial I}{\partial x} - x\frac{\partial I}{\partial y} - (1+z^2)\frac{\partial I}{\partial z} = 0. \quad (9.26)$$

The characteristic system (9.24) for (9.26) is

$$\frac{dx}{y} = \frac{dy}{-x} = \frac{dz}{-(1+z^2)}.$$

The first equation reduces to a simple separable ordinary differential equation $dy/dx = -x/y$, with general solution $x^2 + y^2 = c_1$, for c_1 a constant of integration; therefore, the cylindrical radius $r = \sqrt{x^2 + y^2}$ provides one invariant. To solve the second characteristic equation, we replace x by $\sqrt{r^2 - y^2}$ and treat r as constant. The general solution is found to be $\tan^{-1} z - \sin^{-1}(y/r) = \tan^{-1} z - \tan^{-1}(y/x) = c_2$, and hence $\tan^{-1} z - \tan^{-1}(y/x)$ is a second invariant. A more convenient choice is provided by the tangent of this invariant. We deduce that, for $yz + x \neq 0$, the functions $r = \sqrt{x^2 + y^2}$, $w = (xz - y)/(yz + x)$ form a complete system of functionally independent invariants.

For multi-parameter groups, the invariants are simultaneous solutions to an overdetermined system of linear, homogeneous, first order partial differential equations. One solution method is to look for invariants of each generator in turn and try to re-express subsequent generators in terms of the invariants. However, this can become quite complicated to implement in practice; see [**168, 169**] for examples.

We now specialize these general results to our topic at hand: the invariants and covariants of binary forms. We begin with the simple case of binary quadratics.

Example 9.31. Consider the action of SL(2) on the space $\mathcal{Q}$ of binary quadratic forms $Q(x,y) = ax^2 + 2bxy + cy^2$. (The extension to GL(2) will be treated shortly.) We use the coefficients a, b, c as coordinates on $\mathcal{Q} \simeq \mathbb{R}^3$. The infinitesimal generators will be found by differentiating the transformation rules $Q(e^{-tJ}\mathbf{x})$ with respect to t and setting $t = 0$. The first three one-parameter subgroups (9.11) correspond to the transformations $(x + ty, y)$, $(e^t x, e^{-t} y)$, and $(x, y + tx)$, and hence, by (1.12), induce the following transformation rules on the coefficients:

$$
(a, b, c) \quad \longmapsto \quad
\begin{cases}
(a - 2tb + t^2 c, b - tc, c), \\
(e^{-2t} a, b, e^{2t} c), \\
(a, b - ta, c - 2tb + t^2 a),
\end{cases}
$$

respectively. Differentiating with respect to t and setting $t = 0$ produces the three infinitesimal generators

$$
\mathbf{v}_- = a\partial_b + 2b\partial_c, \qquad \mathbf{v}_0 = 2a\partial_a - 2c\partial_c, \qquad \mathbf{v}_+ = 2b\partial_a + c\partial_b. \qquad (9.27)
$$

These differential operators are readily seen to annihilate the discriminant $\Delta = ac - b^2$, re-proving that it forms an SL(2)–invariant.

The generalization of this example to higher degree forms is straightforward, and the result dates back to Cayley, [**41**].

Lemma 9.32. *The action of* SL(2) *on the space* $\mathcal{P}^{(n)}$ *of homogeneous polynomials of degree* n *has the following infinitesimal generators:*

$$
\mathbf{v}_- = na_1 \frac{\partial}{\partial a_0} + (n-1)a_2 \frac{\partial}{\partial a_1} + \cdots + 2a_{n-1} \frac{\partial}{\partial a_{n-2}} + a_n \frac{\partial}{\partial a_{n-1}},
$$

$$
\mathbf{v}_0 = -na_0 \frac{\partial}{\partial a_0} + (2-n)a_1 \frac{\partial}{\partial a_1} + \cdots + (n-2)a_{n-1} \frac{\partial}{\partial a_{n-1}} + na_n \frac{\partial}{\partial a_n},
$$

$$
\mathbf{v}_+ = a_0 \frac{\partial}{\partial a_1} + 2a_1 \frac{\partial}{\partial a_2} + \cdots + (n-1)a_{n-2} \frac{\partial}{\partial a_{n-1}} + na_{n-1} \frac{\partial}{\partial a_n}. \qquad (9.28)
$$

Proposition 9.33. *A function* $I(\mathbf{a})$ *is an invariant of the induced action of* SL(2) *on the space of binary forms if and only if it satisfies the infinitesimal invariance criteria*

$$
\mathbf{v}_-(I) = \mathbf{v}_0(I) = \mathbf{v}_+(I) = 0. \qquad (9.29)
$$

Thus, the classification of classical invariants can be now viewed as the study of the solution space to a particular linear system of partial differential equations (9.29). The method of characteristics is available here, but the computations become too complicated for it to be effective. On the other hand, if we restrict our attention to invariants of a *fixed* order, then the infinitesimal invariance conditions (9.29) reduce to a (large) system of *linear* equations for the coefficients, which can be directly solved with the aid of computer algebra.

In terms of the coordinates $(a_0, \ldots, a_n)$ on $\mathcal{P}^{(n)}$, the vector fields corresponding to the infinitesimal generators (9.28) are given by

$$\xi_-(\mathbf{a}) = (na_1, (n-1)a_2, \ldots, 2a_{n-1}, a_n, 0),$$

$$\xi_0(\mathbf{a}) = (-na_0, (2-n)a_1, \ldots, (n-2)a_{n-1}, -na_n), \qquad (9.30)$$

$$\xi_+(\mathbf{a}) = (0, a_0, 2a_1, \ldots, (n-1)a_{n-1}, na_n).$$

According to Proposition 9.26, the dimension $0 \le k \le 3$ of the subspace spanned by $\xi_-(\mathbf{a})$, $\xi_0(\mathbf{a})$, $\xi_+(\mathbf{a})$, when evaluated on the coefficients of a given binary form $Q \in \mathcal{P}^{(n)}$, equals the dimension of the SL(2) orbit containing Q. Moreover, the symmetry (or isotropy) group of Q forms a subgroup of SL(2) of dimension $3 - k$. In particular, if the vector fields are linearly independent, then Q admits at most a discrete group of unimodular symmetries. This provides us with an alternative method for recognizing the symmetry groups of binary forms, classified earlier in Theorem 8.66. In particular, a binary form Q is equivalent to a monomial if and only if the three vectors (9.30) are linearly dependent at Q. The reader might enjoy proving this result directly.

There is a corresponding infinitesimal criterion for covariants. In this case, we must also take infinitesimal changes in the x, y coordinates into account.

Theorem 9.34. *Let $Q(\mathbf{x})$ be a binary form as in (2.1). A function $J(\mathbf{a}, \mathbf{x})$ depending on the coefficients of Q and the coordinates $\mathbf{x} = (x, y)$ is an SL(2)–covariant if and only if it satisfies the infinitesimal invariance conditions*

$$\mathbf{v}_-(J) = \mathbf{v}_0(J) = \mathbf{v}_+(J) = 0, \qquad (9.31)$$

where the infinitesimal generators are

$$\mathbf{v}_- = na_1 \frac{\partial}{\partial a_0} + (n-1)a_2 \frac{\partial}{\partial a_1} + \cdots + a_n \frac{\partial}{\partial a_{n-1}} - y \frac{\partial}{\partial x},$$

$$\mathbf{v}_0 = -na_0 \frac{\partial}{\partial a_0} + (2-n)a_1 \frac{\partial}{\partial a_1} + \cdots + na_n \frac{\partial}{\partial a_n} - x \frac{\partial}{\partial x} + y \frac{\partial}{\partial y},$$

$$\mathbf{v}_+ = a_0 \frac{\partial}{\partial a_1} + 2a_1 \frac{\partial}{\partial a_2} + \cdots + na_{n-1} \frac{\partial}{\partial a_n} - x \frac{\partial}{\partial y}. \qquad (9.32)$$

Exercise 9.35. Show that the differential operators (9.32) annihilate the binary form Q itself, as in (4.26). Further verify that they annihilate the cubic and quartic Hessians (2.24) and (2.31).

Exercise 9.36. Prove that if $J(\mathbf{a}, \mathbf{x})$ is any polynomial function, then $(\mathbf{v}_-)^N(J) = 0 = (\mathbf{v}_+)^N(J)$ for $N \gg 0$ sufficiently large. Can you determine the minimal N in advance?

Infinitesimal Multipliers and Relative Invariants

Multiplier representations and relative invariants can also be effectively treated using the infinitesimal calculus. Let G be a connected Lie group acting on a manifold X, and let $\mu(g, x)$ be a scalar multiplier. As before, we look at the infinitesimal change in a function $F(x)$ under the associated multiplier representation. According to (4.20), given $g \in G$, the transformed function $\overline{F} = g \cdot F$ is given by

$$\overline{F}(x) = \mu(g, g^{-1} \cdot x)\, F(g^{-1} \cdot x) = \frac{F(g^{-1} \cdot x)}{\mu(g^{-1}, x)},$$

the second equality resulting from the multiplier equation (4.21). Therefore, given $J \in \mathfrak{g}$, the associated *infinitesimal generator* is a first order differential operator $\mathcal{D}_J$, defined so that

$$\mathcal{D}_J(F(x)) = \frac{d}{dt} \left. \frac{F\left(e^{-tJ} \cdot x\right)}{\mu\left(e^{-tJ}, x\right)} \right|_{t=0} = -\frac{d}{dt} \left. \frac{F\left(e^{tJ} \cdot x\right)}{\mu\left(e^{tJ}, x\right)} \right|_{t=0} \tag{9.33}$$

for every analytic function $F \colon X \to \mathbb{R}$. A straightforward chain rule computation demonstrates that

$$\mathcal{D}_J = \mathbf{v}_J + h_J(x) = \sum_{i=1}^{p} \xi_J^i(x)\, \frac{\partial}{\partial x^i} + h_J(x), \tag{9.34}$$

where $\mathbf{v}_J$ is the infinitesimal generator (9.14) of the original action on X, and

$$h_J(x) = -\frac{d}{dt} \left. \frac{1}{\mu(e^{-tJ}, x)} \right|_{t=0} = \frac{d}{dt} \left. \mu(e^{tJ}, x) \right|_{t=0}. \tag{9.35}$$

The map taking an element J to the infinitesimal generator $\mathcal{D}_J$ is a Lie algebra homomorphism, and so

$$\mathcal{D}_{[J,K]} = [\mathcal{D}_J, \mathcal{D}_K] = \mathcal{D}_J \cdot \mathcal{D}_K - \mathcal{D}_K \cdot \mathcal{D}_J. \tag{9.36}$$

Therefore, the infinitesimal generators of a multiplier representation form a Lie algebra of first order differential operators. Conversely, any such Lie algebra forms the generator of some multiplier representation. Consequently, the problem of classifying multiplier representations of connected Lie groups can be reduced to the problem of classifying Lie algebras of first order differential operators. We refer the reader to the survey paper [**85**] for extensive results, including new applications to quantum mechanics via the theory of "quasi-exactly solvable" Schrödinger operators, cf. [**220**].

Example 9.37. Applying (9.35), the fundamental multiplier $\mu_{n,k}$ corresponding to the action of GL(2) on the space of weight k functions given in (4.22) is readily seen to have associated infinitesimal generators

$$\mathcal{D}_- = -\partial_p, \quad \mathcal{D}_0 = -2p\partial_p + n, \quad \mathcal{D}_+ = p^2\partial_p - np, \quad \mathcal{D}_e = -n - 2k. \tag{9.37}$$

For example, under the subgroup $p \mapsto p/(-tp + 1)$ generated by J^+, we have

$$h_+(p) = \frac{d}{dt}\left.\mu_{n,k}(e^{tJ^+}, p)\right|_{t=0} = \frac{d}{dt}\left.(-tp + 1)^{-n}\right|_{t=0} = -np,$$

and so $\mathcal{D}_+ = \mathbf{v}_+ + h_+(p)$ is as given. In accordance with the general theory, the differential operators (9.37) form a Lie algebra of differential operators isomorphic to $\mathfrak{gl}(2)$.

Theorem 9.38. *Let G be a connected group of transformations acting on X. A function $R(x)$ is a relative invariant for an associated multiplier representation if and only if it is annihilated by all the infinitesimal generators:*

$$\mathcal{D}_J(R) = 0 \qquad \text{for every} \qquad J \in \mathfrak{g}. \tag{9.38}$$

Consider the action of GL(2) on the space of homogeneous polynomials of degree n and weight m. A covariant of weight k must be annihilated by the infinitesimal generators (9.32), as well as the scaling generator

$$\mathcal{D} = (n + 2m)\left\{a_0\frac{\partial}{\partial a_0} + a_1\frac{\partial}{\partial a_1} + \cdots + a_n\frac{\partial}{\partial a_n}\right\} - x\frac{\partial}{\partial x} - y\frac{\partial}{\partial y} - 2k, \tag{9.39}$$

which corresponds to the one-parameter center $\{\lambda\mathbb{1}\}$. According to Euler's formula (2.13), a function $F(\mathbf{a}, \mathbf{x})$ is annihilated by the scaling generator $\mathcal{D}$ if and only if it is homogeneous of degree $l = 2k/(n + 2m)$ under the scaling $\mathbf{a} \mapsto \lambda\mathbf{a}$, $\mathbf{x} \mapsto \lambda^{-1/(n+2m)}\mathbf{x}$.

Exercise 9.39. Prove that if $n \geq 3$, the generic orbits of GL(2) on the space $\mathcal{P}^{(n)}$ of binary forms of degree n are four-dimensional.

Isobaric and Semi-invariants

The determination of the invariants for a complicated transformation group can often be effectively handled inductively by a preliminary analysis of a simpler subgroup. If G is any group acting on a space X, and

$H \subset G$ is a subgroup, then any G-invariant function $I(x)$ is automatically H-invariant. Thus, if we know a complete system of invariants for H, say, $J_1(x), \ldots, J_s(x)$, then we can write all the G-invariants as functions, $I(x) = F(J_1(x), \ldots, J_s(x))$, thereof. In favorable situations, this will reduce the number of variables, changing the direct problem of determining G-invariants into several simpler subproblems.

In the case of $GL(2) = GL(2, \mathbb{R})$, there are five important subgroups, forming a chain[†]

$$\{e\} \subset S \subset U \subset SL(2) \subset GL(2)^+ \subset GL(2). \tag{9.40}$$

Here

$$S = \left\{ \begin{pmatrix} \lambda & 0 \\ 0 & \lambda^{-1} \end{pmatrix} \right\}, \qquad U = \left\{ \begin{pmatrix} \lambda & \beta \\ 0 & \lambda^{-1} \end{pmatrix} \right\}, \tag{9.41}$$

are, respectively, the unimodular or *isobaric* scaling subgroup, and the upper triangular unimodular subgroup. In the real case, we require S and U to be connected, and so we restrict to $\lambda > 0$. Note that S is a one-parameter subgroup, while according to Example 9.18, $U \simeq A(1)$ is isomorphic to the one-dimensional affine group. We shall be interested in the action of these subgroups on the space $\mathcal{P}^{(n)}$ of binary forms, although the inductive approach is applicable to more general actions of $GL(2)$. Classical invariants are absolutely invariant under the unimodular subgroup $SL(2)$, and relatively invariant under the full group. Consequently, any classical invariant must also be invariant under the subgroups S and U as well.

Definition 9.40. A function $F(\mathbf{a}, \mathbf{x})$ is called *isobaric* if it is invariant under the action of the unimodular scaling subgroup S, and is a *semi-invariant* if it is invariant under the unimodular upper triangular subgroup U.

In particular, semi-invariants are isobaric, and invariants are semi-invariant. We can thus try to determine classical invariants by first determining the isobaric invariants, and then using these to construct semi-invariants, with the final step in the procedure being the construction of the full invariant from semi-invariants.

The subgroup chain (9.40) implies a chain of Lie algebras

$$\begin{array}{ccccccc} \{0\} \subset & \mathbb{R} & \subset & \mathfrak{a}(1) & \subset & \mathfrak{sl}(2) & \subset & \mathfrak{gl}(2) \\ \{0\} \subset & \{J^0\} & \subset & \{J^0, J^-\} & \subset & \{J^0, J^-, J^+\} & \subset & \{J^0, J^-, J^+, \mathbb{1}\}. \end{array} \tag{9.42}$$

[†] In the complex case, the penultimate group does not appear.

(Both $GL(2)^+$ and $GL(2)$ have the same Lie algebra.) Moreover, according to Theorem 9.28, any function which is invariant under one of these subgroups must be annihilated by the corresponding infinitesimal generators. Thus, $F(\mathbf{a}, \mathbf{x})$ is isobaric if and only if it satisfies the differential equation

$$\mathbf{v}_0(F) = 0, \tag{9.43}$$

where $\mathbf{v}_0$ is given in (9.32). For example, in the case of quadratic polynomials, the $\mathbf{x}$-independent isobaric functions are those satisfying the first order partial differential equation $aF_a = cF_c$. Using the method of characteristics — or by inspection — we find that the general isobaric function has the form $I = F(ac, b)$.

Since the subgroup S acts on monomials by scaling $a_i \mapsto \lambda^{2i-n} a_i$, the construction of polynomial isobaric invariants is straightforward.

Theorem 9.41. *Every isobaric polynomial depending on the coefficients $a_0, \ldots, a_n$ of a binary form is a sum of monomials of the form*

$$a_{i_1} a_{i_2} \cdots a_{i_k}, \qquad \text{where} \qquad i_1 + \cdots + i_k = \tfrac{1}{2} nk. \tag{9.44}$$

Exercise 9.42. Prove that if F is isobaric, so are the functions $\mathbf{v}_+ \mathbf{v}_-(F)$ and $\mathbf{v}_- \mathbf{v}_+(F)$. Prove that these functions are identical.

Exercise 9.43. Extend Theorem 9.41 to the case of isobaric covariants $F(\mathbf{a}, \mathbf{x})$. What about rational isobaric invariants?

Similarly, the infinitesimal criterion for semi-invariants is

$$\mathbf{v}_-(F) = 0 = \mathbf{v}_0(F). \tag{9.45}$$

For example, a function $F(a, b, c)$ depending on the coefficients of a binary quadratic is a semi-invariant if and only if it satisfies the two partial differential equations

$$\mathbf{v}_0(F) = a\frac{\partial F}{\partial a} - c\frac{\partial F}{\partial c} = 0, \qquad \mathbf{v}_-(F) = a\frac{\partial F}{\partial b} + 2b\frac{\partial F}{\partial c} = 0.$$

First, F must be isobaric, so $F = F(s, b)$, where $s = ac$. Then $\mathbf{v}_-(F) = a\big(2bF_s + F_b\big)$, and hence $F = F(s - b^2) = F(ac - b^2)$ is a function of the discriminant.

There is an important general correspondence between covariants and semi-invariants. Any nonzero covariant of degree n has the form

$$J(\mathbf{a}, x, y) = K(\mathbf{a})\, x^n + M(\mathbf{a})\, x^{n-1} y + \cdots. \tag{9.46}$$

We call $L(\mathbf{a}, x) = J(\mathbf{a}, x, 0) = K(\mathbf{a})\, x^n$ the *leading term* of J, and $K(\mathbf{a}) = J(\mathbf{a}, 1, 0)$ its *leading coefficient*. Note that the leading term of an invariant $I(\mathbf{a})$ is the invariant itself.

Lemma 9.44. *The linear map taking a polynomial covariant to its leading term is invertible.*

Proof: We need only check that a nonzero covariant (9.46) cannot have a zero leading term, i.e., $K(\mathbf{a}) \not\equiv 0$. Note that the unimodular shearing transformation $(x, y) \mapsto (x, \gamma x + y)$ does not alter a covariant. On the other hand, we can always arrange a value of γ so that any function $J(\mathbf{a}, x, \gamma x + y)$ will have a nonzero leading coefficient. *Q.E.D.*

Proposition 9.45. *The leading term of any polynomial covariant is a semi-invariant.*

Proof: Since $J(\mathbf{a}, \mathbf{x})$ is a covariant, it is also invariant under the upper triangular subgroup U, and so

$$J(\mathbf{a}, \lambda x + \beta y, \lambda^{-1} y) = J(\overline{\mathbf{a}}, x, y)$$

for all x, y, where $\overline{\mathbf{a}}$ denote the transformed coefficients of the binary form. Setting $y = 0$ in this identity demonstrates that

$$J(\mathbf{a}, \lambda x, 0) = L(\mathbf{a}, \lambda x) = L(\overline{\mathbf{a}}, x) = J(\overline{\mathbf{a}}, x, 0),$$

which proves the semi-invariance of L. Therefore, the leading coefficient is a relative semi-invariant of weight n, so $K(\overline{\mathbf{a}}) = \lambda^{-n} K(\mathbf{a})$. *Q.E.D.*

Proposition 9.45 shows that one can effectively work with semi-invariants instead of covariants. A particular consequence of Lemma 9.44 is that one can use semi-invariants to detect syzygies and other algebraic identities among the more complicated covariants.

Theorem 9.46. *The relation* $P(J_1, \ldots, J_m) = 0$ *defines a syzygy among covariants if and only if* $P(K_1, \ldots, K_m) = 0$ *gives the identical relation among their leading coefficients.*

Proof: It suffices to note that the leading coefficient of a polynomial combination $P(J_1, \ldots, J_m) = 0$ of covariants is precisely $P(K_1, \ldots, K_m)$. Therefore, by Lemma 9.44, P defines a syzygy if and only if it has zero leading term. *Q.E.D.*

Example 9.47. For a binary cubic Q, the leading coefficients of its fundamental covariants $Q, H = \frac{1}{2}(Q, Q)^{(2)}$, $T = (Q, H)^{(1)}$ are given by $\widehat{Q} = a_0$, $\frac{1}{36}\widehat{H} = a_1 a_3 - a_2^2$, and $\frac{1}{108}\widehat{T} = a_0 a_3^2 - 3 a_1 a_2 a_3 + 2 a_2^3$, respectively. The discriminant Δ, cf. (2.22), is its own leading term since it is an invariant. The evident relation $\widehat{T}^2 + \widehat{H}^3 = 2^4 3^6 \widehat{Q}^2 \Delta$ implies the fundamental syzygy (2.47) for the cubic.

Remark: See Cushman and Sanders, [57], for applications of semi-invariants to the determination of normal forms for systems of ordinary differential equations.

The Hilbert Operator

As we have seen, isobaric functions and polynomials are relatively easy to construct. Hilbert, [106], made the remarkable discovery of a differential operator which converts isobaric functions into genuine invariants. The Hilbert operators therefore play a role analogous to the symmetrization operator for a finite group, or invariant integration for compact groups. The only caveat is that one needs to start with isobaric invariants instead of general functions. In the case of $GL(n)$, $n \geq 3$, Hilbert's construction was simplified by Story, [202]. To keep things reasonably simple, though, we shall concentrate on the case of binary forms.

The starting point is the commutation relations (9.19) between the infinitesimal generators of the unimodular group, which we rewrite in the equivalent form

$$\mathbf{v}_0\mathbf{v}_- = \mathbf{v}_-(\mathbf{v}_0 - 2), \qquad \mathbf{v}_0\mathbf{v}_+ = \mathbf{v}_+(\mathbf{v}_0 + 2), \qquad \mathbf{v}_+\mathbf{v}_- = \mathbf{v}_-\mathbf{v}_+ + \mathbf{v}_0.$$
$$(9.47)$$

The first two immediately imply corresponding commutation formulae between $\mathbf{v}_0$ and the kth powers of the other two generators:

$$\mathbf{v}_0(\mathbf{v}_-)^k = (\mathbf{v}_-)^k(\mathbf{v}_0 - 2k), \qquad \mathbf{v}_0(\mathbf{v}_+)^k = (\mathbf{v}_+)^k(\mathbf{v}_0 + 2k). \quad (9.48)$$

A simple induction establishes the next important formulae:

$$\mathbf{v}_+(\mathbf{v}_-)^k = (\mathbf{v}_-)^k\mathbf{v}_+ + (\mathbf{v}_-)^{k-1}[k\mathbf{v}_0 - k(k-1)],$$
$$\mathbf{v}_-(\mathbf{v}_+)^k = (\mathbf{v}_+)^k\mathbf{v}_- + (\mathbf{v}_+)^{k-1}[-k\mathbf{v}_0 - k(k-1)].$$
$$(9.49)$$

All these formulae just depend on the commutation relations and so work whenever $\mathbf{v}_-, \mathbf{v}_0, \mathbf{v}_+$ are the infinitesimal generators of any $SL(2)$ action; the reader can explicitly verify them directly in the case of (9.27) or, more generally, (9.32). The general commutator between powers of the generators is a bit too complicated, but the following special case is needed.

Lemma 9.48. *Let F be an isobaric function. Then*

$$\mathbf{v}_+(\mathbf{v}_-)^k F = (\mathbf{v}_-)^k \mathbf{v}_+ F - k(k-1)(\mathbf{v}_-)^{k-1} F,$$
$$\mathbf{v}_-(\mathbf{v}_+)^k F = (\mathbf{v}_+)^k \mathbf{v}_- F - k(k-1)(\mathbf{v}_+)^{k-1} F, \qquad (9.50)$$
$$(\mathbf{v}_+)^k (\mathbf{v}_-)^k F = (\mathbf{v}_-)^k (\mathbf{v}_+)^k F.$$

Proof: The first two equations follow immediately from (9.49) and the infinitesimal isobaric condition (9.43). To prove the third condition, the case $k = 1$ appears in Exercise 9.42. We then use induction and apply the first two formulae as follows:

$$(\mathbf{v}_+)^k (\mathbf{v}_-)^k F = (\mathbf{v}_+)^{k-1} (\mathbf{v}_-)^k \mathbf{v}_+ F - k(k-1)(\mathbf{v}_+)^{k-1}(\mathbf{v}_-)^{k-1} F$$
$$= (\mathbf{v}_-)^{k-1} (\mathbf{v}_+)^{k-1} \mathbf{v}_- \mathbf{v}_+ F - k(k-1)(\mathbf{v}_-)^{k-1}(\mathbf{v}_+)^{k-1} F$$
$$= (\mathbf{v}_-)^{k-1} (\mathbf{v}_+)^k \mathbf{v}_- F - k(k-1)(\mathbf{v}_-)^{k-1}(\mathbf{v}_+)^{k-1} F$$
$$= (\mathbf{v}_-)^k (\mathbf{v}_+)^k F.$$

The second equality follows since $\mathbf{v}_- \mathbf{v}_+ F = \mathbf{v}_+ \mathbf{v}_- F$ is isobaric whenever F is — as is $(\mathbf{v}_+)^k (\mathbf{v}_-)^k F$. $\qquad$ *Q.E.D.*

We shall first apply these formulae to provide a mechanism for reconstructing covariants from their leading terms, in accordance with Lemma 9.44.

Theorem 9.49. *Suppose $L(\mathbf{a}, x) = K(\mathbf{a}) x^n$ is a y-independent semi-invariant. Then*

$$J(\mathbf{a}, \mathbf{x}) = \sum_{k=0}^{n} \frac{(\mathbf{v}_+)^k K}{k!} \, x^{n-k} y^k \qquad (9.51)$$

is the unique covariant which has L as its leading term.

Proof: We must show that (9.51) is annihilated by the three infinitesimal generators (9.32). Note that since K is a polynomial semi-invariant of weight n, and $\mathbf{v}_+$ lowers the weight of a semi-invariant by 1, we have $(\mathbf{v}_+)^{n+1} K = 0$. Now, proving $\mathbf{v}_+(J) = 0$ is easy, since

$$\mathbf{v}_+[\{(\mathbf{v}_+)^k K\} x^{n-k} y^k] = \{(\mathbf{v}_+)^{k+1} K\} x^{n-k} y^k - k\{(\mathbf{v}_+)^k K\} x^{n-k+1} y^{k-1}.$$

For the scaling generator $\mathbf{v}_0$, we use the fact that $\mathbf{v}_0(K) = -nK$, and hence by (9.48),

$$\mathbf{v}_0 \big[\{(\mathbf{v}_+)^k K\} \, x^{n-k} y^k \big] =$$
$$= \{(\mathbf{v}_+)^k (\mathbf{v}_0 + 2k) K\} \, x^{n-k} y^k + (n - 2k)\{(\mathbf{v}_+)^k K\} \, x^{n-k} y^k = 0,$$

so $\mathbf{v}_0(J) = 0$. Finally, we use (9.49) and the fact that $\mathbf{v}_-(K) = 0$ to analyze the third generator:

$$\mathbf{v}_-\big[\{(\mathbf{v}_+)^k K\}\, x^{n-k} y^k\big] =$$
$$= \{(\mathbf{v}_+)^k \mathbf{v}_- K + (\mathbf{v}_+)^{k-1}[-k\mathbf{v}_0 - k(k-1)]K\}\, x^{n-k} y^k -$$
$$- (n-k)\{(\mathbf{v}_+)^k K\}\, x^{n-k-1} y^{k+1}$$
$$= k(n-k+1)\{(\mathbf{v}_+)^{k-1} K\}\, x^{n-k} y^k - (n-k)\{(\mathbf{v}_+)^k K\}\, x^{n-k-1} y^{k+1}.$$

Substituting into (9.51) demonstrates that $\mathbf{v}_-(J) = 0$.   *Q.E.D.*

Example 9.50.   Consider the Hessian (2.24) of a binary cubic. The leading coefficient of $\frac{1}{36}H$ is $K = a_1 a_3 - a_2^2$. We compute $\mathbf{v}_+(K) = a_0 a_3 - a_1 a_2$, $(\mathbf{v}_+)^2 K = 2(a_0 a_2 - a_1^2)$. Equation (9.51) then recovers the Hessian covariant: $\frac{1}{36}H = K\, x^2 + [\mathbf{v}_+(K)]\, xy + \frac{1}{2}[(\mathbf{v}_+)^2 K]\, y^2$.

The Hilbert operators, which are more sophisticated versions of the summation formulae in (9.51), provide a direct route from isobaric invariants to covariants.

Definition 9.51.   The *Hilbert operators* are the formal infinite series differential operators

$$\mathcal{H}_+ = 1 - \frac{\mathbf{v}_+\mathbf{v}_-}{1!\,2!} + \frac{(\mathbf{v}_+)^2(\mathbf{v}_-)^2}{2!\,3!} - \frac{(\mathbf{v}_+)^3(\mathbf{v}_-)^3}{3!\,4!} + \cdots ,$$
$$\mathcal{H}_- = 1 - \frac{\mathbf{v}_-\mathbf{v}_+}{1!\,2!} + \frac{(\mathbf{v}_-)^2(\mathbf{v}_+)^2}{2!\,3!} - \frac{(\mathbf{v}_-)^3(\mathbf{v}_+)^3}{3!\,4!} + \cdots .$$
$$(9.52)$$

Again, when applying $\mathcal{H}_+$ or $\mathcal{H}_-$ to a polynomial, only finitely many terms in the summation will be nonzero. One can also prove the convergence of the series when applied to an analytic function.

Theorem 9.52.   *If $F(\mathbf{a}, \mathbf{x})$ is any isobaric function, then*

$$I = \mathcal{H}_+(F) = \mathcal{H}_-(F) \qquad (9.53)$$

is an SL(2)–covariant.

Proof: We must verify the infinitesimal invariance criterion (9.31). The fact that $\mathcal{H}_+(F) = \mathcal{H}_-(F)$ for isobaric F is an immediate consequence of (9.50). Indeed, each summand $(\mathbf{v}_+)^k(\mathbf{v}_-)^k F = (\mathbf{v}_-)^k(\mathbf{v}_+)^k F$ is isobaric, and hence I satisfies $\mathbf{v}_0(I) = 0$. Next, we apply (9.49) and then (9.48) to compute

$$\mathbf{v}_-(\mathbf{v}_+)^k(\mathbf{v}_-)^k =$$
$$= (\mathbf{v}_+)^k(\mathbf{v}_-)^{k+1} - k(\mathbf{v}_+)^{k-1}\mathbf{v}_0(\mathbf{v}_-)^k - k(k-1)(\mathbf{v}_+)^{k-1}(\mathbf{v}_-)^k$$
$$= (\mathbf{v}_+)^k(\mathbf{v}_-)^{k+1} - k(\mathbf{v}_+)^{k-1}(\mathbf{v}_-)^k\mathbf{v}_0 + k(k+1)(\mathbf{v}_+)^{k-1}(\mathbf{v}_-)^k.$$

Therefore, for isobaric F, the sum in $\mathbf{v}_-(\mathcal{H}_+(F))$ telescopes:

$$\mathbf{v}_-(I) = \sum_{k=0}^{\infty} \frac{(-1)^k}{k!(k+1)!} \, \mathbf{v}_-(\mathbf{v}_+)^k(\mathbf{v}_-)^k F$$

$$= \sum_{k=0}^{\infty} (-1)^k \left\{ \frac{(\mathbf{v}_+)^k(\mathbf{v}_-)^{k+1}F}{k!\,(k+1)!} + \frac{(\mathbf{v}_+)^{k-1}(\mathbf{v}_-)^k F}{(k-1)!\,k!} \right\} = 0.$$

The proof that $\mathbf{v}_+(I) = \mathbf{v}_+(\mathcal{H}_- I) = 0$ is similar, using the second form $\mathcal{H}_-$ of the Hilbert operator. *Q.E.D.*

Example 9.53. Let Q be a binary cubic, (2.21). Consider the isobaric function $F(a_0, a_1, a_2, a_3, x, y) = a_1 a_3 x^2$. Applying the infinitesimal operators (9.32), we find

$$\tfrac{1}{2}\mathbf{v}_+\mathbf{v}_-(F) = (3a_2^2 + 3a_1 a_3)x^2 - (3a_1 a_2 + a_0 a_3)xy$$

$$\tfrac{1}{12}(\mathbf{v}_+)^2(\mathbf{v}_-)^2(F) = (7a_2^2 + 5a_1 a_3)x^2 - (14a_1 a_2 + 2a_0 a_3)xy +$$
$$+ (a_1^2 + a_0 a_2)y^2$$

$$\tfrac{1}{144}(\mathbf{v}_+)^3(\mathbf{v}_-)^3(F) = (6a_2^2 + 4a_1 a_3)x^2 - (18a_1 a_2 + 2a_0 a_3)xy +$$
$$+ (3a_1^2 + 2a_0 a_2)y^2$$

$$\tfrac{1}{2880}(\mathbf{v}_+)^4(\mathbf{v}_-)^4(F) = \tfrac{1}{5}\big[\,(9a_2^2 + 6a_1 a_3)x^2 - (36a_1 a_2 + 4a_0 a_3)xy +$$
$$+ (9a_1^2 + 6a_0 a_2)y^2\,\big]$$

$$(\mathbf{v}_+)^n(\mathbf{v}_-)^n(F) = 0, \qquad n \geq 5.$$

Summing, the resulting covariant $\mathcal{H}_+(F) = \mathcal{H}_-(F)$ is exactly $\frac{1}{180}H$, where $H = \frac{1}{2}(Q, Q)^{(2)}$ is the Hessian, given by (2.24).

Proof of the Hilbert Basis Theorem

The Hilbert operators will now be applied to complete the proof of Hilbert's Basis Theorem 2.42. In outline, the proof proceeds in two steps. One first demonstrates that every polynomial covariant can be written as a linear combination, with polynomial coefficients, of a finite system of covariants. In commutative algebra, cf. [13], such a system is referred to as a "basis". This term should *not* be confused with the eventual Hilbert basis, but is rather the algebraic analog of a basis for a subspace in linear algebra. In fact, the existence of a finite linear basis is not particular to the set of polynomial invariants or covariants of a system of forms but is a completely general property that is valid for *any* system of polynomials!

Definition 9.54. Let $\mathcal{I}$ be a collection of polynomials $P(x)$ depending on n variables $x = (x_1, \ldots, x_n)$. A *finite* set of polynomials $P_1, \ldots, P_m \in \mathcal{I}$ is said to form a *basis* for $\mathcal{I}$ if every other polynomial $P \in \mathcal{I}$ can be written as a linear combination

$$P = Q_1 P_1 + \cdots + Q_m P_m, \qquad (9.54)$$

for polynomials $Q_1(x), \ldots, Q_m(x)$ — which are not necessarily in $\mathcal{I}$.

The fact that any system of polynomials admits a finite basis was first recognized by Hilbert, [**106, 107**], who, because he was primarily concerned with its invariant theoretic consequences, named this important result a mere "Hilfsatz" (lemma). Hilbert's lemma has since been recognized to be one of the most fundamental results in modern commutative algebra. In fact, the Hilfsatz itself has become known as the "Hilbert Basis Theorem", leading to some confusion in terminology!

Theorem 9.55. *Every collection of polynomials in n variables admits a finite basis.*

Remark: In the modern formulation of the Hilbert Basis Theorem Theorem 9.55, one usually restricts attention to polynomial ideals. In general, a linear subspace of the space of polynomials forms an *ideal* whenever $P \in \mathcal{I}$ implies that $Q \cdot P \in \mathcal{I}$ for any polynomial Q. Theorem 9.55 is easily seen to be equivalent to the ideal version; one merely replaces $\mathcal{I}$ by the smallest polynomial ideal containing it.

Proof: The proof proceeds by an induction on the number of variables. At first, it might seem logical to begin with the univariate case. However, it is possible to start with an even simpler case — that of polynomials that depend on zero variables, i.e., constant polynomials! Clearly, the result is completely trivial when $\mathcal{I}$ contains only constant polynomials.

Proceeding to the induction step, we assume that we have already proved the result for systems of polynomials in $n - 1$ variables, where $n \geq 1$. Any n-variate polynomial $P(x_1, \ldots, x_n)$ can be written in the form $P = S(x_1, \ldots, x_{n-1})(x_n)^k + \cdots$, where k is the degree of P in x_n, and $S(x_1, \ldots, x_{n-1})$ is its *leading coefficient*. Let S denote the set of all leading coefficients of polynomials in $\mathcal{I}$. Applying our induction hypothesis, since S consists of polynomials in $n - 1$ variables, it admits a basis $S_1, \ldots, S_m$. Let $P_1, \ldots, P_m \in \mathcal{I}$ be the corresponding n-variate polynomials, each of the form

$$P_\nu(x_1, \ldots, x_n) = S_\nu(x_1, \ldots, x_{n-1})(x_n)^{k_\nu} + \cdots, \qquad \nu = 1, \ldots, m.$$

Let $k = \max k_\nu$. Suppose first that $P = S(x_n)^l + \cdots \in \mathcal{I}$ has degree $l \geq k$ in x_n. Using our basis for the set of leading coefficients, we can write $S = L_1 S_1 + \cdots + L_m S_m$. Then the polynomial

$$\tilde{P} = P - \left[\, L_1(x_n)^{l-k_1} P_1 + \cdots + L_m(x_n)^{l-k_m} P_m \,\right]$$

has degree (in x_n) strictly less than that of P. Continuing in this manner, we can express every $P \in \mathcal{I}$ in the form

$$P = K_1 P_1 + \cdots + K_m P_m + T(x_1, \ldots, x_{n-1})\,(x_n)^{k-1} + \cdots, \qquad (9.55)$$

where the remaining terms are of degree $\leq k-2$ in x_n. Let $\mathcal{T}$ denote the set of all order $k-1$ "remainder coefficients" $T(x_1, \ldots, x_{n-1})$. As before, our induction hypothesis says that $\mathcal{T}$ admits a finite basis $T_1, \ldots, T_q$. Let

$$P_{\kappa+m} = K_{1,\kappa} P_1 + \cdots + K_{m,\kappa} P_m + T_\kappa\,(x_n)^{k-1} + \cdots, \qquad \kappa = 1, \ldots, q,$$

be the corresponding polynomials in $\mathcal{I}$. For any other $P \in \mathcal{I}$, we can write its remainder coefficient $T = M_1 T_1 + \cdots + M_q T_q$ in terms of the basis of $\mathcal{T}$. Therefore,

$$
\begin{aligned}
P = \tilde{K}_1 P_1 + \cdots + \tilde{K}_m P_m + \tilde{K}_{m+1} P_{m+1} + \cdots + \tilde{K}_{m+q} P_{m+q} + \\
+ U(x_1, \ldots, x_{n-1})\,(x_n)^{k-2} + \cdots,
\end{aligned}
\qquad (9.56)
$$

where

$$
\tilde{K}_\nu = \begin{cases} K_\nu - \displaystyle\sum_\kappa M_\kappa K_{\nu,\kappa}\,(x_n)^{k-1}, & \nu = 1, \ldots, m, \\[2mm] M_{\nu-m}, & \nu = m+1, \ldots, m+q. \end{cases}
$$

By this procedure, we have reduced the degree of the remainder in x_n from $k-1$ to $k-2$ by introducing q further polynomials into our basis. Applying this procedure recursively will complete the proof. *Q.E.D.*

We are now in a position to complete the proof of the invariant theoretic Hilbert Basis Theorem 2.42. Let $\mathcal{I}$ denote the set of all non-constant invariants (or covariants) for a system of forms. According to Theorem 9.55, $\mathcal{I}$ admits a (linear) basis of invariants $I_1, \ldots, I_m \in \mathcal{I}$, which means that every other invariant can be written in the form

$$I = Q_1 I_1 + \cdots + Q_m I_m. \qquad (9.57)$$

Now, the coefficient polynomials Q_ν are *not* necessarily invariant. But since I and the I_ν's are isobaric, we can assume, without loss of generality, that the Q_ν's are also isobaric. (Just throw away any non-isobaric terms in the Q_ν; this will not affect the equation.) The goal is to produce *invariant* coefficient polynomials. To accomplish this, we apply the Hilbert operator $\mathcal{H}_+$ to both sides of (9.57). Since the differential operators $\mathbf{v}_-$ and $\mathbf{v}_+$ both annihilate invariants, we find

$$\mathcal{H}_+ I = I = (\mathcal{H}_+ Q_1) I_1 + \cdots + (\mathcal{H}_+ Q_m) I_m = J_1 I_1 + \cdots + J_m I_m, \quad (9.58)$$

where each $J_\nu = \mathcal{H}_+ Q_\nu$ is an invariant. Moreover, each J_ν has degree strictly less than I. Therefore, we can use an evident induction to conclude that the linear basis $I_1, \ldots, I_m$ in fact provides a Hilbert basis, and so we can write $I = F(I_1, \ldots, I_m)$ as a polynomial function of the linear basis invariants. This completes the proof of Theorem 2.42.   *Q.E.D.*

Remark: Clearly this method of proof generalizes to any group with an operator (of an appropriate form) that projects to the space of invariants. For the general linear group, a second such operator, based on the omega process, appears in Hilbert's second proof of the Basis Theorem, cf. [**105, 204**]. For finite groups, averaging via the symmetrization operator (4.8) plays this role, while for compact groups one can use the method of invariant integration based on the Haar measure. For semi-simple Lie groups, the more sophisticated approach is Weyl's "unitarian trick", [**231**], which does not construct the operator directly, but relies on its existence for a "maximal" compact subgroup, and then uses analytic continuation to conclude that any polynomial identity which holds on the compact subgroup also holds for the entire group.

Nullforms

As we have seen, the invariants do not typically distinguish the different equivalence classes of binary forms. The extreme examples are forms having all zero invariants, which certainly do not lie on regular orbits.

Definition 9.56. A binary form for which all the invariants vanish is known as a *nullform*.

Nullforms were introduced in Hilbert's second, deeper proof of the Basis Theorem, [**106**]. (In the same paper, Hilbert proves his fundamental Nullstellensatz, cf. [**155**], as a Hilfsatz for the analysis of the nullforms!) They play a pivotal role in the advanced Hilbert–Mumford approach to geometric invariant theory, [**156**]. We do not have space to dwell on these important special polynomials, but we can prove one basic classification result, [**107**; p. 159].

Theorem 9.57. *Let* $Q(p)$ *be a binary form of even degree* $n = 2m$ *or odd degree* $n = 2m + 1$. *Then* Q *is a nullform if and only if it has a root of multiplicity at least* $m + 1$.

Proof: Suppose first that Q has a root of multiplicity at least $m+1$. Because it is such a high multiplicity, there can only be one such root, and hence if Q is a real nullform, the distinguished root is also real. Therefore, in either the real or complex case, we can send the root to ∞ by a suitable projective transformation, and hence place Q in the semi-canonical form $Q(p) = \binom{n}{n-m-1}a_{n-m-1}p^{n-m-1} + \cdots + a_0$. In other words, the first $m+1$ coefficients of Q vanish: $a_n = a_{n-1} = \cdots = a_{n-m} = 0$. But, according to Exercise 2.31, every term in an invariant $I = \sum c_M \mathbf{a}^M$ must contain at least one of the coefficients $\mathbf{a}_+ = (a_{n-l}, \ldots, a_n)$, and hence I must vanish for the nullform.

To prove the converse, we begin by forming the transvectants $S_k = (Q, Q)^{(2k)}$, $k = 1, \ldots, m$, as in (6.54), which are of degree $2n - 2k$. If $n = 2m$ is even, then the last transvectant S_m is an invariant, which vanishes by hypothesis. Set $l = m - 1$ if n is even, while $l = m$ if n is odd. Let L be the least common multiple of the numbers $n, 2n-2, 2n-4, \ldots, 2n-2l$. Write $L = j_0 n = j_k(2n - 2k)$, and consider the polynomials

$$U = y_0 Q^{j_0} + y_1(S_1)^{j_1} + \cdots + y_l(S_l)^{j_l},$$
$$V = z_0 Q^{j_0} + z_1(S_1)^{j_1} + \cdots + z_l(S_l)^{j_l},$$

where the y's and z's are indeterminate parameters. The resultant of U, V can be written in the form

$$R(U, V) = \sum_{\kappa} I_\kappa M_\kappa(y, z),$$

where the $M_\kappa(y, z)$ are the various monomials in y, z. Since the I_κ are invariants of U, V, they are invariants for Q and hence, by hypothesis, must vanish. This implies that the polynomials U, V have a common root for every possible assignment of the parameters y, z.

We conclude that the common root must be independent of y and z and hence is a common root of $Q, S_1, \ldots, S_l$. Let us place the root at ∞, and therefore $a_n = 0$. The leading term in $S_1 = (Q, Q)^{(2)}$ is $(a_n a_{n-2} - a_{n-1}^2)p^{2n-2}$, which (assuming $n > 1$) must vanish, and so $a_{n-1} = 0$. Proceeding to S_2, we compute its leading coefficient, and use this to conclude that $a_{n-2} = 0$. In general, if $a_n = \cdots = a_{n-k+1} = 0$ and the leading coefficient of S_k vanishes, due to the presence of an infinite root, then the explicit formula implies $a_{n-k} = 0$. When n is even, the last step uses the fact that S_m is an invariant and hence vanishes by assumption. Therefore, we conclude that $a_n = \cdots = a_{n-m-1} = 0$ and hence ∞ is a root of multiplicity at least $m + 1$.   Q.E.D.

Chapter 10

Multivariate Polynomials

We have now reached the culmination of our brief tour of the most note-worthy features in the invariant theory of binary forms and stand ready to launch into new and less well charted territories. Unfortunately, our space limitations prevent us from realizing this grander aim in any depth, but we do have room in this final chapter for a brief introduction to the invariant theory of homogeneous polynomials in more than two variables (along with their inhomogeneous counterparts). Not surprisingly, almost all of the classical — and not so classical — methods that have been developed for binary forms can be straightforwardly adapted to the multivariate situation. On the other hand, the required computations are considerably more involved, and much less is known about explicit bases for invariants and covariants, solution to equivalence problems, canonical forms, structure of symmetries, and so on.

Given a system of multivariate homogeneous polynomials or, more generally, functions, we may construct a wide variety of invariants and covariants by a combination of interrelated techniques, the most useful being modeled on the transvectant processes in conjunction with symbolic or transform methods. Since many of the tools are straightforward extensions of the bivariate versions that were developed in depth in the preceding chapters, we can pass very quickly through them. A signifi-cant complication is that, in addition to classical invariants and covari-ants, there exist nontrivial "contravariants" and "mixed concomitants", which, unlike the case of binary forms, cannot be directly obtained by a simple determinantal reweighting. Furthermore, our graph-theoretic methods are not nearly as pleasant to devise or use, and so one must, typically, revert to the more involved (albeit more computer-friendly) algebraic manipulations. On the other hand, the moving frame and Lie algebraic methods remain pertinent, but, to date, they have been much less systematically developed in the multivariate context.

Additional areas that are amenable to methods from the classical theory have been suggested from a variety of applications. In the final section, we discuss a hybrid case — the invariant theory of "biforms" — which arises in the study of linear elasticity and the calculus of variations. Historically, the lion's share of the attention has been devoted to symmetric forms, although some work has been done on their antisymmetric counterparts, which are of importance in differential geometry and topology. The investigation of more general symmetry classes of tensorial objects is uncharted territory for the would-be explorer. However, at this point, our modest resources have been exhausted, and we must bid the traveler bon voyage as she or he sets sail for the modern general theory.

Polynomials and Algebraic Curves

As in the case of binary forms, the primary focus for the classical theory is on polynomial functions. A general homogeneous polynomial of degree n depending on m variables $\mathbf{x} = (x_1, \ldots, x_m)$ has the form

$$Q(\mathbf{x}) = \sum_I \binom{n}{I} a_I x^I, \quad \text{where} \quad x^I = (x_1)^{i_1} \cdots (x_m)^{i_m}. \quad (10.1)$$

The summation is over all multi-indices $I = (i_1, \ldots, i_m)$ with $0 \leq i_\nu$ and $\#I = i_1 + \cdots + i_m = n$. The form (10.1) is classically referred to as a binary, ternary, quaternary, and, generally, m-ary form in accordance with the number of (homogeneous) variables it depends on. Both real and complex forms are of interest, of course, but for specificity we shall concentrate on the complex case throughout this chapter.

We have already encountered the basic geometry that underlies the passage from homogeneous functions on a vector space to their inhomogeneous counterparts on the associated projective space. When $X = \mathbb{C}^m$, the corresponding projective space $\mathbb{P}(X) = \mathbb{CP}^{m-1}$ can be identified as the set of lines through the origin in $\mathbb{C}^m$. The standard coordinates $\mathbf{x} = (x_1, \ldots, x_m)$ on $\mathbb{C}^m$ induce homogeneous coordinates $[\mathbf{x}] = [x_1, \ldots, x_m]$ on $\mathbb{CP}^{m-1}$; not all the x_i's are allowed to simultaneously vanish, and we identify points that are nonzero scalar multiples of each other. In particular, the dense open subset consisting of all lines that are not contained in the coordinate hyperplane $\mathcal{H}_m = \{x_m = 0\}$, say, forms a coordinate chart $\mathbb{C}^{m-1} \subset \mathbb{CP}^{m-1}$ whose coordinates $p_1 = x_1/x_m, \ldots, p_{m-1} = x_{m-1}/x_m$ correspond to the canonical homogeneous coordinates $[p_1, \ldots, p_{m-1}, 1]$. Instead of merely adjoining a point at infinity, as was the case for the projective line, the remainder $\mathbb{CP}^{m-1} \setminus \mathbb{C}^{m-1} \simeq \mathbb{CP}^{m-2}$ forms an entire projective space, of one lower dimension, at infinity.

Example 10.1. The case of the complex projective plane is instructive. The homogeneous coordinates $[x, y, z]$ on $\mathbb{C}^3$ induce the projective coordinates $p = x/z$ and $q = y/z$ on the chart $\mathbb{C}^2 \subset \mathbb{CP}^2$ associated with all non-horizontal lines in $\mathbb{C}^3$. The lines at ∞ are those contained in the (x, y)–plane, which form thus a projective line $\mathbb{CP}^1 \simeq \mathbb{CP}^2 \setminus \mathbb{C}^2$. The other two basic coordinate charts consist of lines not contained in one of the other two coordinate planes. For instance, the projective coordinates $\tilde{p} = x/y$, $\tilde{q} = z/y$ apply to lines not in the (x, z)–plane. Note that the changes of coordinates $\tilde{p} = p/q$, $\tilde{q} = 1/q$ are analytic on the overlap of the two charts, proving that $\mathbb{CP}^2$ is an analytic manifold, in accordance with the general definition.

Example 10.2. The real projective plane, $\mathbb{RP}^2 = \mathbb{P}(\mathbb{R}^3)$, is the space of lines through the origin in $\mathbb{R}^3$ and has similar homogeneous and projective coordinates. Each line intersects the unit sphere $S^2 \subset \mathbb{R}^3$ in two antipodal points, and hence $\mathbb{RP}^2$ can be regarded as the space obtained by identifying antipodal points on a sphere. However, the projective plane is *not* a sphere! Indeed, $\mathbb{RP}^2$ is a non-orientable surface, like the Möbius strip and Klein bottle. One way of seeing this is to note that the antipodal map $\iota \colon S^2 \to S^2$ reverses orientation, and hence one cannot endow $\mathbb{RP}^2$ with a consistent, global, oriented coordinate frame, cf. [**199**; p. 120]. (In general, even-dimensional real projective spaces are not orientable, whereas odd-dimensional ones are.) We refer the reader to [**161**] for additional details on the topology of higher dimensional projective spaces.

Exercise 10.3. The (complex) *Grassmann space* $\mathcal{G}(k, m)$ is defined as the set of all k-dimensional subspaces in the m-dimensional vector space $\mathbb{C}^m$. Show that $\mathcal{G}(k, m)$ can be given the structure of an analytic manifold, and determine its dimension and local coordinates. What is the corresponding formula for the induced action of $\mathrm{GL}(m)$ on $\mathcal{G}(k, m)$? Prove the natural identification $\mathcal{G}(k, m) \simeq \mathcal{G}(m - k, m)$; in particular, the Grassmann space of hyperplanes $\mathcal{G}(m - 1, m) \simeq \mathcal{G}(1, m) = \mathbb{CP}^{m-1}$ can be identified with ordinary projective space.

For the m-ary form (10.1), the inhomogeneous representative associated with the mth coordinate chart in $\mathbb{CP}^{m-1}$ is obtained by setting $x_m = 1$, and so $Q(p_1, \ldots, p_{m-1}) = Q(p_1, \ldots, p_{m-1}, 1)$, where, in accordance with our general convention, we use the same letter Q to denote both polynomials. (As in the binary form case, other representatives are obtained on alternative coordinate charts, or by applying suitable inversions.) To recover the homogeneous version, we adapt formula (2.3) to deduce

$$Q(x_1, \ldots, x_m) = (x_m)^n \, Q\left(\frac{x_1}{x_m}, \ldots, \frac{x_{m-1}}{x_m}\right) . \qquad (10.2)$$

For a form depending on three or more homogeneous variables, the roots are no longer isolated points, but define general *algebraic varieties* inside projective space. In other words, the zero locus $\mathcal{Z}_Q = \{Q(\mathbf{x}) = 0\}$ of the inhomogeneous representative $Q \colon \mathbb{CP}^{m-1} \to \mathbb{C}$ will be a certain algebraic subset — typically a submanifold of dimension $m - 2$, but which may contain singularities.

Example 10.4. Consider a ternary quadratic form

$$Q(x, y, z) = ax^2 + 2bxy + cy^2 + dxz + eyz + fz^2. \qquad (10.3)$$

The corresponding inhomogeneous polynomial is obtained by setting $z = 1$, so

$$Q(p, q) = ap^2 + 2bpq + cq^2 + dp + eq + f. \qquad (10.4)$$

The zero locus $Q(p, q) = 0$ defines a conic section $\mathcal{C} \subset \mathbb{CP}^2$. Note that a complex conic section will include points at "infinity", namely, any nonzero solution to the homogeneous quadratic $ax^2 + 2bxy + cy^2 = 0$, which is obtained by setting $z = 0$ in (10.3).

Transformations and Covariants

Next we review the fundamentals of classical covariants and invariants, now adapted to functions and polynomials in more than two variables. As in the bivariate case, we do not need to assume that our functions on the original vector space are actually homogeneous.

The general linear group $\mathrm{GL}(m)$ acts on $\mathbb{C}^m$ via the usual linear maps $\mathbf{x} \mapsto A\mathbf{x}$, where the $m \times m$ matrix A has entries $a_{i,j}$, $1 \leq i, j \leq m$. The action preserves lines and so induces a projective action on $\mathbb{CP}^{m-1}$. In terms of the projective coordinates $p_i = x_i/x_m$ on $\mathbb{C}^{m-1}$, this action is given by

$$\bar{p}_i = \frac{a_{i,1}p_1 + \cdots + a_{i,m-1}p_{m-1} + a_{i,m}}{a_{m,1}p_1 + \cdots + a_{m,m-1}p_{m-1} + a_{m,m}}, \qquad i = 1, \ldots, m - 1.$$

$$(10.5)$$

For example, projective transformations on $\mathbb{CP}^2$ take the explicit form

$$(p,q) \longmapsto \left(\frac{\alpha p + \beta q + \gamma}{\rho p + \sigma q + \tau}, \frac{\lambda p + \mu q + \nu}{\rho p + \sigma q + \tau} \right), \qquad \det \begin{vmatrix} \alpha & \beta & \gamma \\ \lambda & \mu & \nu \\ \rho & \sigma & \tau \end{vmatrix} \neq 0,$$

$$(10.6)$$

in the coordinate system $p = x/z$ and $q = y/z$.

For the linear action of $\mathrm{GL}(m)$ on $\mathbb{C}^m$, a complete system of inequivalent multipliers is provided by the powers $\rho_k(A) = (\det A)^k$ of the determinantal multiplier. Classical covariants appear as relative invariants for these multiplier representations.

Definition 10.5. A function $Q(\mathbf{x})$ has *weight* k if it transforms according to the determinantal rule

$$Q(\mathbf{x}) = (\det A)^k \, \overline{Q}(\overline{\mathbf{x}}), \qquad \overline{\mathbf{x}} = A\,\mathbf{x}, \quad A \in \mathrm{GL}(m). \qquad (10.7)$$

Usually k is taken to be a non-negative integer, although extensions to more general real and even complex weights are straightforwardly implemented. The basic equivalence and canonical form problems ask when two multivariate forms can be mapped to each other by such a transformation (10.7).

Exercise 10.6. Determine the multiplier and transformation rules for inhomogeneous forms of degree n and weight k on $\mathbb{CP}^{m-1}$.

Example 10.7. Let us consider the case of ternary quadratics (10.3). Sylvester's Theorem 3.56 implies that we can find a linear transformation that diagonalizes Q, with the (complex) canonical form determined by its rank. Therefore, under a general projective transformation, we can place *any* conic section into one of three simple canonical forms:

$$p^2 + q^2 = 1, \qquad p^2 = q^2, \qquad p^2 = 0. \qquad (10.8)$$

In other words, every nontrivial complex conic section is projectively equivalent to either a (complex) circle, two (projective) lines, or a single line with multiplicity 2. Note that the solutions to (10.8) include points on the projective line at infinity. Indeed, if we use the canonical homogeneous form $Q = z^2$ in the third case, we end up with the inhomogeneous equation $1 = 0$. This is not a contradiction; it merely means that the solution set is the entire projective line at ∞.

Exercise 10.8. A nondegenerate planar, real conic is either an ellipse, a parabola, or a hyperbola. Determine how many points at infinity each one has. Explain why they are all equivalent under a complex projective transformation. Which ones are equivalent under a real projective transformation on $\mathbb{RP}^2$? Find a complete list of canonical forms for real conic sections.

Exercise 10.9. Show how the elliptic curve $y^2 = x^3 + ax + b$, which plays a key role in analytic function theory, cf. [8], can be identified as the algebraic variety corresponding to a certain homogeneous ternary cubic. Discuss as far as you can the canonical forms of ternary cubics, cf. [65; §229]. See [236, 185] for the remarkable connections between elliptic curves, modular forms, and Fermat's Last Theorem.

Consider a collection $Q_1(\mathbf{x}), \ldots, Q_l(\mathbf{x})$ of one or more analytic functions on $\mathbb{C}^m$. In general, the functions can be assigned different weights $k_1, \ldots, k_l$, the most important case, of course, being when they are all of weight 0.

Definition 10.10. A differential polynomial $\Delta[\mathbf{x}, Q_1, \ldots, Q_l]$ defines a *covariant* of the system if it transforms with weight k under the action of GL(m).

In particular, if each $Q_\nu(\mathbf{x}) = \sum_J a_J^\nu x^J$ is a homogeneous polynomial, then, just as in the bivariate case, any function $J(\mathbf{a}, \mathbf{x})$ depending on their coefficients $\mathbf{a} = (a_J^\nu)$, and, possibly, the variables $\mathbf{x}$, can always be written (in many ways) as a differential polynomial $\Delta[\mathbf{x}, Q^1, \ldots, Q^k]$, and so our definition automatically incorporates the classical concepts of invariant — where $J(\mathbf{a})$ does not depend on $\mathbf{x}$ — and covariant. As before, we discover that all covariants for homogeneous functions are obtained as constant coefficient differential polynomials — see below.

The simplest example is the *Jacobian determinant*

$$J[Q_1, \ldots, Q_m] = \frac{\partial(Q_1, \ldots, Q_m)}{\partial(x_1, \ldots, x_m)} = \det\left(\frac{\partial Q_i}{\partial x_j}\right), \qquad (10.9)$$

which depends linearly on m functions of m variables. If each Q_ν has weight k_ν, then the Jacobian is a joint covariant of weight $k_1 + \cdots + k_m + 1$. The *Hessian* covariant

$$H[Q] = \det\left(\frac{\partial^2 Q}{\partial x_i \partial x_j}\right) \qquad (10.10)$$

of a single function Q of weight k is a covariant of order m and has weight $mk + 2$. This particular covariant has an interesting history. The name dates back to the original papers [103, 104], in which Hesse claimed to have proven the following result, generalizing Proposition 2.23 on the vanishing of the Hessian of a binary form.

Hesse's "Theorem". Let $Q(x_1, \ldots, x_m)$ be a homogeneous polynomial in m variables. The Hessian of Q is identically zero, $H[Q] \equiv 0$, if and only if there is a linear change of variables $\mathbf{y} = A\mathbf{x}$ such that $Q(\mathbf{y}) = Q(y_1, \ldots, y_{m-1})$ depends on at most $m - 1$ variables.

In other words, Hesse claimed that an m-ary form should have zero Hessian if and only if it is in fact a function of fewer than m variables in disguise, i.e., a k-ary form for some $k < m$. That such functions have zero Hessian is readily seen: the Hessian matrix has its last row and last column all zero and hence trivially has zero determinant. Remarkably, Hesse's "Theorem" was accepted for 25 years until Gordan and Nöther, [90], showed that the theorem is true if $n \leq 4$ but is *false* if the number of variables is 5 or more! The simplest counterexample occurs in dimension 5 and is the cubic form

$$Q(x_1, \ldots, x_5) = x_1^2 x_3 + x_1 x_2 x_4 + x_2^2 x_5, \qquad (10.11)$$

whose Hessian matrix has its lower right 3×3 block identically zero, but which, nevertheless, cannot be written in fewer than 5 variables. See also [110] for some further developments.

Transvectants

Both the Jacobian and Hessian covariants are particular cases of the general concept of a transvectant, which, in the multivariate version, will be a certain multilinear differential polynomial depending on m functions of the m variables $\mathbf{x}$. We introduce m distinct sets of variables $\mathbf{x}^\alpha = (x_1^\alpha, \ldots, x_m^\alpha)$, $\alpha = 1, \ldots, m$, and consider the joint action of $\mathrm{GL}(m)$ on the m-fold Cartesian product $\mathbb{C}^{m^2} = \mathbb{C}^m \times \cdots \times \mathbb{C}^m$. Given m functions (some of which may coincide), we employ our tensor product notation $Q_1 \otimes \cdots \otimes Q_m$ to denote the m-fold joint product $Q_1(\mathbf{x}^1) Q_2(\mathbf{x}^2) \cdots Q_m(\mathbf{x}^m)$.

The m-dimensional Cayley *omega process* is the mth order partial differential operator

$$\Omega = \det \begin{vmatrix} \dfrac{\partial}{\partial x_1^1} & \cdots & \dfrac{\partial}{\partial x_m^1} \\ \vdots & \ddots & \vdots \\ \dfrac{\partial}{\partial x_1^m} & \cdots & \dfrac{\partial}{\partial x_m^m} \end{vmatrix}. \qquad (10.12)$$

Definition 10.11. The rth *transvectant* of functions $Q_1, \ldots, Q_m$ is

$$(Q_1, \ldots, Q_m)^{(r)} = \operatorname{tr} \Omega^r (Q_1 \otimes \cdots \otimes Q_m), \qquad (10.13)$$

where tr means to set all variables equal: $\mathbf{x}^1 = \cdots = \mathbf{x}^m = \mathbf{x}$.

For example, the first transvectant is just the Jacobian determinant (10.9), while the Hessian (10.10) is a multiple of the second transvectant $H[Q] = \frac{1}{m!} (Q, \ldots, Q)^{(2)}$. (The general second transvectant is the fully polarized form of the Hessian.)

Under a linear change of variables $\mathbf{x} \mapsto A \mathbf{x}$, the multivariate omega process obeys the same transformation rule (5.3) as its bivariate version. As a consequence, the rth transvectant (10.13) of functions of respective weights $k_1, \ldots, k_m$ forms a joint covariant of weight $k_1 + \cdots + k_m + r$.

Exercise 10.12. Discuss the symmetry properties of the rth transvectant under permutations of the functions. Write out the particular case $(Q, Q, Q)^{(4)}$ when $Q(x, y, z)$ is a ternary form.

Generalizations of the omega calculus to the multivariate situation are straightforward and left to the reader. For any collection of functions, one can form many different joint covariants via partial transvectant processes by using various products of omega processes. Additional covariants are constructed using the scaling process $\sigma = \sum_i x_i \partial/\partial x_i$, although Euler's formula (2.13) says that these are just multiples of the partial transvectant covariants in the case of homogeneous functions or forms. As a consequence of the multivariate First Fundamental Theorem (see below), every polynomial covariant of a system of forms can be written as a linear combination of partial transvectants, and, in the inhomogeneous case, scaling processes. However, the analog of Theorem 5.14 does not appear to be true — even in the special case of homogeneous polynomials, not every partial transvectant can be written as a linear combination of ordinary transvectants!

Remark: Transvectants are particular cases of a general class of differential polynomials called *hyperjacobians*, which appear in the calculus of variations, cf. [15, 163]. Both concepts have their origins in Cayley's hyperdeterminant theory, [38, 39]. There are also interesting connections with the less well-known theory of higher dimensional determinants, cf. [66, 76, 162].

Exercise 10.13. Determine the formula for the rth transvectant in projective coordinates. Does this formula help simplify the study of forms with vanishing Hessian?

Tensorial Invariants

Let us now restrict the discussion to multivariate forms, that is, homogeneous polynomials. We can regard a form and its classical covariants as symmetric covariant tensors on the underlying space (manifold) $X = \mathbb{C}^m$. As before, the elements of X transform according to the identity representation of $\mathrm{GL}(m)$, with $\mathbf{x} \mapsto A\mathbf{x}$, and are classically known as *contravariant vectors*. A covariant defines a polynomial function $J(\mathbf{a}, \mathbf{x})$ depending on the coefficients of one or more forms and the contravariant vectors. An element $\mathbf{u} = (u^1, \ldots, u^m)$ of the dual space X^* is subject to the contragredient transformation rules $\mathbf{u} \mapsto A^{-T}\mathbf{u}$, and is known as a *covariant vector*. Thus, one can consider more general invariant objects $K(\mathbf{a}, \mathbf{x}, \mathbf{u})$ that depend on both co- and contravariant variables. The key difference between binary forms and their multivariate analogs is that, for $m \geq 3$, there is no natural identification of the dual space with a determinantally weighted version of the original space. Therefore, one cannot reduce these more general invariant quantities to suitably weighted covariants.

For ternary forms, there is a very pretty geometric interpretation of the dual, covariant coordinates. In general, a homogeneous linear equation

$$ax + by + cz = 0 \qquad (10.14)$$

defines a plane P passing through the origin in $\mathbb{C}^3$. The coefficients $\mathbf{a} = (a, b, c)$ that define the plane can be invariantly interpreted as coordinates in the dual space $(\mathbb{C}^3)^*$, i.e., as covariant coordinates. Two nonzero points $\mathbf{a}, \widetilde{\mathbf{a}} \in (\mathbb{C}^3)^*$ define the same plane if and only if they are scalar multiples of each other, $\widetilde{\mathbf{a}} = \lambda \mathbf{a}$. In other words, the different planes are parametrized by points in the dual projective space $\mathbb{P}((\mathbb{C}^3)^*) = (\mathbb{CP}^2)^*$, subject to the dual transformation rules.

A plane P in $\mathbb{C}^3$ projects to a line L in the corresponding projective space $\mathbb{CP}^2$. If we use $p = x/z$, $q = y/z$ as the projective coordinates on $\mathbb{CP}^2$, and $l = a/c$, $m = b/c$ as "dual" coordinates on $(\mathbb{CP}^2)^*$, then the line $L \subset \mathbb{CP}^2$ corresponding to the plane (10.14) is given by the inhomogeneous equation

$$lp + mq + 1 = 0 \qquad (10.15)$$

in the given coordinate chart. There is a single additional point at infinity, so by a line in $\mathbb{CP}^2$ we really mean a (linear) embedding of a projective line $\mathbb{CP}^1 \hookrightarrow \mathbb{CP}^2$.

Consider a (nonsingular) algebraic curve $\mathcal{C} \subset \mathbb{CP}^2$ defined by the vanishing of an inhomogeneous polynomial $Q(p, q) = 0$. At each point $\mathbf{p} = (p, q) \in \mathcal{C}$, let $L_\mathbf{p}$ denote the tangent line to $\mathcal{C}$ at $\mathbf{p}$; we identify $L_\mathbf{p}$ with a point $\mathbf{a} \in (\mathbb{CP}^2)^*$ in the dual projective plane. The set of all such points defines the *dual curve* $\widehat{\mathcal{C}} \subset (\mathbb{CP}^2)^*$, classically known as the "line equation" for the given curve, [92]. Analytic continuation allows us to extend this correspondence through singularities.

Remark: Duality is closely related with the identification of the hyperplane Grassmann space with ordinary projective space, as in Exercise 10.3. In higher dimensions, duality takes each point $\mathbf{x} \in \mathcal{C} \subset \mathbb{CP}^{m-1}$ in an algebraic variety to its tangent hyperplane.

Given an m-ary form $Q(\mathbf{x})$, by a *contravariant* we mean a polynomial function $C(\mathbf{a}, \mathbf{u})$ depending on the coefficients of the form and a single covariant variable which is unaffected by linear transformations up to a determinantal power. (The classical terminology is becoming quite confusing at this point!) More generally, a *mixed concomitant* is a function $K(\mathbf{a}, \mathbf{x}, \mathbf{u})$ depending on both co- and contravariant vectors that is similarly unaffected. Specifically, if $\bar{\mathbf{a}}$ denotes the coefficients of the transformed form (10.2), then we require

$$K(\mathbf{a}, \mathbf{x}, \mathbf{u}) = (\det A)^k\, K(\bar{\mathbf{a}}, \bar{\mathbf{x}}, \bar{\mathbf{u}}), \quad \begin{matrix} \bar{\mathbf{x}} = A\,\mathbf{x}, \\ \bar{\mathbf{u}} = A^{-T}\,\mathbf{u}, \end{matrix} \quad A \in \mathrm{GL}(m),$$

(10.16)

where k denotes the weight of the concomitant.

Example 10.14. A trivial example of a mixed concomitant is the *identity tensor*

$$\mathbb{1} = \mathbf{u} \cdot \mathbf{x} = \sum_{i=1}^{n} u^i x_i,$$

(10.17)

depending on both covariant and contravariant variables. It trivially satisfies mixed concomitant condition (10.16) for weight 0 — independently of any actual form.

Example 10.15. The simplest invariant of a ternary quadratic form (10.3) is its Hessian or discriminant

$$\Delta = \frac{1}{8} H[Q] = \det \begin{vmatrix} a & b & d \\ b & c & e \\ d & e & f \end{vmatrix},$$

(10.18)

238 Multivariate Polynomials

which is a multiple of the determinant of its coefficient matrix. The
simplest contravariant of a ternary quadratic is the dual quadratic form

$$R(u,v,w) = \det \begin{vmatrix} a & b & d & u \\ b & c & e & v \\ d & e & f & w \\ u & v & w & 0 \end{vmatrix}, \qquad (10.19)$$

whose coefficient matrix is the adjunct or cofactor matrix to Q. The
dual form is a contravariant of weight 2 and defines the dual conic sec-
tion, $R(\mathbf{u}) = 0$. It can be proved that Q, R, Δ, and $\mathbb{1}$ provide a complete list of concomitants of the ternary quadratic, meaning that any
other concomitant can be written as a suitably homogeneous polynomial therein, cf. [229; §15]. The (inhomogeneous) canonical forms for
complex quadratics are invariantly distinguished in the following table.

Canonical Forms for Complex Ternary Quadratics

I.	$p^2 + q^2 + 1$	$\Delta \neq 0$	conic
II.	$p^2 + 1$	$\Delta = 0,\ R \not\equiv 0$	two lines
III.	1	$R \equiv 0,\ Q \not\equiv 0$	double line
IV.	0	$Q \equiv 0$	

Exercise 10.16. Determine an invariant classification of the possible canonical forms of real ternary quadratics.

According to [92, 226], complete systems of invariants, covariants,
contravariants, and mixed concomitants are known only in the cases of a
single ternary quadratic, a ternary cubic, two or three ternary quadratics, and two quaternary quadratics. The case of a ternary quartic was
given to Noether by her thesis advisor Gordan; the calculations were
horrendous, [160], and motivated her, in reaction, to establish the foundations of modern abstract (and nonconstructive) algebra! As far as I
know, there is no analog of the Stroh–Hilbert Theorem 6.32 for rationally
independent covariants and concomitants in the multivariate context.

Symbolic Methods

The extension of the transform and symbolic methods for binary forms
to multivariate functions and polynomials is straightforward. As before,

one has a choice whether to base the construction on a multilinear analog of the Fourier transform — which is the method preferred here since it applies equally to polynomials and general analytic functions — or to base it symbolically on an "artificial factorization" of the form of degree n as the nth power of a linear form. The latter version, which appears in the standard literature, is left to the interested reader to construct. As in the bivariate Theorem 6.23, there is a simple rule for changing one version to the other. For simplicity, we shall just formulate the construction for a single function $Q(\mathbf{x}) = Q(x_1, \ldots, x_m)$, the most important subcase being that of an m-ary form. Extending the method to several functions of m variables is straightforward.

Consider a differential polynomial $\Delta[\mathbf{x}, Q]$ which is homogeneous of degree k in Q. We introduce k distinct symbolic letters $\alpha, \beta, \ldots, \varepsilon$, and associated contravariant variables $\mathbf{x}^\alpha = (x_1^\alpha, \ldots, x_m^\alpha)$. We define the "tensor product" $\mathbf{Q} = Q \otimes \cdots \otimes Q = Q(\mathbf{x}^\alpha) \cdots Q(\mathbf{x}^\varepsilon)$. The trace $\operatorname{tr} \mathbf{Q} = Q(\mathbf{x})^k$ sets all the contravariant variables equal. Let $\nabla_\alpha = (\partial/\partial x_1^\alpha, \ldots, \partial/\partial x_m^\alpha)$ denote the gradient operator with respect to $\mathbf{x}^\alpha$, and let $\boldsymbol{\xi}_\alpha = (\xi_\alpha^1, \ldots, \xi_\alpha^m)$ be the corresponding symbolic variables, both of which transform covariantly under linear maps.

Definition 10.17. Any polynomial $T(\mathbf{x}, \boldsymbol{\xi}_\alpha, \ldots, \boldsymbol{\xi}_\varepsilon)$ which reduces to Δ upon evaluation,

$$\Delta[\mathbf{x}, Q] = \operatorname{tr} T(\mathbf{x}, \nabla_\alpha, \ldots, \nabla_\varepsilon) \, \mathbf{Q} \,, \tag{10.20}$$

is called a *transform* of Δ.

Theorem 10.18. *The transform $\mathcal{F}$ determines a linear isomorphism from the space of homogeneous differential polynomials to the space of multi-symmetric polynomials in the symbolic letters.*

In the case of invariant differential polynomials, the omega and scaling processes are rewritten as multivariate bracket factors. A *bracket factor of the first kind* is a pairing between a covariant (symbolic) variable and a contravariant variable:

$$(\alpha \, \mathbf{x}) = \xi_\alpha^1 \, x_1 + \cdots + \xi_\alpha^m \, x_m. \tag{10.21}$$

A *bracket factor of the second kind* is an $m \times m$ determinant of symbolic variables

$$[\alpha_1 \, \alpha_2 \, \ldots \, \alpha_m] = \det\left(\xi_{\alpha_j}^i\right). \tag{10.22}$$

There are also bracket factors of the third kind depending on m contravariant variables. The First Fundamental Theorem states that any covariant has transform (symbolic form) given by a bracket polynomial in the first and second kind of bracket factors. The Second Fundamental Theorem gives a complete list of syzygies among the bracket factors:

$$\sum_{j=1}^{p+1} [\alpha_1 \ldots \alpha_{j-1} \alpha_{j+1} \ldots \alpha_{p+1}] [\alpha_j \beta_1 \ldots \beta_{p-1}] = 0, \qquad (10.23)$$

$$\sum_{j=1}^{p+1} [\alpha_1 \ldots \alpha_{j-1} \alpha_{j+1} \ldots \alpha_{p+1}] (\alpha_j \mathbf{x}) = 0. \qquad (10.24)$$

See [97, 229, 231] for details.

Remark: It is possible to assign a graph-theoretic interpretation to the bracket algebra, but one must rely on the less intuitive theory of hypergraphs, [44], and the simplifications are not as dramatic as in the bivariate case.

Exercise 10.19. Let $Q(\mathbf{x})$ be a ternary cubic. Prove that the bracket monomial $[\alpha\beta\gamma][\alpha\beta\delta][\alpha\gamma\delta][\beta\gamma\delta]$ represents the simplest nonzero invariant. Determine a second independent invariant. It can be proved that two invariants suffice to form a Hilbert basis for the invariants of a ternary cubic, cf. [204; p. 166], [229; §17]. The complete system of 34 fundamental concomitants of the ternary cubic was established by Gordan, [87]; see also [50, 65] for additional details.

Contravariants and mixed concomitants are also included in the symbolic bracket algebra. Since the symbolic variables represent derivatives with respect to the contravariant variables, they transform as covariant vectors. Therefore, we can replace symbolic variables by covariant variables in bracket factors without affecting their transformation properties. Thus, for example, the dual quadratic form (10.19) has a transform given by a multiple of the bracket monomial $[\alpha\beta\mathbf{u}]^2$. (The multiple depends on whether we are viewing this as the transform or the symbolic form of R.) Indeed, since the covariant variables are not necessarily equated upon reverting to the explicit form, one can also construct interesting types of "joint concomitants" depending on several different covariant vectors, as in the following example.

Exercise 10.20. Let $Q(\mathbf{x}) = Q(x, y, z, w)$ be a quaternary quadratic form. Its inhomogeneous counterpart $Q(p, q, r) = Q(p, q, r, 1)$ defines a *quadric surface* $\mathcal{Q} = \{Q(p, q, r) = 0\} \subset \mathbb{CP}^3$. Use Sylvester's Theorem 3.56 to determine the real and complex canonical forms for quadric surfaces under projective transformations. Discuss the following invariants and their geometrical significance, as indicated by their symbolic forms:

(a) $[\alpha\beta\gamma\delta]^2$ — the discriminant,

(b) $[\alpha\beta\gamma\mathbf{u}]^2 = 0$ — the equation for the dual quadric, i.e., the planes tangent to $\mathcal{Q}$,

(c) $[\alpha\beta\mathbf{u}\mathbf{v}]^2 = 0$ — equation for the lines tangent to $\mathcal{Q}$. Each line is given as the intersection of the two planes corresponding to the two covariant variables $\mathbf{u}$ and $\mathbf{v}$,

(d) $[\alpha\mathbf{u}\mathbf{v}\mathbf{w}]^2 = 0$ — equation for the points on $\mathcal{Q}$. Each point is characterized as the intersection of three planes.

The classical theory concentrates almost exclusively on homogeneous, symmetric forms. These are a very special type of covariant tensor, and one might reasonably investigate the invariant theory of other types of tensorial objects, cf. [97]. A particularly important class are the anti-symmetric tensors, which are the algebraic counterparts of differential forms on manifolds, [25, 168]. The invariant theory of anti-symmetric forms has been studied by Weitzenböck, [229], Gurevich, [97], and Grosshans, Rota, and Stein, [94, 187]. It is interesting that, while the symbolic method for symmetric forms leads to the anti-symmetric bracket algebra, the symbolic method for anti-symmetric forms leads to an algebra of symmetric brackets! The latter references develop a theory of "superalgebras" that incorporate both objects and can be used to investigate joint invariants of both symmetric and anti-symmetric tensors. Another classical case is that of an "affinor", which is just a fancy name for a matrix or linear transformation under the conjugation action $S \mapsto ASA^{-1}$ of $\mathrm{GL}(m)$. Affinors can be viewed as elements of the tensor product space $X \otimes X^*$, and the Jordan canonical form provides a complete classification. See [97] for lists of invariants.

In tensor analysis, the symmetric and anti-symmetric tensors are but two of a wide variety of possible symmetry classes of tensors, which are governed by the decomposition of the representation of the general linear group $\mathrm{GL}(V)$ on the tensor powers $\otimes^n V$ into various irreducible summands, cf. Example 4.9. For example, the Riemannian metric on a manifold defines a covariant 2-tensor. The Riemann curvature tensor has three irreducible components — the Weyl and Ricci tensors and the Ricci scalar curvature — each with its own intrinsic geometrical significance. See [74, 133, 169] for details and applications to relativity. Despite its potential in a wide variety of applications, the invariant theory of more general tensors remains completely undeveloped.

Biforms

Classical invariant theory concentrates almost exclusively on the standard action of the general linear group $GL(m)$ on the associated m-dimensional vector space $\mathbb{R}^m$ or $\mathbb{C}^m$. However, the methods readily extend to other actions, e.g., the joint action of $GL(m)$ on Cartesian products of copies of $\mathbb{C}^m$. Yet another extension that is amenable to the classical techniques is that of the Cartesian product of general linear groups acting on a Cartesian product space. In this final section, we discuss the most interesting special case, first analyzed by Peano, [**178**].

Definition 10.21. A *biform* of *bidegree* (m, n) is a polynomial function $Q(\mathbf{x}, \mathbf{w})$ depending on $(\mathbf{x}, \mathbf{w}) \in \mathbb{C}^p \times \mathbb{C}^q$, which, for fixed $\mathbf{w}$, is a homogeneous polynomial of degree m in $\mathbf{x}$, and, for fixed $\mathbf{x}$, is a homogeneous polynomial of degree n in $\mathbf{w}$.

In other words, Q satisfies the homogeneity requirement

$$Q(\lambda \mathbf{x}, \mu \mathbf{w}) = \lambda^m \mu^n \, Q(\mathbf{x}, \mathbf{w}).$$

The most general such biform is

$$Q(\mathbf{x}, \mathbf{w}) = \sum_{\substack{\#I=m \\ \#J=n}} \binom{m}{I} \binom{n}{J} a_{IJ} \, x^I u^J. \tag{10.25}$$

The biform (10.25) is said to have *biweight* (k, l) if, under the simultaneous linear transformations $\mathbf{x} \mapsto A \cdot \mathbf{x}$, $\mathbf{w} \mapsto B \cdot \mathbf{w}$, where $A \in GL(p)$, $B \in GL(q)$, it satisfies the transformation rule

$$Q(\mathbf{x}, \mathbf{w}) = (\det A)^k (\det B)^l \, \overline{Q}(\overline{\mathbf{x}}, \overline{\mathbf{w}}), \qquad \begin{aligned} \overline{\mathbf{x}} &= A \cdot \mathbf{x}, \\ \overline{\mathbf{w}} &= B \cdot \mathbf{w}. \end{aligned} \tag{10.26}$$

As with forms, two biforms are *equivalent* if they can be mapped to each other by a suitable linear transformation.

A *covariant* of biweight (k', l') is a polynomial function $J(\mathbf{a}, \mathbf{x}, \mathbf{w})$, depending on the coefficients $\mathbf{a} = (a_{IJ})$ of the biform (10.25), and the variables $\mathbf{x}, \mathbf{w}$, which transforms according to the multiplier rule (10.26). As with forms, an *invariant* is just a covariant which does not depend on the variables $\mathbf{x}$ or $\mathbf{w}$. Covariants can be constructed by successive (partial) transvectants, but now there are two different omega processes available: one with respect to the $\mathbf{x}$ variables and the other with respect to the $\mathbf{w}$ variables.

For specificity, let us consider the first nontrivial case: a binary biquadratic

$$Q(\mathbf{x}, \mathbf{w}) = a_{11}^{11} x^2 u^2 + 2a_{12}^{11} xyu^2 + a_{22}^{11} y^2 u^2 + 2a_{11}^{12} x^2 uv + 4a_{12}^{12} xyuv +$$
$$+ 2a_{22}^{12} y^2 uv + a_{11}^{22} x^2 v^2 + 2a_{12}^{22} xyv^2 + a_{22}^{22} y^2 v^2, \quad (10.27)$$

depending on $\mathbf{x} = (x, y)$ and $\mathbf{w} = (u, v)$, which we take of biweight $(0, 0)$. We define the (r, s)-*transvectant* of two such biforms by the formula

$$(Q, R)^{(r,s)} = \operatorname{tr} \Omega^r \Theta^s \, (Q \otimes R), \quad (10.28)$$

where $Q \otimes R = Q(\mathbf{x}_\alpha, \mathbf{w}_\alpha) R(\mathbf{x}_\beta, \mathbf{w}_\beta)$ and the two omega processes are given by

$$\Omega = \det \begin{vmatrix} \dfrac{\partial}{\partial x_\alpha} & \dfrac{\partial}{\partial y_\alpha} \\[2mm] \dfrac{\partial}{\partial x_\beta} & \dfrac{\partial}{\partial y_\beta} \end{vmatrix}, \qquad \Theta = \det \begin{vmatrix} \dfrac{\partial}{\partial u_\alpha} & \dfrac{\partial}{\partial v_\alpha} \\[2mm] \dfrac{\partial}{\partial u_\beta} & \dfrac{\partial}{\partial v_\beta} \end{vmatrix}. \quad (10.29)$$

If Q, R have respective biweights $(k, l), (\widetilde{k}, \widetilde{l})$, then (10.28) is a covariant of biweight $(k + \widetilde{k} + r, l + \widetilde{l} + s)$.

For each fixed $\mathbf{w}$, the biform (10.27) is a quadratic in $\mathbf{x}$, and so we can form its discriminant $\Delta_\mathbf{x}(\mathbf{w})$, which is a binary quartic in $\mathbf{w}$. The $\mathbf{x}$ discriminant obeys all the usual properties enjoyed by the discriminant of an ordinary quadratic polynomial; for instance, if $\Delta_\mathbf{x}(\mathbf{w}_0) = 0$, then $Q(\mathbf{x}, \mathbf{w}_0)$ is a perfect square for that particular value of $\mathbf{w}$. A similar construction holds in $\mathbf{w}$, and hence we deduce the first important covariants

$$\Delta_\mathbf{x}(\mathbf{w}) = \tfrac{1}{2} (Q, Q)^{(2,0)} = Q_{xx} Q_{yy} - Q_{xy}^2,$$
$$\Delta_\mathbf{w}(\mathbf{x}) = \tfrac{1}{2} (Q, Q)^{(0,2)} = Q_{uu} Q_{vv} - Q_{uv}^2, \quad (10.30)$$

of respective biweights $(2, 0)$, $(0, 2)$. There is a mixed biquadratic covariant of biweight $(1, 1)$, which has the explicit formula

$$C = \tfrac{1}{2} (Q, Q)^{(1,1)} = Q_{xu} Q_{yv} - Q_{xv} Q_{yu}.$$

The simplest invariant is the quadratic expression

$$I = \tfrac{1}{32} (Q, Q)^{(2,2)} = a_{11}^{11} a_{22}^{22} + a_{22}^{11} a_{11}^{22} - 2a_{12}^{11} a_{12}^{22} - 2a_{11}^{12} a_{22}^{12} + 2(a_{12}^{12})^2,$$

of biweight $(2, 2)$. There is a single cubic invariant of biweight $(3, 3)$,

$$J = \tfrac{1}{1536} (C, Q)^{(2,2)} = a_{11}^{11} a_{12}^{12} a_{22}^{22} - a_{11}^{11} a_{22}^{12} a_{12}^{22} - a_{12}^{11} a_{11}^{12} a_{22}^{22} +$$
$$+ a_{12}^{11} a_{22}^{12} a_{11}^{22} + a_{22}^{11} a_{11}^{12} a_{12}^{22} - a_{22}^{11} a_{12}^{12} a_{11}^{22}.$$

The technique of composition can be used to produce further interesting covariants. Thus, since the discriminant $\Delta_\mathbf{w}(\mathbf{x})$ of a binary biquadratic is a binary quartic in the variables $\mathbf{x} = (x, y)$, all the standard quartic covariants yield, under composition, covariants of the original biquadratic Q. Thus, we have the Hessian of the $\mathbf{w}$-discriminant

$$H_{\mathbf{w}}(\mathbf{x}) = H \circ \Delta_{\mathbf{w}} = (\Delta_{\mathbf{w}}, \Delta_{\mathbf{w}})^{(2)},$$

which is again a binary quartic in $\mathbf{x}$, and a covariant of biweight $(2, 4)$, as well as the two invariants

$$i_{\mathbf{w}} = i \circ \Delta_{\mathbf{w}} = (\Delta_{\mathbf{w}}, \Delta_{\mathbf{w}})^{(4)}, \qquad j_{\mathbf{w}} = j \circ \Delta_{\mathbf{w}} = (\Delta_{\mathbf{w}}, H_{\mathbf{w}})^{(4)},$$

of respective biweights $(4, 4)$ and $(6, 6)$. Alternatively, the Hessian and the two invariants of the $\mathbf{x}$ discriminant

$$H_{\mathbf{x}}(\mathbf{w}) = (\Delta_{\mathbf{x}}, \Delta_{\mathbf{x}})^{(2)}, \qquad i_{\mathbf{x}} = (\Delta_{\mathbf{x}}, \Delta_{\mathbf{x}})^{(4)}, \qquad j_{\mathbf{x}} = (\Delta_{\mathbf{x}}, H_{\mathbf{x}})^{(4)},$$

have respective biweights $(4, 2)$, $(4, 4)$, and $(6, 6)$. The two Hessians $H_{\mathbf{w}}$ and $H_{\mathbf{x}}$ are easily seen to be different quartic polynomials in general (even if one identifies the variables $\mathbf{x}$ and $\mathbf{w}$). Remarkably, the invariants of the two discriminants are the *same*!

Proposition 10.22. *Let Q be a binary biquadratic form, and let $\Delta_{\mathbf{x}}(\mathbf{w})$, $\Delta_{\mathbf{w}}(\mathbf{x})$ be its two discriminants. Then the quartic invariants of the discriminants are identical:*

$$i_{\mathbf{x}} = i \circ \Delta_{\mathbf{x}} = i_{\mathbf{w}} = i \circ \Delta_{\mathbf{w}}, \qquad j_{\mathbf{x}} = j \circ \Delta_{\mathbf{x}} = j_{\mathbf{w}} = j \circ \Delta_{\mathbf{w}}.$$

Consequently, $\Delta_{\mathbf{x}}(\mathbf{w})$ and $\Delta_{\mathbf{w}}(\mathbf{x})$ either both have all simple roots, or both have multiple roots.

The most direct way to prove Proposition 10.22 is by a modification of our graph-theoretic method, where the arcs in the digraphs now come in two different colors to represent the two different available omega processes (10.29). The syzygy rules (7.2), (7.3), (7.6), are only allowed on arcs having the same color. Details are left as an exercise for the reader.

The final statement in Proposition 10.22 immediately follows from the fact that the invariants of a quartic determine its discriminant, whose vanishing indicates the presence of repeated roots. Note that it is *not* asserted that $\Delta_{\mathbf{x}}(\mathbf{w})$ and $\Delta_{\mathbf{w}}(\mathbf{x})$ have identical root multiplicities! For example, the biquadratic form $Q = x^2 u^2 + xy v^2$ has $\mathbf{w}$-discriminant $\Delta_{\mathbf{w}} = -4x^3 y$, which has a triple root at 0 and a simple root at ∞, whereas the $\mathbf{x}$-discriminant $\Delta_{\mathbf{x}} = v^4$ has a quadruple root at ∞. Thus, the two discriminants of a biquadratic cannot be completely arbitrary but have subtle interrelationships.

In a series of papers, Peano, [**178**], and Turnbull, [**215–218**], give a complete classification and geometric interpretation for the Hilbert basis for covariants of the binary biquadratic. The complete system consists of the following 15 covariants and 3 invariants, [**216**; p. 49]:

$$Q, \quad C = (Q, Q)^{(1,1)}, \quad I = (Q, Q)^{(2,2)},$$
$$\Delta_{\mathbf{x}} = (Q, Q)^{(2,0)}, \quad \Delta_{\mathbf{w}} = (Q, Q)^{(0,2)}, \quad (10.31)$$
$$D = (Q, \Delta_{\mathbf{x}})^{(0,2)} = (Q, \Delta_{\mathbf{w}})^{(2,0)},$$
$$J = (Q, C)^{(2,2)}, \quad i_{\mathbf{x}} = i_{\mathbf{w}} = (\Delta_{\mathbf{x}}, \Delta_{\mathbf{x}})^{(0,4)} = (\Delta_{\mathbf{w}}, \Delta_{\mathbf{w}})^{(4,0)},$$
$$R_{\mathbf{x}} = (Q, \Delta_{\mathbf{x}})^{(0,1)}, \quad R_{\mathbf{w}} = (Q, \Delta_{\mathbf{w}})^{(0,1)}, \quad S_{\mathbf{x}} = (Q, D)^{(1,0)},$$
$$S_{\mathbf{w}} = (Q, D)^{(0,1)}, \quad T_{\mathbf{x}} = (Q, D)^{(2,0)}, \quad T_{\mathbf{w}} = (Q, D)^{(0,2)},$$
$$U_{\mathbf{x}} = (Q, T_{\mathbf{x}})^{(0,1)}, \quad U_{\mathbf{w}} = (Q, T_{\mathbf{w}})^{(1,0)},$$
$$V_{\mathbf{x}} = (T_{\mathbf{x}}, \Delta_{\mathbf{x}})^{(0,1)}, \quad V_{\mathbf{w}} = (T_{\mathbf{w}}, \Delta_{\mathbf{w}})^{(1,0)}.$$

A complete list of canonical forms for complex binary biquadratics follows. The different canonical forms are distinguished by the root configurations of the associated quartic discriminants (10.30), based on the restrictions in Proposition 10.22. See [**165**] for details.

Canonical Forms for Complex Binary Biquadratics

1.	$x^2 u^2 + y^2 v^2 + \alpha(x^2 v^2 + y^2 u^2) + 2\beta x y u v$
2.	$x^2 u^2 + y^2 v^2 + y^2 u^2 + 2\beta x y u v$
3.	$x^2 u^2 - x^2 v^2 + x y u^2 + x y u v^2$
4.	$x^2 u^2 - y^2 u^2 + x^2 u v + y^2 u v$
5.	$x^2 u^2 + y^2 v^2 + y^2 u v + 2 x y u v$
6.	$x^2 u^2 + y^2 u v$
7.	$x^2 u^2 + x y u v^2$
8.	$x^2 u v + x y u^2$
9.	$x^2 u^2 + y^2 u^2 + x y u v$
10.	$x^2 u^2 + x^2 v^2 + x y u v$
11.	$x^2 u^2 + y^2 u^2$
12.	$x^2 u^2 + x^2 v^2$
13.	$x^2 u^2 + x y u v$
14.	$x^2 u^2$
15.	$x y u v$
16.	0

See [**165, 166**] for the corresponding real classification and applications to the study of canonical forms of quadratic variational problems and conservation laws in linear elasticity.

Exercise 10.23. Determine the formula for $j_{\mathbf{x}} = j_{\mathbf{w}}$ in terms of the Hilbert basis (10.31).

Exercise 10.24. Use the covariants (10.31) to produce an invariant characterization of each canonical form.

References

[1] Abraham, R., and Marsden, J.E., *Foundations of Mechanics*, 2nd ed., Benjamin–Cummings Publ. Co., Reading, Mass., 1978.

[2] Ackerman, M., and Hermann, R., *Sophus Lie's 1880 Transformation Group Paper*, Lie Groups: History, Frontiers and Applications, vol. 1, Math Sci Press, Brookline, Mass., 1975.

[3] Ackerman, M., and Hermann, R., *Sophus Lie's 1884 Differential Invariant Paper*, Lie Groups: History, Frontiers and Applications, vol. 3, Math Sci Press, Brookline, Mass., 1976.

[4] Ackerman, M., and Hermann, R., *Hilbert's Invariant Theory Papers*, Lie Groups: History, Frontiers and Applications, vol. 8, Math Sci Press, Brookline, Mass., 1978.

[5] Adem, A., Maginnis, J., and Milgram, R.J., Symmetric invariants and cohomology of groups, *Math. Ann.* **287** (1990), 391–411.

[6] Adler, M., On a trace functional for formal pseudo-differential operators and the symplectic structure of the Korteweg–deVries type equations, *Invent. Math.* **50** (1979), 219–248.

[7] Ado, I.D., The representations of Lie algebras by matrices, *Uspekhi Mat. Nauk* **2** (1947), 159–173; *Amer. Math. Soc. Transl.* #2, New York, 1949.

[8] Ahlfors, L., *Complex Analysis*, McGraw–Hill, New York, 1966.

[9] Anderson, I.M., and Pohjanpelto, J., Variational principles for differential equations with symmetries and conservation laws I. Second order scalar equations, *Math. Ann.* **299** (1994), 191–222.

[10] Anderson, I.M., and Pohjanpelto, J., Variational principles for differential equations with symmetries and conservation laws II. Polynomial differential equations, *Math. Ann.* **301** (1995), 627–653.

[11] Antman, S.S., *Nonlinear Problems of Elasticity*, Appl. Math. Sci., vol. 107, Springer–Verlag, New York, 1995.

[12] Aronhold, S., Theorie der homogenen Functionen dritten Grades von drei Veränderlichen, *J. Reine Angew. Math.* **55** (1858), 97–191.

[13] Atiyah, M.F., and Macdonald, I.G., *Introduction to Commutative Algebra*, Addison–Wesley Publ. Co., Reading, Mass., 1969.

[14] Ball, J.M., Differentiability properties of symmetric and isotropic functions, *Duke Math. J.* **51** (1984), 699–728.

[15] Ball, J.M., Currie, J.C., and Olver, P.J., Null Lagrangians, weak continuity, and variational problems of arbitrary order, *J. Func. Anal.* **41** (1981), 135–174.

[16] Barbançon, G., Théorème de Newton pour les fonctions de class C^r, *Ann. Sci. École Norm. Sup.* (4) **5** (1972), 435–457.

248   *References*

[17] Bargmann, V., Irreducible unitary representations of the Lorentz group, *Ann. Math.* **48** (1947), 568–640.

[18] Bass, H., Connell, E.H., and Wright, D., The Jacobian conjecture: reduction of degree and formal expansion of the inverse, *Bull. Amer. Math. Soc.* **7** (1982), 287–330.

[19] Bayen, F., Flato, M., Fronsdal, C., Lichnerowicz, A., and Sternheimer, D., Deformation theory and quantization. I. Deformations of symplectic structures, *Ann. Physics* **111** (1978), 61–110.

[20] Beardon, A.F., *The Geometry of Discrete Groups*, Graduate Texts in Mathematics, vol. 91, Springer–Verlag, New York, 1983.

[21] Berezin, F.A., *The Method of Second Quantization*, Academic Press, New York, 1966.

[22] Bleecker, D., *Gauge Theory and Variational Principles*, Addison–Wesley Publ. Co., Reading, Mass., 1981.

[23] Bôcher, M., *Introduction to Higher Algebra*, Macmillan Co., New York, 1907.

[24] Boole, G., Exposition of a general theory of linear transformations, *Camb. Math. J.* **3** (1841–2), 1–20, 106–119.

[25] Bott, R., and Tu, L.W., *Differential Forms in Algebraic Topology*, Springer–Verlag, New York, 1982.

[26] Boyer, C.B., *A History of Mathematics*, J. Wiley & Sons, New York, 1968.

[27] Bryant, R.L., Chern, S.-S., Gardner, R.B., Goldschmidt, H.L., and Griffiths, P.A., *Exterior Differential Systems*, Math. Sci. Res. Inst. Publ., vol. 18, Springer–Verlag, New York, 1991.

[28] Buchberger, B., Applications of Gröbner bases in non-linear computational geometry, in: *Scientific Software*, J.R. Rice, ed., IMA Volumes in Mathematics and Its Applications, vol. 14, Springer–Verlag, New York, 1988.

[29] Calabi, E., Olver, P.J., Shakiban, C., Tannenbaum, A., and Haker, S., Differential and numerically invariant signature curves applied to object recognition, *Int. J. Computer Vision* **26** (1998), 107–135.

[30] Calabi, E., Olver, P.J., and Tannenbaum, A., Affine geometry, curve flows, and invariant numerical approximations, *Adv. in Math.* **124** (1996), 154–196.

[31] Capelli, A., Ueber die Zurückführung der Cayley'schen Operation Ω auf gewöhnliche Polar-Operationen, *Math. Ann.* **29** (1887), 331–338.

[32] Cartan, É., *Sur la Structure des Groupes de Transformations Finis et Continus*, Thèse, Paris, Nony, 1894; also *Oeuvres Complètes*, part. I, vol. 1, Gauthier–Villars, Paris, 1952, pp. 137–287.

[33] Cartan, É., *La Méthode du Repère Mobile, la Théorie des Groupes Continus, et les Espaces Généralisés*, Exposés de Géométrie, no. 5, Hermann, Paris, 1935.

[34] Cartan, É., *La Topologie des Groupes de Lie*, Exposés de Géométrie, no. 8, Hermann, Paris, 1936.

[35] Cartan, É., Les problèmes d'équivalence, in: *Oeuvres Complètes*, part. II, vol. 2, Gauthiers-Villars, Paris, 1952, pp. 1311–1334.

References 249

[36] Cayley, A., *The Collected Mathematical Papers*, vols. 1–12, Cambridge Univ. Press, Cambridge, England, 1889–1898.

[37] Cayley, A., *The Collected Mathematical Papers*, vol. 1, Cambridge Univ. Press, Cambridge, England, 1889.

[38] Cayley, A., On the theory of linear transformations, *Camb. Math. J.* **4** (1845), 193–209; also *The Collected Mathematical Papers*, vol. 1, Cambridge Univ. Press, Cambridge, England, 1889, pp. 80–94.

[39] Cayley, A., On linear transformations, *Camb. and Dublin Math. J.* **1** (1846), 104–122; also *The Collected Mathematical Papers*, vol. 1, Cambridge Univ. Press, Cambridge, England, 1889, pp. 95–112.

[40] Cayley, A., An introductory memoir upon quantics, *Phil. Trans. Roy. Soc. London* **144** (1854), 244–258; also *The Collected Mathematical Papers*, vol. 2, Cambridge Univ. Press, Cambridge, England, 1889, pp. 221–234.

[41] Cayley, A., A second memoir upon quantics, *Phil. Trans. Roy. Soc. London* **146** (1856), 101–126; also *The Collected Mathematical Papers*, vol. 2, Cambridge Univ. Press, Cambridge, England, 1889, pp. 250–275.

[42] Cayley, A., On Tschirnhausen's transformation, *Phil. Trans. Roy. Soc. London* **151** (1861), 561–578; also *The Collected Mathematical Papers*, vol. 4, Cambridge Univ. Press, Cambridge, England, 1891, pp. 375–394.

[43] Chakravarty, S., Ablowitz, M.J., and Clarkson, P.A., Reductions of self-dual Yang–Mills fields and classical systems, *Phys. Rev. Lett.* **65** (1990), 1085–1087.

[44] Chartrand, G., and Lesniak, L., *Graphs & Digraphs*, 3rd ed., Chapman & Hall, London, 1996.

[45] Chazy, J., Sur les équations différentielles du troisième ordre et d'ordre supérieur dont l'intégrale générale a ses points critiques fixes, *Acta Math.* **34** (1911), 317–385.

[46] Chen, J.Q., *Group Representation Theory for Physicists*, World Scientific, Singapore, 1989.

[47] Clarkson, P.A., and Olver, P.J., Symmetry and the Chazy equation, *J. Diff. Eq.* **124** (1996), 225–246.

[48] Clebsch, A., Ueber symbolische Darstellung algebraischer Formen, *J. Reine Angew. Math.* **59** (1861), 1–62.

[49] Clebsch, A., *Theorie der Binären Algebraischen Formen*, B.G. Teubner, Leipzig, 1872.

[50] Clebsch, A., and Gordan, P., Ueber cubische ternäre Formen, *Math. Ann.* **6** (1873), 436–512.

[51] Clifford, W., Extract of a letter to Mr. Sylvester from Prof. Clifford of University College, London, *Amer. J. Math.* **1** (1878), 126–128.

[52] Cohen, H., Sums involving the values at negative integers of L-functions of quadratic characters, *Math. Ann.* **217** (1975), 271–285.

[53] Coleman, A.J., The greatest mathematical paper of all time, *Math. Intelligencer* **11** (1989), 29–38.

250   *References*

[54] Cox, D., Little, J., and O'Shea, D., *Ideals, Varieties, and Algorithms*, 2nd ed., Springer–Verlag, New York, 1996.

[55] Crilly, T., The rise of Cayley's invariant theory (1841–1862), *Hist. Math.* **13** (1986), 241–254.

[56] Curtis, C.W., and Reiner, I., *Representation Theory of Finite Groups and Associative Algebras*, Interscience, New York, 1962.

[57] Cushman, R., and Sanders, J.A., A survey of invariant theory applied to normal forms of vectorfields with nilpotent linear part, *in:* *Invariant Theory and Tableaux*, D. Stanton, ed., IMA Volumes in Mathematics and Its Applications, vol. 19, Springer–Verlag, New York, 1990, pp. 82–106.

[58] Dickson, L.E., *Modern Algebraic Theories*, Benj. H. Sanborn & Co., Chicago, 1926.

[59] Dirac, P.A.M., *Lectures on Quantum Mechanics*, Belfer Graduate School of Science Monographs Series, no. 2, Yeshiva Univ., New York, 1964.

[60] Dixmier, J., and Lazard, D., Le nombre minimum d'invariants fondamentaux pour les formes binaires de degré 7, *Portugal. Math.* **43** (1986), 377–392.

[61] Dixmier, J., and Lazard, D., Minimum number of fundamental invariants for the binary form of degree 7, *J. Symb. Comp.* **6** (1988), 113–115.

[62] Ðoković, D., and Hofmann, K.H., The exponential in real Lie groups: A status report, *J. Lie Theory* **7** (1997), 171–199.

[63] Edelman, P., personal communication, 1992.

[64] Ehresmann, C., Introduction à la théorie des structures infinitésimales et des pseudo-groupes de Lie, *in:* *Géométrie Différentielle*, Colloq. Inter. du Centre Nat. de la Rech. Sci., Strasbourg, 1953, pp. 97–110.

[65] Elliott, E.B., *An Introduction to the Algebra of Quantics*, Oxford Univ. Press, Oxford, 1913.

[66] Escherich, G.V., Die Determinanten hoheren Ranges und ihre Verwendung zur Bildung von Invarianten, *Denkschr. Kais. Aka. Wiss. Wien* **43** (1882), 1–12.

[67] Faà di Bruno, F., *Einleitung in die Theorie der Binären Formen*, B.G. Teubner, Leipzig, 1881.

[68] Fels, M., and Olver, P.J., On relative invariants, *Math. Ann.* **308** (1997), 701–732.

[69] Fels, M., and Olver, P.J., Moving coframes. I. A practical algorithm, *Acta Appl. Math.* **51** (1998), 161–213.

[70] Fels, M., and Olver, P.J., Moving coframes. II. Regularization and theoretical foundations, *Acta Appl. Math.*, to appear.

[71] Fine, B., and Rosenberger, G., *The Fundamental Theorem of Algebra*, Undergraduate Texts in Mathematics, Springer–Verlag, New York, 1997.

[72] Fisher, C.S., The death of a mathematical theory: a study in the sociology of knowledge, *Arch. Hist. Exact Sci.* **3** (1966), 137–159.

[73] Frobenius, G., Über das Pfaffsche Problem, *J. Reine Angew. Math.* **82** (1877), 230–315.

[74] Fulling, S.A., King, R.C., Wybourne, B.G., and Cummins, C.J., Normal forms for tensor polynomials: I. The Riemann tensor, *Class. Quantum Grav.* **9** (1992), 1151–1197.

[75] Gantmacher, F.R., *The Theory of Matrices*, Chelsea Publ. Co., New York, 1959.

[76] Gegenbauer, L., Über Determinanten hoheren Ranges, *Denkschr. Kais. Aka. Wiss. Wien* **43** (1882), 17–32.

[77] Gel'fand, I.M., and Dikii, L.A., Asymptotic behavior of the resolvent of Sturm-Liouville equations and the algebra of the Korteweg–deVries equation, *Russian Math. Surveys* **30**:5 (1975), 77–113; also *Collected Papers*, vol. 1, Springer-Verlag, New York, 1987, pp. 647–683.

[78] Gel'fand, I.M., and Dikii, L.A., A family of Hamiltonian structures related to integrable nonlinear differential equations, preprint, Acad. Sci. USSR, no. 136, 1978; also *Collected Papers*, vol. 1, Springer-Verlag, New York, 1987, pp. 625–646.

[79] Gel'fand, I.M., Graev, M.I., and Piatetski-Shapiro, I.I., *Representation Theory and Automorphic Functions*, Academic Press, Boston, 1990.

[80] Gessel, I.M., Enumerative applications of symmetric functions, *in*: *Actes 17e Séminaire Lotharingien*, Publ. I.M.R.A., 348/S–17, Strasbourg, 1988, pp. 5–21.

[81] Glaeser, G., Fonctions composées différentiables, *Ann. Math.* **77** (1963), 193–209.

[82] Golubitsky, M., Stewart, I., and Schaeffer, D.G., *Singularities and Groups in Bifurcation Theory*, vol. II, Appl. Math. Sci. #69, Springer–Verlag, New York, 1988.

[83] González–López, A., Hernández, R., and Marí-Beffa, G., Invariant differential equations and the Adler–Gel'fand–Dikii bracket, *J. Math. Phys.* **38** (1997), 5720–5738.

[84] González–López, A., Kamran, N., and Olver, P.J., Lie algebras of differential operators in two complex variables, *Amer. J. Math.* **114** (1992), 1163–1185.

[85] González–López, A., Kamran, N., and Olver, P.J., Quasi-exact solvability, *Contemp. Math.* **160** (1994), 113–140.

[86] Gordan, P., Beweis, dass jede Covariante und Invariante einer binären Form eine ganz Funktion mit numerischen Coefficienten einer endlichen Anzahl solcher Formen ist, *J. Reine Angew. Math.* **69** (1868), 323–354.

[87] Gordan, P., Ueber ternäre Formen dritten Grades, *Math. Ann.* **1** (1869), 90–128.

[88] Gordan, P., Ueber die Bildung der Resultante zweier Gleichungen, *Math. Ann.* **3** (1869), 355–414.

[89] Gordan, P., *Vorlesungen über Invariantentheorie*, Chelsea Publ. Co., New York, 1987.

252   References

[90] Gordan, P., and Nöther, M., Ueber die algebraischen Formen, deren Hesse'sche Determinante identisch verschwindet, *Math. Ann.* **10** (1876), 547–568.

[91] Gorenstein, D., *Finite Simple Groups*, Plenum Press, New York, 1982.

[92] Grace, J.H., and Young, A., *The Algebra of Invariants*, Cambridge Univ. Press, Cambridge, 1903.

[93] Griffiths, P.A., On Cartan's method of Lie groups and moving frames as applied to uniqueness and existence questions in differential geometry, *Duke Math. J.* **41** (1974), 775–814.

[94] Grosshans, F.D., Rota, G.-C., and Stein, J.A., *Invariant Theory and Superalgebras*, CBMS Regional Conference Series, no. 69, American Math. Soc., Providence, R.I., 1987.

[95] Guggenheimer, H.W., *Differential Geometry*, McGraw–Hill, New York, 1963.

[96] Gundelfinger, S., Zur Theorie der binären Formen, *J. Reine Angew. Math.* **100** (1886), 413–424.

[97] Gurevich, G.B., *Foundations of the Theory of Algebraic Invariants*, P. Noordhoff Ltd., Groningen, Holland, 1964.

[98] Hale, J.K., *Ordinary Differential Equations*, Wiley–Interscience, New York, 1969.

[99] Halphen, G.-H., Sur les invariants différentiels, *in: Oeuvres*, vol. 2, Gauthier–Villars, Paris, 1913, pp. 197–253.

[100] Hamermesh, M., *Group Theory and Its Application to Physical Problems*, Addison–Wesley Publ. Co., Reading, Mass., 1962.

[101] Hawkins, T., Jacobi and the birth of Lie's theory of groups, *Arch. Hist. Exact Sci.* **42** (1991), 187–278.

[102] Hermite, M., Sur la theorie des fonctions homogenes à deux indéterminées, *Camb. and Dublin Math. J.* **9** (1854), 172–217.

[103] Hesse, O., Über die Bedingung, unter welcher eine homogene ganze Function von n unabhängigen Variabeln durch lineäre Substitutionen von n andern unabhängigen Variabeln auf ein homogene Function sich zurückführen lässt, die eine Variable weniger enthält, *J. Reine Angew. Math.* **42** (1851), 117–124.

[104] Hesse, O., Zur Theorie der ganzen homogenen Functionen, *J. Reine Angew. Math.* **56** (1859), 263–269.

[105] Hilbert, D., Über die Theorie der algebraischen Formen, *Math. Ann.* **36** (1890), 473–534; also *Gesammelte Abhandlungen*,vol. 2, Springer–Verlag, Berlin, 1933, pp. 199–257; see [4; pp. 143–224] for an English translation.

[106] Hilbert, D., Über die vollen Invariantensystem, *Math. Ann.* **42** (1893), 313–373; also *Gesammelte Abhandlungen*, vol. 2, Springer–Verlag, Berlin, 1933, pp. 287–344; see [4; pp. 225–301] for an English translation.

[107] Hilbert, D., *Theory of Algebraic Invariants*, Cambridge Univ. Press, New York, 1993.

[108] Hille, E., *Ordinary Differential Equations in the Complex Domain*, John Wiley & Sons, New York, 1976.

[109] Hirota, R., Direct method for finding exact solutions of nonlinear evolution equations, in: *Bäcklund Transformations*, R.M. Miura, ed., Lecture Notes in Math., vol. 515, Springer–Verlag, New York, 1976, pp. 40–68.

[110] Ho, S.–M., On the isotropic group of a homogeneous polynomial, *Trans. Amer. Math. Soc.* **183** (1973), 495–498.

[111] Hochster, M., and Roberts, J.L., Rings of invariants of reductive groups acting on regular rings are Cohen–Macaulay, *Adv. in Math.* **13** (1974), 115–175.

[112] Howe, R., Remarks on classical invariant theory, *Trans. Amer. Math. Soc.* **313** (1989), 539–570.

[113] Hughston, L.P., and Ward, R.S. (eds.), *Advances in Twistor Theory*, Research Notes in Math., vol. 37, Pitman Publ., San Francisco, 1979.

[114] Hurwitz, A., Zur Invariantentheorie, *Math. Ann.* **45** (1894), 381–404.

[115] Iachello, F., and Levine, R.D., *Algebraic Theory of Molecules*, Oxford Univ. Press, Oxford, 1995.

[116] Ince, E.L., *Ordinary Differential Equations*, Dover Publ., New York, 1956.

[117] Jacobson, N., *Lie Algebras*, Interscience Publ., New York, 1962.

[118] Jacquet, H., and Langlands, R.P., *Automorphic Forms on* GL(2), Lecture Notes in Math., vol. 114, Springer–Verlag, New York, 1970.

[119] Janson, S., and Peetre, J., A new generalization of Hankel operators (the case of higher weights), *Math. Nachr.* **132** (1987), 313–328.

[120] Jensen, G.R., *Higher Order Contact of Submanifolds of Homogeneous Spaces*, Lecture Notes in Math., vol. 610, Springer–Verlag, New York, 1977.

[121] Junker, F., Ueber symmetrischen Functionen von mehren Reihen von Veränderlichen, *Math. Ann.* **43** (1893), 225–270.

[122] Junker, F., Die symmetrischen Functionen und die Relationen zwischen den Elementarfunctionen derselben, *Math. Ann.* **45** (1894), 1–84.

[123] Kaplansky, I., *An Introduction to Differential Algebra*, 2nd ed., Hermann, Paris, 1976.

[124] Kauffman, L.H., *Knots and Physics*, 2nd ed., World Scientific, Singapore, 1993.

[125] Kempe, A.B., On the application of Clifford's graphs to ordinary binary quantics, *Proc. London Math. Soc.* **17** (1885), 107–121.

[126] Killing, W., Die Zusammensetzung der stetigen, endlichen Transformationgruppen, *Math. Ann.* **31** (1888), 355–414; **33** (1889), 1–48; **33** (1889), 57–122; **33** (1890), 161–189.

[127] Killing, W., Erweiterung der Begriffes der Invarianten von Transformationgruppen, *Math. Ann.* **35** (1890), 423–432.

[128] Klein, F., *Vergleichende Betrachtungen über neuere geometrische Forschungen*, A. Deichert, Erlangen, Germany, 1872; also *Gesammelte Mathematische Abhandlungen*, vol. 1, Julius Springer, Berlin, 1921, pp. 460–497; see *Bull. New York Math. Soc.* **2** (1893) 215–249 for an English translation.

[**129**] Klein, F., and Lie, S., Über diejenigen ebenen Curven, welche durch ein geschlossenes System von einfach unendlich vielen vertauschbaren linearen Transformationen in sich übergeben, *Math. Ann.* **4** (1871), 50–84.

[**130**] Knox, R.S., and Gold, A., *Symmetry in the Solid State*, W.A. Benjamin, New York, 1964.

[**131**] Knuth, D.E., Two notes on notation, *Amer. Math. Monthly* **99** (1992), 403–422.

[**132**] Kostant, B., An editorial perspective, *in: Izrail M. Gel'fand Collected Papers*, vol. 3, Springer-Verlag, New York, 1989, pp. 1025–1026.

[**133**] Kramer, D., Stephani, H., MacCallum, M., and Herlt, E., *Exact Solutions of Einstein's Field Equations*, VEB Deutscher Verlag Wissen., Berlin, 1980.

[**134**] Kung, J.P.S., Canonical forms for binary forms of even degree, *in: Invariant Theory*, S.S. Koh, ed., Lecture Notes in Math, vol. 1278, Springer–Verlag, New York, 1987, pp. 52–61.

[**135**] Kung, J.P.S., and Rota, G.–C., The invariant theory of binary forms, *Bull. Amer. Math. Soc.* **10** (1984), 27–85.

[**136**] Landau, L.D., and Lifshitz, E.M., *Quantum Mechanics* (*Non-relativistic Theory*), Course of Theoretical Physics, vol. 3, Pergamon Press, New York, 1977.

[**137**] Lickorish, W.B.R., *An Introduction to Knot Theory*, Graduate Texts in Mathematics, vol. 175, Springer–Verlag, New York, 1997.

[**138**] Lie, S., Theorie der Transformationsgruppen I, *Math. Ann.* **16** (1880), 441–528; also *Gesammelte Abhandlungen*, vol. 6, B.G. Teubner, Leipzig, 1927, pp. 1–94; see [**2**] for an English translation.

[**139**] Lie, S., Über Differentialinvarianten, *Math. Ann.* **24** (1884), 537–578; also *Gesammelte Abhandlungen*, vol. 6, B.G. Teubner, Leipzig, 1927, pp. 95–138; see [**3**] for an English translation.

[**140**] Lie, S., *Vorlesungen über Differentialgleichungen mit Bekannten Infinitesimalen Transformationen*, B.G. Teubner, Leipzig, 1891.

[**141**] Lie, S., and Scheffers, G., *Vorlesungen über Continuierliche Gruppen mit Geometrischen und Anderen Anwendungen*, B.G. Teubner, Leipzig, 1893.

[**142**] Lindeberg, T., *Scale-Space Theory in Computer Vision*, Kluwer Acad. Publ., Boston, 1994.

[**143**] Macdonald, I.G., *Symmetric Functions and Hall Polynomials*, Clarendon Press, Oxford, 1979.

[**144**] Mackey, G.W., *The Theory of Unitary Group Representations*, Univ. of Chicago Press, Chicago, 1976.

[**145**] Mackey, G.W., *Unitary Group Representations in Physics, Probability, and Number Theory*, Benjamin Cummings Publ. Co., Reading, Mass., 1978.

[**146**] MacMahon, P.A., The perpetuant invariants of binary quantics, *Proc. London Math. Soc.* **26** (1895), 262–284.

[**147**] MacMahon, P.A., *Combinatory Analysis*, Cambridge Univ. Press, Cambridge, 1915, 1916.

[148] Magid, A.R., *Lectures on Differential Galois Theory*, Univ. Lecture Series, vol. 7, American Math. Soc., Providence, R.I., 1994.

[149] Maschke, H., A new method for determining the differential parameters and invariants of quadratic differential quantics, *Trans. Amer. Math. Soc.* 1 (1900), 197–204.

[150] Maschke, H., A symbolic treatment of the theory of invariants of quadratic differential quantics of n variables, *Trans. Amer. Math. Soc.* 4 (1903), 445–469.

[151] Meyer, F., Bericht über den gegenwärtigen Stand der Invariantentheorie, *Jahresberichte Deutschen Math.-Verein.* 1 (1890), 79–290.

[152] Miller, W., Jr., *Lie Theory and Special Functions*, Academic Press, New York, 1968.

[153] Miller, W., Jr., *Symmetry Groups and Their Applications*, Academic Press, New York, 1972.

[154] Moyal, J.E., Quantum mechanics as a statistical theory, *Proc. Camb. Phil. Soc.* 45 (1949), 99–124.

[155] Mumford, D., *Algebraic Geometry I Complex Projective Varieties*, Springer–Verlag, New York, 1976.

[156] Mumford, D., *Geometric Invariant Theory*, 3rd ed., Springer–Verlag, New York, 1994.

[157] Mundy, J.L., and Zisserman, A., Projective geometry for machine vision, appendix in: *Geometric Invariance in Computer Vision*, J.L. Mundy and A. Zisserman, eds., MIT Press, Cambridge, Mass., 1992, pp. 463–519.

[158] Netto, E., *Vorlesungen über Algebra*, vol. 2, B.G. Teubner, Leipzig, 1900.

[159] Newell, A., *Solitons in Mathematics and Physics*, CBMS–NSF Conf. Series in Applied Math., vol. 48, SIAM, Philadelphia, 1985.

[160] Noether, E., Über die Bildungen des Formensystems der ternären biquadratischen Form, *J. Reine Angew. Math.* 134 (1908), 23–90.

[161] Novikov, S.P. (ed.), *Topology I*, Encyclopedia of Mathematical Sciences, vol. 12, Springer–Verlag, New York, 1996.

[162] Oldenburger, R., Higher dimensional determinants, *Amer. Math. Monthly* 47 (1940), 25–33.

[163] Olver, P.J., Hyperjacobians, determinantal ideals and weak solutions to variational problems, *Proc. Roy. Soc. Edinburgh* 95A (1983), 317–340.

[164] Olver, P.J., Invariant theory and differential equations, in: *Invariant Theory*, S.S. Koh, ed., Lecture Notes in Mathematics, vol. 1278, Springer–Verlag, New York, 1987, 62–80.

[165] Olver, P.J., The equivalence problem and canonical forms for quadratic Lagrangians, *Adv. Appl. Math.* 9 (1988), 226–257.

[166] Olver, P.J., Canonical elastic moduli, *J. Elasticity* 19 (1988), 189–212.

[167] Olver, P.J., Classical invariant theory and the equivalence problem for particle Lagrangians. I. Binary forms, *Adv. in Math.* 80 (1990), 39–77.

[168] Olver, P.J., *Applications of Lie Groups to Differential Equations*, 2nd ed., Graduate Texts in Mathematics, vol. 107, Springer–Verlag, New York, 1993.

[169] Olver, P.J., *Equivalence, Invariants, and Symmetry*, Cambridge Univ. Press, Cambridge, 1995.

[170] Olver, P.J., Singularities of prolonged group actions on jet bundles, preprint, Univ. of Minnesota, 1997.

[171] Olver, P.J., and Shakiban, C., Graph theory and classical invariant theory, *Adv. in Math.* **75** (1989), 212–245.

[172] Olver, P.J., and Shakiban, C., Dissipative decomposition of partial differential equations, *Rocky Mountain J. Math.* **22** (1992), 1483–1510.

[173] Onishchik, A.L. (ed.), *Lie Groups and Lie Algebras I*, Encyclopedia of Mathematical Sciences, vol. 20, Springer–Verlag, New York, 1993.

[174] Ovsienko, V., Exotic deformation quantization, *J. Diff. Geom.* **45** (1997), 390–406.

[175] Ovsienko, O.D., and Ovsienko, V.Y., Lie derivatives of order n on the line. Tensor meaning of the Gelfand–Dikii bracket, *Adv. Soviet Math.* **2** (1991), 221–231.

[176] Palais, R.S., *A Global Formulation of the Lie Theory of Transformation Groups*, Memoirs Amer. Math. Soc., no. 22, Providence, R.I., 1957.

[177] Pascal, E., Sopra le relazioni che possono sussistere identicamente fra formacioni simbolische del tipo invariantivo nella teoria generale delle forme algebriche, *Rom. Acc. Lincei Mem.* **5** (1888), 375–387.

[178] Peano, G., Formazioni invariantive delle corrispondenze, *Giornale Mat. Univ. Ital.* **20** (1882), 79–100.

[179] Peter, F., and Weyl, H., Die Vollständigkeit der primitiven Darstellungen einer geschlossenen kontinuerlichen Gruppe, *Math. Ann.* **97** (1927), 737–755.

[180] Pochhammer, L., Ueber hypergeometrische Functionen n^{ter} Ordnung, *J. Reine Angew. Math.* **71** (1870), 316–352.

[181] Popov, V.L., *Groups, Generators, Syzygies, and Orbits in Invariant Theory*, Transl. Math. Monographs, vol. 100, Amer. Math. Soc., Providence, R.I., 1992.

[182] Rankin, R.A., The construction of automorphic forms from the derivatives of a given form, *J. Indian Math. Soc.* **20** (1956), 103–116.

[183] Reichstein, B., On symmetric operators of higher degree and their application, *Linear Algebra Appl.* **75** (1986), 155–172.

[184] Reznick, B.A., *Sums of Even Powers of Real Linear Forms*, Memoirs Amer. Math. Soc., vol. 96, no. 463, Providence, R.I., 1992.

[185] Ribet, K.A., Galois representations and modular forms, *Bull. Amer. Math. Soc.* **32** (1995), 375–402.

[186] Richman, D.R., The fundamental theorems of vector invariants, *Adv. in Math.* **73** (1989), 43–78.

[187] Rota, G.–C., and Stein, J.A., Symbolic method in invariant theory, *Proc. Nat. Acad. Sci.* **83** (1986), 844–847.

[188] Sanders, J.A., and Wang, J.P., On the integrability of homogeneous scalar evolution equations, *J. Diff. Eq.* **147** (1998), 410–434.

[189] Sartori, G., Geometric invariant theory: a model-independent approach to spontaneous symmetry and/or supersymmetry breaking, *Rivista Nuovo Cimento* **14**, #11 (1991), 1–120.

[190] Sattinger, D.H., *Group Theoretic Methods in Bifurcation Theory*, Lecture Notes in Math., vol. 762, Springer–Verlag, New York, 1979.

[191] Schrödinger, E., *Collected Papers on Wave Mechanics*, Chelsea Publ. Co., New York, 1982.

[192] Shakiban, C., A resolution of the Euler operator II, *Math. Proc. Camb. Phil. Soc.* **89** (1981), 501–510.

[193] Shakiban, C., An invariant theoretic characterization of conservation laws, *Amer. J. Math.* **104** (1982), 1127–1152.

[194] Shioda, T., On the graded ring of invariants of binary octavics, *Amer. J. Math.* **89** (1967), 1022–1046.

[195] Sibirsky, K.S., *Introduction to the Algebraic Theory of Invariants of Differential Equations*, Manchester Univ. Press, New York, 1988.

[196] Smith, L., Polynomial invariants of finite groups. A survey of recent results, *Bull. Amer. Math. Soc.* **34** (1997), 211–250.

[197] Spencer, A.J.M., Theory of invariants, in: *Continuum Physics*, A.C. Eringen, ed., vol. 1, Academic Press, New York, 1971.

[198] Spivak, M., *Calculus on Manifolds*, W.A. Benjamin, Menlo Park, Calif., 1965.

[199] Spivak, M., *A Comprehensive Introduction to Differential Geometry*, vol. 1, 2nd ed., Publish or Perish, Inc., Wilmington, Del., 1979.

[200] Springer, T.A., *Invariant Theory*, Lecture Notes in Math., vol. 585, Springer–Verlag, New York, 1977.

[201] Stanley, R.P., Invariants of finite groups and their applications to combinatorics, *Bull. Amer. Math. Soc.* **1** (1979), 475–511.

[202] Story, W.E., On the covariants of systems of quantics, *Math. Ann.* **41** (1893), 469–490.

[203] Stroh, E., Ueber eine fundamentale Eigenschaft des Ueberschiebungsprocesses und deren Verwerthung in der Theorie der binären Formen, *Math. Ann.* **33** (1889), 61–107.

[204] Sturmfels, B., *Algorithms in Invariant Theory*, Springer–Verlag, New York, 1993.

[205] Sturmfels, B., and White, N., Computing combinatorial decompositions of rings, *Combinatorica* **11** (1991), 275–293.

[206] Sylvester, J.J., On an application of the new atomic theory to the graphical representation of the invariants and covariants of binary quantics, with three appendices, *Amer. J. Math.* **1** (1878), 64–125; also *The Collected Mathematical Papers*, vol. 3, Cambridge Univ. Press, Cambridge, England, 1909, pp. 148–206.

[207] Sylvester, J.J., Sur les actions mutuelles des formes invariantives dérivées, *J. Reine Angew. Math.* **85** (1878), 89–114; also *The Collected Mathematical Papers*, vol. 3, Cambridge Univ. Press, Cambridge, England, 1909, pp. 218–240.

[208] Sylvester, J.J., Tables of the generating functions and groundforms for the binary quantics of the first ten orders, *Amer. J. Math.* **2** (1879), 223–251; also *The Collected Mathematical Papers*, vol. 3, Cambridge Univ. Press, Cambridge, England, 1909, pp. 283–311.

[209] Sylvester, J.J., Tables of the generating functions and groundforms for the binary duadecimic, with some general remarks, and tables of the irreducible syzygies of certain quantics, *Amer. J. Math.* **4** (1881), 41–61; also *The Collected Mathematical Papers*, vol. 3, Cambridge Univ. Press, Cambridge, England, 1909, pp. 489–508.

[210] Sylvester, J.J., On subinvariants, that is, semi-invariants to binary quantics of an unlimited order, *Amer. J. Math.* **5** (1882), 79–136; also *The Collected Mathematical Papers*, vol. 3, Cambridge Univ. Press, Cambridge, England, 1909, pp. 568–622.

[211] Takhtajan, L.A., A simple example of modular-forms as tau-functions for integrable equations, *Theor. Math. Phys.* **93** (1993), 1308–1317.

[212] Talman, J.D., *Special Functions. A Group Theoretic Approach*, W.A. Benjamin,, New York, 1968.

[213] Tannenbaum, A., *Invariance and System Theory: Algebraic and Geometric Aspects*, Lecture Notes in Math., vol. 845, Springer–Verlag, New York, 1981.

[214] Thomas, T.Y., *The Differential Invariants of Generalized Spaces*, Chelsea Publ. Co., New York, 1991.

[215] Turnbull, H.W., Gordan's theorem for double binary forms, *Proc. Edinburgh Math. Soc.* **41** (1922/23), 116–127.

[216] Turnbull, H.W., Double binary forms III, *Proc. Roy. Soc. Edinburgh* **43** (1922/23), 43–50.

[217] Turnbull, H.W., Double binary forms IV, *Proc. Edinburgh Math. Soc.* **42** (1923/24), 69–80.

[218] Turnbull, H.W., A geometrical interpretation of the complete system of the double binary (2,2) form V, *Proc. Roy. Soc. Edinburgh* **44** (1923/24), 23–50.

[219] Turnbull, H.W., and Aitken, A.C., *An Introduction to the Theory of Canonical Matrices*, Blackie and Sons, London, 1932.

[220] Ushveridze, A.G., *Quasi-Exactly Solvable Models in Quantum Mechanics*, Inst. of Physics Publ., Bristol, England, 1994.

[221] van der Waerden, B.L., Über die fundamentalen Identitäten der Invariantentheorie, *Math. Ann.* **95** (1926), 706–735.

[222] van der Waerden, B.L., *Algebra*, vol.1, Frederick Ungar, New York, 1970.

[223] Varadarajan, V.S., *Lie Groups, Lie Algebras, and Their Representations*, Springer–Verlag, New York, 1984.

[224] Vey, J., Déformation du crochet de Poisson sur une variété symplectique, *Comment. Math. Helv.* **50** (1975), 421–454.

[225] Vilenkin, N.J., and Klimyk, A.U., *Representation of Lie Groups and Special Functions*, Kluwer Acad. Publ., Dordrecht, 1991.

[226] Vinberg, E.B., Effective invariant theory, *Amer. Math. Soc. Transl.* **137** (1987), 15–19.

[227] Warner, F.W., *Foundations of Differentiable Manifolds and Lie Groups*, Springer–Verlag, New York, 1983.

[228] Warner, G., *Harmonic Analysis on Semi-simple Lie Groups*, Springer–Verlag, New York, 1972.

[229] Weitzenböck, R., *Invariantentheorie*, P. Noordhoff, Groningen, 1923.

[230] Weyl, H., Invariants, *Duke Math. J.* **5** (1939), 489–502.

[231] Weyl, H., *Classical Groups*, Princeton Univ. Press, Princeton, N.J., 1946.

[232] Weyl, H., *The Theory of Groups and Quantum Mechanics*, Dover Publ., New York, 1950.

[233] Weyl, H., *Space–Time–Matter*, Dover Publ., New York, 1952.

[234] Wilczynski, E.J., An application of group theory to hydrodynamics, *Trans. Amer. Math. Soc.* **1** (1900), 339–352.

[235] Wilczynski, E.J., *Projective Differential Geometry of Curves and Ruled Surfaces*, B.G. Teubner, Leipzig, 1906.

[236] Wiles, A., Modular elliptic curves and Fermat's Last Theorem, *Ann. Math.* **141** (1995), 443–551.

[237] Wussing, H., *The Genesis of the Abstract Group Concept*, MIT Press, Cambridge, Mass., 1984.

[238] Wybourne, B.G., *Classical Groups for Physicists*, John Wiley, New York, 1974.

[239] Yaglom, I.M., *Felix Klein and Sophus Lie*, Birkhäuser, Boston, 1988.

[240] Yale, P.B., *Geometry and Symmetry*, Holden–Day, San Francisco, 1968.

Author Index

Subject Index

Printed in the United States
By Bookmasters